Third Canadian Edition

Electrical Wiring Commercial

Ray C. Mullin

Gary Miller

Paul Stephenson

Based on the 2002 Canadian Electrical Code

THOMSON
NELSON

Australia   Canada   Mexico   Singapore   Spain   United Kingdom   United States

**Electrical Wiring: Commercial
Third Canadian Edition**

by Ray C. Mullin, Gary Miller,
and Paul Stephenson

Editorial Director and Publisher: Evelyn Veitch	**Production Editor:** Wendy Yano	**Cover Design:** Pedro Gaudenz
Acquisitions Editor: Anthony Rezek	**Production Coordinator:** Julie Preston	**Cover Photo:** Ryan McVay/PhotoDisc
Marketing Manager: Kevin Smulan	**Copy Editor/Proofreader:** Edie Franks	**Compositor:** Erich Falkenberg
Senior Developmental Editor: Edward Ikeda	**Creative Director:** Angela Cluer	**Printer:** Webcom

COPYRIGHT © 2003 by Nelson, a division of Thomson Canada Limited.

Printed and bound in Canada
3 4 05 04

For more information contact Nelson, 1120 Birchmount Road, Scarborough, Ontario, M1K 5G4. Or you can visit our Internet site at http://www.nelson.com

Notice to the Reader
Publisher does not warrant or guarantee any of the products described herein or perform any independent analysis in connection with any product information contained herein. Publisher does not assume, and expressly disclaims, any obligation to obtain and include information other than that provided to it by the manufacturer.

The reader is expressly warned to consider and adopt all safety precautions that might be indicated by the activities described herein and to avoid all potential hazards. By following the instructions contained herein, the reader willingly assumes all risks in connection with such instructions.

The publisher makes no representations or warranties of any kind, including but not limited to, the warranties of fitness for particular purpose or merchantability, nor are any such representations implied with respect to the material set forth herein, and the publisher shall not be liable for any special, consequential or exemplary damages resulting, in whole or in part, from the reader's use of, or reliance upon, this material.

ALL RIGHTS RESERVED. No part of this work covered by the copyright hereon may be reproduced, transcribed, or used in any form or by any means—graphic, electronic, or mechanical, including photocopying, recording, taping, Web distribution, or information storage and retrieval systems—without the written permission of the publisher.

For permission to use material from this text or product, contact us by
Tel 1-800-730-2214
Fax 1-800-730-2215
www.thomsonrights.com

Every effort has been made to trace ownership of all copyrighted material and to secure permission from copyright holders. In the event of any question arising as to the use of any material, we will be pleased to make the necessary corrections in future printings.

National Library of Canada Cataloguing in Publication Data

Mullin, Ray C
  Electrical wiring : commercial / Ray C. Mullin, Gary Miller, Paul Stephenson. — 3rd Canadian ed.

Includes index.
Based on 2002 Canadian Electrical Code.
Second Canadian ed. written by Ray C. Mullen ... [et al.].
ISBN 0-17-622370-3

1. Electrical wiring, Interior.
2. Commercial buildings — Electric equipment. I. Miller, Gary, 1955–
II. Stephenson, Paul, 1946–

TK3285.M83 2002   621.319'24
C2002-901960-5

CONTENTS

Introduction	ix
Preface	xi
Acknowledgments	xii

Unit 1 COMMERCIAL BUILDING PLANS AND SPECIFICATIONS 1

Objectives	1
Contracts	1
Specifications	1
Supplementary General Conditions	4
Commercial Building Plans	8
Estimating	14
Job Management	14
Change Orders	16
Approval of Equipment	16
Organizations	16
Electrical Inspection	19
Review	20

Unit 2 THE ELECTRIC SERVICE 23

Objectives	23
Liquid-Filled Transformers	23
Dry-Type Transformers	25
Transformer Overcurrent Protection	27
Transformer Connections	27
The Service Entrance	30
Metering	30
Service-Entrance Equipment	32
Service Disconnecting Means	33
Grounding	33
Bonding	39
Ground-Fault Protection (*Rule 14–102*)	45
Review	48

Unit 3 BRANCH CIRCUIT AND FEEDERS 51

Objectives	51
Branch-Circuit Calculation	51
General Lighting Loads	52
Motor Loads (Appliances and Air-Conditioning and Refrigeration Equipment)	53
Branch Circuits	54
Considerations When Sizing Wire	55
Determining Conductor Size and Type	59

	Correction Factors for Ambient Temperature	60
	Derating Factors for More Than Three Current-Carrying Conductors in One Raceway	60
	Feeders	62
	Overcurrent Protection and Circuit Rating	62
	Voltage Drop	63
	Determining Voltage Drop (Volt Loss) Using Tables	64
	Energy Saving Considerations	66
	Aluminum Conductors	67
	Proper Installation Procedures	68
	Review	69
Unit 4	**HOUSE CIRCUITS (OWNER'S CIRCUITS)**	**73**
	Objectives	73
	Loading Schedule	73
	Lighting Circuits	73
	Sump Pump Control	75
	Boiler Control	77
	Emergency and Exit Lighting	79
	Central Supply	80
	Unit Equipment	80
	Review	80
Unit 5	**EMERGENCY POWER SYSTEMS**	**82**
	Objectives	82
	Sources of Power (*Rules 46–200 to 46–210*)	82
	Special Service Arrangements	82
	Emergency Generator Source (*Rule 46–202(3)*)	83
	Review	90
Unit 6	**FIRE ALARM AND SAFETY SYSTEMS**	**92**
	Objectives	92
	Special Terminology	92
	Four Stages of a Fire	93
	Four Classes of Fire	93
	Fire Triangle	94
	Purpose of Fire Alarm Systems	95
	Types of Occupancy Requiring a Fire Alarm	96
	Classification of Major Occupancies	96
	Areas Where Fire Detectors Are to Be Installed	96
	Areas Where Heat Detectors Are to Be Installed	96
	Areas Where Smoke Detectors Are to Be Installed	97
	Types of Systems	97
	Smoke Detectors	99
	Zoning Requirements of Fire Alarm Systems	105
	Fire Codes and Standards	105
	C.E.C., Part I Requirements (*Section 32*)	106
	Fire Pumps	106
	Review	107
Unit 7	**HAZARDOUS MATERIALS IN THE WORKPLACE**	**109**
	Objectives	109

	Workplace Hazardous Materials Information System	109
	Health	110
	Typical Hazards	110
	Workplace Controls	110
	Personal Protective Equipment	110
	Material Safety Data Sheet	111
	Supplier Label	111
	Workplace Label	115
	Hazardous Materials	116
	Review	116
Unit 8	**PANELBOARD SELECTION AND INSTALLATION**	**118**
	Objectives	118
	Panelboards	118
	The Feeder	121
	Neutral Sizing (*Rule 4–022*)	125
	Feeder Loading Schedule	125
	Review	126
Unit 9	**THE COOLING SYSTEM**	**127**
	Objectives	127
	Refrigeration	127
	Evaporator	128
	Compressor	129
	Condenser	129
	Expansion Valve	130
	Hermetic Compressors	131
	Cooling System Control	132
	Cooling System Installation	132
	Electrical Requirements for Air-Conditioning and Refrigeration Equipment	134
	Special Terminology	135
	Review	140
Unit 10	**READING ELECTRICAL DRAWINGS—BAKERY, AND SWITCHES AND RECEPTACLES**	**141**
	Part I: Reading Electrical Drawings—Bakery	**141**
	Objectives	141
	Prints	141
	The Bakery Prints	141
	Part II: Switches and Receptacles	**143**
	Objectives	143
	Receptacles	143
	Hospital-Grade Receptacles	145
	Electronic Equipment Receptacles	146
	Ground-Fault Circuit Interrupter Receptacles (*Rule 26–700(13)*)	147
	Receptacles in Electric Baseboard Heaters	149
	Switches (*Rule 14–500*)	149
	Switch and Receptacle Covers	151
	Review	153

Unit 11 BRANCH-CIRCUIT INSTALLATION ... 156
Objectives ... 156
Branch-Circuit Installation ... 156
Rigid Metal Conduit (*Rules 12–1000* to *12–1014*) ... 156
Electrical Metallic Tubing (*Rules 12–1400* to *12–1412*) ... 158
Electrical Nonmetallic Tubing (*Rules 12–1500* to *12–1516*) ... 159
Flexible Connections (*Rules 12–1010* and *12–1300* to *12–1306*) ... 159
Rigid PVC Conduit (*Rule 12–1100*) ... 162
Raceway Sizing ... 166
Special Considerations ... 167
Box Styles and Sizing ... 167
Raceway Support ... 175
Review ... 177

Unit 12 APPLIANCE CIRCUITS (BAKERY) ... 180
Objectives ... 180
Appliances ... 180
The Exhaust Fan (Connected to a Lighting Branch Circuit) ... 180
The Basics of Motor Circuits ... 181
Several Motors on a Single Branch Circuit ... 186
Conductors Supplying Several Motors ... 186
Several Motors on One Feeder ... 186
Grounding ... 186
Overcurrent Protection ... 187
Typical Bakery Equipment ... 187
The Doughnut Machine ... 188
The Bake Oven ... 189
Review ... 190

Unit 13 LUMINAIRES AND LAMPS ... 193
Part I: Luminaires ... 193
Objectives ... 193
Definitions ... 193
Installation ... 193
Labelling ... 197
Loading Calculations ... 201
Location of Luminaires in Clothes Closets ... 208

Part II: Lamps for Lighting ... 209
Objectives ... 209
Lighting Terminology ... 209
Lamp Energy Efficiency ... 210
Incandescent Lamps ... 211
Low-Voltage Incandescent Lamps ... 214
Fluorescent Lamps ... 214
High-Intensity Discharge (HID) Lamps ... 221
Review ... 223

Unit 14 READING ELECTRICAL DRAWINGS—INSURANCE OFFICE/BEAUTY SALON ... 225
Objectives ... 225
Prints ... 225
Loading Schedule ... 225
Appliance Circuits ... 225
Review ... 229

Unit 15 SPECIAL SYSTEMS ... 233
Objectives ... 233
Surface Raceways ... 233
Multioutlet Assemblies ... 234
Communications Systems ... 237
Floor Outlets ... 238
Computer Room Circuits ... 238
Review ... 244

Unit 16 READING ELECTRICAL AND ARCHITECTURAL DRAWINGS (PRINTS)—DRUGSTORE ... 245
Objectives ... 245
The Drugstore Prints ... 245
Review ... 247

Unit 17 OVERCURRENT PROTECTION: FUSES AND CIRCUIT BREAKERS ... 248
Objectives ... 248
Fuses and Circuit Breakers ... 250
Types of Fuses ... 252
Cartridge Fuses ... 255
Testing Fuses ... 266
Time–Current Characteristic Curves and Peak Let-Through Charts ... 268
Circuit Breakers ... 273
Series-Tested Combinations ... 278
Current-Limiting Breakers ... 279
Cost Considerations ... 280
Review ... 280

Unit 18 SHORT-CIRCUIT CALCULATIONS AND COORDINATION OF OVERCURRENT PROTECTIVE DEVICES ... 285
Objectives ... 285
Short-Circuit Calculations ... 286
Short-Circuit Current Variables ... 292
Coordination of Overcurrent Protective Devices ... 295
Single Phasing ... 298
Review ... 300

Unit 19 EQUIPMENT AND CONDUCTOR SHORT-CIRCUIT PROTECTION ... 303
Objectives ... 303
Conductor Withstand Rating ... 304
Conductor Heating ... 307

	Calculating an Insulated (75°C Thermoplastic) Conductor's Short-Time Withstand Rating	308
	Calculating Withstand Ratings and Sizing for Bare Copper Conductors	310
	Using Charts to Determine a Conductor's Short-Time Withstand Rating	311
	Magnetic Forces	312
	Tap Conductors	313
	Review	314
Unit 20	**LOW-VOLTAGE REMOTE-CONTROL LIGHTING**	**317**
	Objectives	317
	Low-Voltage Remote-Control Lighting	317
	Wiring Methods	319
	Review	322
	APPENDIX	**324**
	Part A: Division 16 Electrical Specifications	324
	Division 15 Mechanical	329
	Part B: Useful Formulas	331
	Part C: Fire Protection Symbols	332
	GLOSSARY	**333**
	INDEX	**335**

Plans for a commercial building (inserted at back of book)

Sheet A1 Site Plan, East Elevation, West Elevation, Index to Drawings
Sheet A2 Basement Floor Plan
Sheet A3 First Floor Plan
Sheet A4 Second Floor Plan
Sheet A5 North and South Elevations
Sheet A6 Transverse Building Section
Sheet A7 Longitudinal Building Section
Sheet E1 Site plan, Legend, List of drawings
Sheet E2 Basement Electrical Plan
Sheet E3 First Floor Electrical Plan
Sheet E4 Second Floor Electrical Plan
Sheet E5 Power Riser Diagrams
Sheet E6 Schedules
Sheet E7 Schedules and Details

INTRODUCTION

The third Canadian edition of *Electrical Wiring: Commercial* is based on the 2002 edition of the *Canadian Electrical Code, Part I*, the safety standard for electrical installations. This Canadian edition thoroughly and clearly explains the *C.E.C., Part I* rules and changes that relate to commercial wiring.

Because *Electrical Wiring: Commercial* is both comprehensible and readable, it is suitable for colleges, technical institutes, and vocational technical schools. The *Canadian Electrical Code, Part I* is the basic standard for the layout and construction of electrical systems in Canada; however, some local and provincial codes may contain specific amendments that must be adhered to in electrical wiring installations within their jurisdictions.

To gain the greatest benefit from this text, the reader must refer to the *Canadian Electrical Code, Part I* on a continuing basis. The authors encourage the reader to develop a detailed knowledge of the layout and content of the *Canadian Electrical Code, Part I*.

This text takes the reader through the essential minimum requirements as set forth in the *Canadian Electrical Code, Part I* for commercial installations, and often provides further information above and beyond the minimum *C.E.C.* requirements.

The commercial electrician is required to work in three common situations: where the work is planned in advance; where there is no advance planning; and where repairs are needed. The first situation exists when the work is designed by a consulting engineer. In this case, the electrician must know the installation procedures, must be able to read plans, and must be able to understand and interpret specifications. The second situation occurs either during or after construction as a result of changes or remodelling. The third situation arises any time after a system is installed. Whenever a problem occurs with an installation, the electrician must understand the operation of all equipment included in the installation in order to solve the problem.

When the electrician is working on the initial installation or is modifying an existing installation, the circuit loads must be determined. Thorough explanations and numerous examples of calculating these loads help prepare the reader for similar problems on the job. The text and assignments make frequent reference to the commercial building drawings at the back of the text.

Readers should be aware that many of the electrical loads used in the building described in the text were contrived in order to create *C.E.C.* problems. The authors' purpose is to demonstrate, and thus enhance, the learning of as many *C.E.C.* problems as possible. For example, there is a single-phase feeder to the doctor's office. This could have been a three-phase feeder similar to those in the other occupancies.

However, using the single-phase feeder allows us to demonstrate additional *C.E.C.* applications.

This is not a typical nor an ideal design for a commercial building. It is a composite to demonstrate a range of *C.E.C.* applications. The authors also carry many calculations to a higher level of accuracy than that required in many actual job situations. This is done to demonstrate the correct method.

Thorough explanations are provided throughout, as the text guides the reader through the steps necessary to become proficient in the techniques and requirements of the *Canadian Electrical Code, Part I*.

PREFACE

This text will provide a valuable resource to instructors and students alike. It includes all 2002 *Canadian Electrical Code, Part I* references and wiring techniques.

Review questions are included at the end of each chapter to summarize the material being covered. The chapters are sequenced to introduce the student to the basic principles and wiring practices contained within a commercial building. Included in the text is information to introduce the reader to WHMIS and to fire safety in commercial applications.

The student requires a reasonable level of mechanical aptitude and skills in order to be successful in the practical application of the techniques discussed.

FEATURES

- The 2002 edition of the *Canadian Electrical Code, Part I*, the safety standard for electrical installation, is the basis for this text.

- The references to the *Canadian Electrical Code, Part I* are shown in *italics* to simplify cross-reference to the actual 2002 *Canadian Electrical Code, Part I*.

- Easy-to-understand explanations are given of the intent of the 2002 *Canadian Electrical Code, Part I* sections and rules.

- A concise explanation of WHMIS legislation and valuable material relating to fire safety is provided.

- Straightforward, easy-to-follow calculations are illustrated with examples in the text.

- Clearly defined objectives are set out for each unit together with questions and problems dealing with unit objectives.

- A large number of diagrams and illustrations further emphasize unit material.

ACKNOWLEDGMENTS

We wish to acknowledge the valuable assistance of Bill Wright and Jeremy Groves in the preparation of the architectural drawings for this book. We also wish to thank the people from Nelson for their encouragement and professional advice.

The following individuals reviewed portions or all of the text:

Mark Schuetzkowski, *Conestoga College*
Butch Carding, *Cambrian College*
Tony Poirier, *Durham College*

Their critical analyses and recommendations helped the authors extensively in the revision of the manuscript.

The authors wish to thank the following companies for their contributions of data, illustrations, and technical information:

AFC/A Nortek Co.
Anchor Electric Division, Sola Basic Industries
Appleton Electric Co.
Arrow-Hart, Inc.
BRK Electronics, A Division of Pittway Corporation
Brazos Technologies, Inc.
Bussmann Division, Cooper Industries
Carlon
Chromalox
Commander
Electri-Flex Co.
General Electric Co.
Gould Shawmut
Halo Lighting Division, Cooper Industries
Heyco Molded Products, Inc.
Honeywell
Hubbell Incorporated, Wiring Devices Division
International Association of Electrical Inspectors
IPEX Juno
KEPTEL
The Kohler Co.
Leviton Manufacturing Co., Inc.
Lightolier Canada

Midwest Electric Products, Inc.
Milbank Manufacturing Co.
Moe Light Division, Thomas Industries
Nutone Inc.
Noma Division, Danbel Industries Inc.
Pass & Seymour, Inc.
Progress Lighting
Rheem Manufacturing Co.
Seatek Co. Inc.
Sierra Electric Division, Sola Basic Industries
SMART HOUSE, L.P.
Square D Co.
Superior Electric Co.
THERM-O-DISC
Thomas & Betts Corporation
Underwriters' Laboratories of Canada (ULC is accredited by the Standards Council of Canada as a Certification Organization, a Testing Organization, a Registration Organization and Standards Organization under the National Standards System of Canada.)
Winegard Company
Wiremold Co.
Wolberg Electrical Supply Co., Inc.
Woodhead Industries, Inc.

With the permission of the Canadian Standards Association, material is reproduced from CSA C22.1—1998 and 2002 editions (*Canadian Electrical Code, Part I,* 17th and 18th editions—Safety Standard for Electrical Installations), which are copyrighted by CSA, 178 Rexdale Boulevard, Rexdale, Ontario, Canada, M9W 1R3. While the use of material taken from this Standard has been authorized by CSA, Canadian Standards Association shall not be responsible for the manner in which the information is presented, nor for any interpretations thereof.

This copy of CSA material will not be updated to reflect amendments made to the original content after January, 2002. For up-to-date information, see the current edition of the CSA *Catalogue of Standards*, or contact CSA.

For any further information or to give feedback on this text, you can contact the authors by e-mail at

gmiller@fanshawec.on.ca
pstephenson@fanshawec.on.ca

UNIT 1

Commercial Building Plans and Specifications

OBJECTIVES

After completing the study of this unit, the student will be able to
- define the job requirements from the contract documents
- explain the reasons for building plans and specifications
- locate specific information on the building plans
- obtain information from industry-related organizations

CONTRACTS

A contract is an agreement, enforceable by law, to supply goods or perform work at a stated price. A construction contract differs from most contracts in that the owner has the right to make changes to the contract as the project progresses. A construction contract is made up of working drawings and specifications. A bound volume of specifications and a set of working drawings are provided to all contractors who undertake to bid a job. The drawings and specifications are often referred to as the contract documents.

SPECIFICATIONS

Specifications are written descriptions of materials, construction systems, and workmanship and are generally laid out in the following manner:
- Notice to bidders
- Instructions to bidders
- Proposal (bid) form
- Owner- contractor agreement form
- Schedule of drawings
- General conditions
- Supplemental general conditions
- Alternates
- Technical sections

Notice to Bidders

This is the advertisement of the project. Electrical contractors may be a subcontractor (they sign a contract with the general contractor) or they may be a prime contractor (they sign a contract with the owner or owner's representative (architect or engineer).

The notice to bidders identifies the project, the project location, the closing date for bids, where the bidding documents may be obtained, and the amount of deposit required for the bidding documents.

Instructions to Bidders

The instruction to bidders provides a brief description of the project and its location. It includes information on the type of payment. The most common types being lump sum, unit price, or cost plus fee. A lump sum agreement stipulates that the contractor will satisfactorily complete the job for a fixed sum, regardless of any difficulties the contractor may encounter. If the unit price agreement is used, all bidders base their bids on a unit price basis (per outlet, per pole, per fixture). When the unit price method of payment is used, the actual quantities and classifications are determined by an engineer or quantity surveyor. Cost plus fee agreements are used where the job is extremely complex or construction must begin before complete drawings and specifications have been prepared.

In addition, bidders are told where and how the plans and specifications can be obtained prior to the preparation of the bid, how to make out the proposal form, where and when to deliver the proposal, the amount of any bid deposits required, and any performance bonds required.

The instructions to bidders normally require that the bidder visits the site and stipulates the time and date of site visits.

The instructions to bidders often require that the bidder provides the owner with information showing that the contractor has the ability to complete the contract. The information requested may include evidence that the bidder is licensed to do business in the jurisdiction where the project will be constructed, a list of work to be subcontracted, the names of the subcontractors, and what work the contractor currently has in progress.

Bonds

A surety bond is a guarantee made by the bonding company to meet the obligations of the contractor. If the contractor is unable to fulfill the terms of the contract, the bonding company must complete the contract (up to the amount of the bond) and then try to recover their money from the contractor. Owners require contractors to be bonded to ensure they will complete the project and pay their bills. Common forms of bonds are bid bonds, performance bonds, and labour and material bonds.

Bid bonds ensure that, if a contractor is awarded a contract, the contractor will proceed with the contract within a stipulated time period. If the contractor does not accept the contract, the bond is forfeit to the owner. Bid bonds are normally 10% of the bidding price.

Performance bonds guarantee an owner that the contractor will complete the contract. Performance bonds are normally 100% of the contract price. The premiums charged by the bonding company will vary according to experience and financial stability of the contractor.

Labour and material bonds are bonds which guarantee the payment of the contractor's bills to third parties which supply material or labour to the project.

Addenda

An addendum is a written addition to the contract documents issued before the bidding is closed and is used to:

- Clarify questions raised by bidders regarding conflicts, errors, omissions, and ambiguities.
- Add or reduce the scope of the work.
- Provide additional information to bidders in the way of explanations.
- Change the provisions of the contract.

Proposal

A (bid) proposal form is included in a set of specifications. This allows the owner to evaluate all the bids on the same basis. A typical proposal form would include the following:

- Name of tendering company
- Name of owner
- Lump sum tender price

- Breakdown of lump sum price
- A schedule of unit prices
- Alternates
- Statutory declaration
- Agreement to bond
- Statement of experience
- Statement of senior supervisory staff

The proposal is a form that is filled out by the contractor and submitted at the proper time and place. The proposal is the contractor's bid on a project. The form is the legal instrument that binds the contractor to the owner provided that the contractor completes the proposal properly, the contractor does not forfeit the bid bond, and the owner accepts the proposal and signs the agreement.

Unless the proposal is specifically marked as irrevocable, it may be revoked by the contractor without penalty at any time until the agreement has been signed by the owner.

If the owner fails to pay the contractor within the terms of the agreement, the contractor has a right to charge this as a lien against the property being upgraded. This type of lien is known as a "mechanic's lien." This means that the owner may not deal with the property, e.g., increase the mortgage or sell it, until the lien has been satisfactorily discharged or the amount in dispute has been paid into court pending settlement.

The proposal may show that alternate bids were requested by the owner. In this case, the electrician on the job should study the proposal and consult with the contractor to learn which of the alternate bids has been accepted in order to determine the extent of the work to be completed.

On occasion, the proposal may include a specified time for the completion of the project. This information is important to the electrician on the job since the work must be scheduled to meet the completion date.

The Owner-Contractor Agreement. If the proposal is acceptable to the owner, then a legally binding agreement must be executed between the parties. This agreement may be included as part of the proposal document. The contractor and the owner sign the agreement and the result is a legal contract. Once the agreement is signed, both parties are bound by the terms and conditions given in the specification.

General Conditions. General conditions outline the conditions which apply to all projects. The following items are normally included under the General Conditions heading of the *General Clauses and Conditions* section. A brief description is presented for each item.

- *General Note:* This item specifies that the general conditions are part of the contract documents.
- *Definition:* As used in the contract documents, this item defines the owner, contractor, architect, engineer, and other people and objects involved in the project.
- *Contract Documents:* This item lists the documents involved in the contract, including plans, specifications, and agreement.
- *Insurance:* This item specifies the insurance a contractor must carry on all employees and on the materials involved in the project.
- *Workmanship and Materials:* This item specifies that the work must be done by skilled workers and that the materials must be new and of good quality.
- *Substitutions:* Materials used must be as specified or equivalent materials must be shown to have the required properties.
- *Shop Drawings:* This item identifies the drawings that must be submitted by the contractor to show how the specific pieces of equipment are to be installed.
- *Payments:* This item specifies the method of paying the contractor during the construction.
- *Coordination of Work:* This item specifies that each contractor on the job must cooperate with every other contractor to ensure that the final product is complete and functional.
- *Correction of Work:* This section describes how work must be corrected, at no cost to

the owner, if any part of the job is installed improperly by the contractor.

- *Guarantee:* In this item the contractor guarantees the work for a certain length of time, usually one year.

- *Compliance with All Laws and Regulations:* This section specifies that the contractor will perform all work in accordance with all required laws, ordinances, and codes, such as the *Canadian Electrical Code, Part I* and local codes.

- *Others:* These sections are added as necessary by the owner, architect, and engineer when the complexity of the job and other circumstances require them. None of the items listed in the General Conditions has precedence over another item in terms of its effect on the contractor or the electrician on the job. The electrician must study each of the items before taking a position and assuming responsibilities with respect to the job.

SUPPLEMENTARY GENERAL CONDITIONS

Supplementary general conditions are more specific than general conditions. While the General Conditions can be applied to any job or project in almost any location with little change, the Supplementary General Conditions are rewritten for each project. The following list covers the items normally specified by the Supplementary General Conditions.

- The contractor must instruct all crews to exercise caution while digging; any utilities damaged during the digging must be replaced by the contractor responsible.

- The contractor must verify the existing conditions and measurements.

- The contractor must employ qualified individuals to lay out the worksite accurately. A registered land surveyor or engineer may be part of the crew responsible for the layout work.

- Job offices are to be maintained as specified on the site by the contractor; this office space may include space for owner representatives.

- The contractor may be required to provide telephones at the project site for use by the architect, engineer, subcontractor, or owner.

- Temporary toilet facilities and water are to be provided by the contractor for the construction personnel.

- The contractor must supply an electrical service of a specified size to provide temporary light and power at the site.

- It may be necessary for the contractor to supply a specified type of temporary heating to keep the temperature at the level specified for the structure.

- According to the terms of the guarantee, the contractor agrees to replace faulty equipment and correct construction errors for a period of one year.

The above list is by no means a complete catalogue of all of the items that can be included in Supplementary General Conditions.

Other titles may be applied to the Supplementary General Conditions section; these include Special Conditions and Special Requirements. Regardless of the title used, these sections contain the same types of information. All sections of the specifications must be read and studied by all of the construction trades involved. In other words, the electrician must study the heating, plumbing, ventilating, air-conditioning, and general construction specifications to determine if there is any equipment furnished by the other trades that the contract specifies is to be installed and wired by the electrical contractor. The electrician must also study the general construction specifications, since the roughing in of the electrical system will depend on the types of construction that will be encountered in the building.

This overview of the General and Supplementary General Conditions of a specification is intended to show the student that the construction worker on the job is affected by parts of the specification other than the part designated for each particular trade.

Alternates

An alternate is a request by the owner for a price on alternate materials or methods of construction. Alternates are listed on the proposal form. The alternates may be "add price" or "subtract price" meaning the alternate will add to your base bid or subtract from it.

Technical Sections

Specifications are divided into a number of divisions. Each division covers some portion of the work. Technical sections contain descriptions of materials and instructions on methods of construction.

Table 1–1 describes divisions commonly found in specifications.

Schedule of Drawings

This schedule is a list, by number and title, of all of the drawings related to the project. The contractor, the estimator, and the electrician will each use this schedule prior to preparing the bid for the job. The contractor will determine if all the drawings required are at hand, the estimator will do a take-off from the drawings and formulate a bid, and the electrician will determine if all of the drawings necessary to do the installation are available.

The Drawing Set

A set of drawings is made up of a number of sheets. Each sheet is numbered. If there is a large number of sheets in a set, the set will be broken down into subsets of architectural, electrical, mechanical, etc. The electrical drawings will be numbered E1, E2, E3, etc.

A typical set of construction drawings will include:

Table 1–1 Specification Divisions

Division	Title	Information relating to the electrical trade
1	General Requirements	
2	Site Construction	Site Preparation, Utility Services
3	Concrete	
4	Masonry	
5	Metals	
6	Wood and Plastics	
7	Thermal and Moisture Protection	Thermal Protection, Fire and Smoke Protection
8	Doors and Windows	Type of Construction
9	Finishes	
10	Specialties	Pedestrian Control Devices, Scales, Access Flooring, Telephone Enclosures
11	Equipment	Laundry Equipment, Loading Dock Equipment, Medical Equipment, Parking Equipment
12	Furnishings	
13	Special Construction	Incinerators, Instrumentation, Nuclear Reactors, Vaults, Swimming Pools
14	Conveying Systems	Powered Scaffolding, Hoists and Cranes
15	Mechanical	Controls and Instrumentation, Refrigeration Equipment, Heat Generation Equipment
16	Electrical	Division 16 is broken down into subdivisions; each subdivision covers a portion of the work. Examples include 16010 - General Provisions, 16100 - Basic Material and Methods, 16200 - Power Generation, 16300 - Power Transmission, 16400 - Service and Distribution, 16500 - Lighting, 16600 - Special Systems, 16700 - Communications, 16800 - Heating and Cooling, 16900 - Controls and Instrumentation

A Site (Plot) Plan

The site plan will show:

- The dimensions of the lot
- A north-facing arrow
- The location of utilities and their connection points
- Any outdoor lighting

- Easements (rights to the property by someone other than the owner)
- Contour lines to show changes in elevation
- Accessory buildings
- Trees to be saved

Floor Plans

Depending on the complexity of the building, a set of drawings may have only a single floor plan for each storey of the building. As the complexity of the building increases, there will be a number of floor plans for each storey, one for each trade. A logical breakdown would be:

- Architectural
- Electrical
- Mechanical
- Plumbing
- HVAC (Heating, Ventilating, and Air-Conditioning)
- Structural

On very complex installations different electrical systems such as power, lighting, voice/data, fire alarm, and security/access control may each have their own plan for each storey.

Use architectural floor plans to find room measurements, wall thickness, the sizes of doors and windows, and the swing of doors. The floor plans of other trades will show the locations of the equipment associated with that trade.

Elevations

Elevations show the vertical faces of buildings, structures, and equipment. Elevations are used to show both interior and exterior vertical faces. Sheet A1 shows the East and West elevations of the building.

Details and Sections

Details are large-scale drawings that are used when sufficient detail cannot be provided on floor plans and elevations. Sheet A7 shows details of structural elements of the building.

A Legend of Symbols

A legend of symbols provides the contractor or electrician with the information to interpret the symbols shown on the drawing. Refer to Figs. 1–2 to 1–5 for examples of the various symbols normally found on construction drawings.

Schedules

Schedules are lists of equipment. Electrical schedules are made for equipment such as lighting fixtures (luminaires), panelboards, raceways and cables, motors, and mechanical equipment connections.

Diagrams

One-line and riser diagrams are provided to quickly show the flow of power throughout the building. Riser diagrams may be used to show power, telephone, or fire alarm systems.

Shop Drawings

Shop drawings are used by equipment manufacturers to show details of custom-made equipment for a specific job. Manufacturer's drawings of switchgear and MCC's are examples of shop drawings.

The Electrical Drawing Set

A set of electrical drawings will normally include:

- Legend of symbols
- Site plan
- One-line diagram
- Lighting layout
- Power layout
- Electrical details

- Schematic and wiring diagrams
- Schedules

Scale

A drawing of an object that is the same size as the object is said to be drawn *full scale* or at a scale of 1:1.

Construction drawings reduce the size of a building down to fit on a piece of paper by drawing everything proportionally smaller than it actual is. This is called *reduced scale*. A drawing that uses a scale of 1:50 indicates that the building is 50 times larger than the drawing. A drawing with a scale of $\frac{1}{4}"$ = 1 ft has a scale of 1:48.

Enlarged scale would produce a drawing larger than the actual object. Printed circuit boards are drawn using enlarged scale since many of the components used in the manufacture of printed circuit boards are so small. An example of enlarged scale is 2:1 which indicates the drawing of the component is twice as large as the component.

Types of Scales (Measuring Instruments)

There are three types of scales commonly used on construction drawings. When the drawing is in feet and inches, an architectural scale is used.

Figure 1–1 shows a portion of an architect's triangular scale showing the $\frac{1}{8}"$ and $\frac{1}{4}"$ scales.

For metric drawings, a metric scale is used. When working with civil drawings (roads, dams and bridges, etc.) an engineering scale will be used. Table 1–2 lists common construction drawing scales.

Table 1–2

Metric	Architectural	Engineering
1:20	1/2" = 1'0"	1" = 1 ft or mile
1:50	1/4" = 1'0"	1" = 10
1:100	1/8" = 1'0"	1" = 40
1:200	1/16" = 1'0"	1" = 50
1:500	1/32" = 1'0"	1" = 100

Working with the Drawings

- Check that the drawing set is complete.
- Review the floor plans and elevations to get a mental picture of the size and shape of the building.
- Orient the building to the site using the plot plan. Be sure you know which sides are North, South, East, and West. Add this information to your floor plans.
- Check the scale of all drawings.

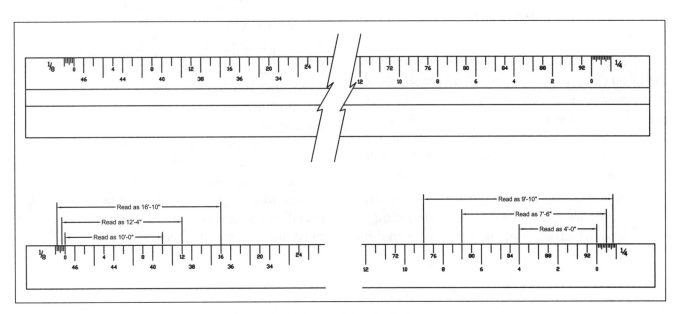

Fig. 1–1 Architect's Scale.

- Identify the type of construction (combustible, noncombustible), materials, and components shown in the drawings. Nonstandard items should be shown in a legend of symbols.
- Read all notes on the drawings carefully.
- Relate details to larger views. The examples below show how details are identified.
- Note multiple or identical drawings. (*Typical* means uniform throughout the building.)

THE COMMERCIAL BUILDING PLANS

A set of fourteen plan sheets is included at the back of the text showing the general and electrical portions of the work specified.

- *Sheet A1—Site Plan, East Elevation, West Elevation, Index to Drawings:* The plot plan shows the location of the commercial building and gives needed elevations. The east elevation is the street view of the building and the west elevation is the back of the building. The index lists the content of all of the plan sheets.
- *Sheet A2—Architectural Floor Plan: Basement.*
- *Sheet A3—Architectural Floor Plan: First Floor.*
- *Sheet A4—Architectural Floor Plan: Second Floor.*

The architectural floor plans give the wall and partition details for the building. These sheets are dimensioned; the electrician can find exact locations by referring to these sheets. The electrician should also check the plans for the material used in the general construction, since these will affect when and how the system will be installed.

- *Sheet A5—Elevations; North and South:* The electrician must study the elevation dimensions, which are given in feet and hundredths of a foot above sea level. For example, the finished second floor, which is shown at 218.33', is 218 feet 4 inches above sea level.
- *Sheets A6 and A7—Sections; Transverse, Longitudinal:* These sheets give detailed drawings of the more important sections of the building. The locations of the sections are indicated on the floor plans. When looking at a section, imagine that you are looking in the direction of the arrows at a building that is cut in two at the place indicated. You should see the section exactly as you view the imaginary building from this point.
- *Sheet E2—Basement Electrical Plan.*
- *Sheet E3—First Floor Electrical Plan.*
- *Sheet E4—Second Floor Electrical Plan.*
- *Sheets E5, E6, and E7—Riser Diagrams, Schedules, and Schedules and Details.*

Sheets E2 to E4 show the detailed electrical work on an outline of the building. Since dimensions usually are not shown on the electrical plans, the electrician must consult the other sheets for this information. It is recommended that the electrician refer frequently to the other plan sheets to ensure that the electrical installation does not conflict with the work of the other construction trades.

- *Sheet E1—Site Plan, Legend, List of Drawings.*

To assist the electrician in recognizing components used by other construction trades, the following illustrations are included: Fig. 1–2, Architectural drafting symbols; Fig. 1–3, Standard symbols for plumbing, piping, and valves; Fig. 1–4, Sheet metal ductwork symbols; and Fig. 1–5, Generic symbols for electrical plans. However, the electrician should be aware that variations of these symbols may be used and the specification and/or plans for a specific project must always be consulted.

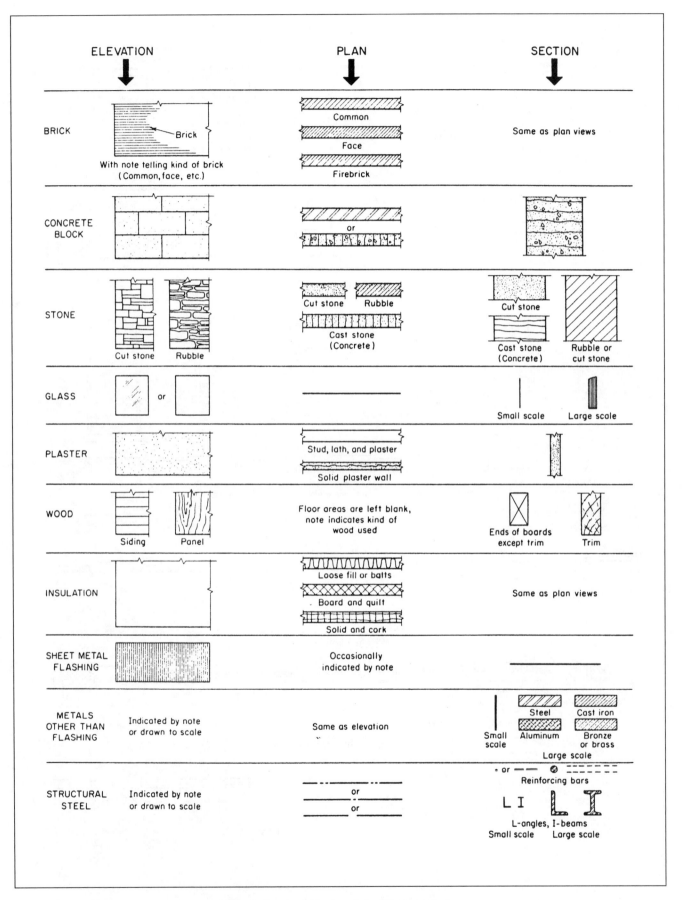

Fig. 1-2 Architectural drafting symbols.

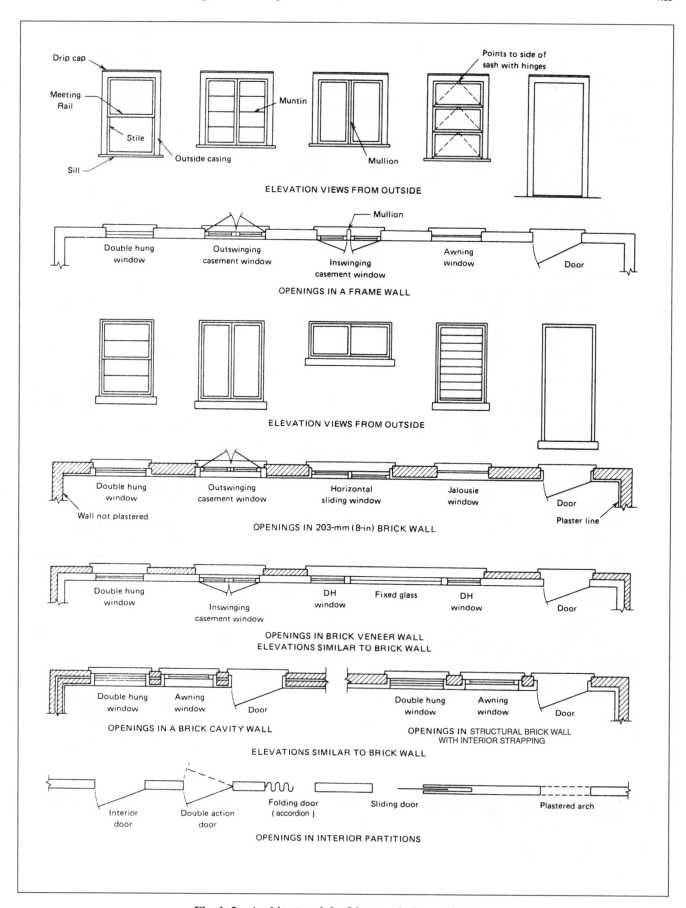

Fig. 1-2 Architectural drafting symbols (continued).

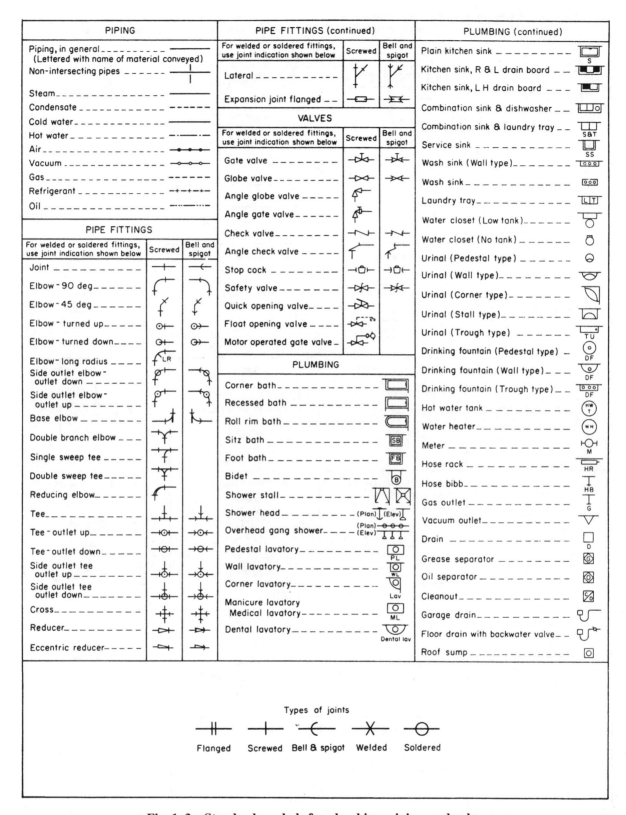

Fig. 1-3 Standard symbols for plumbing, piping, and valves.

12 Unit 1 Commercial Building Plans and Specifications

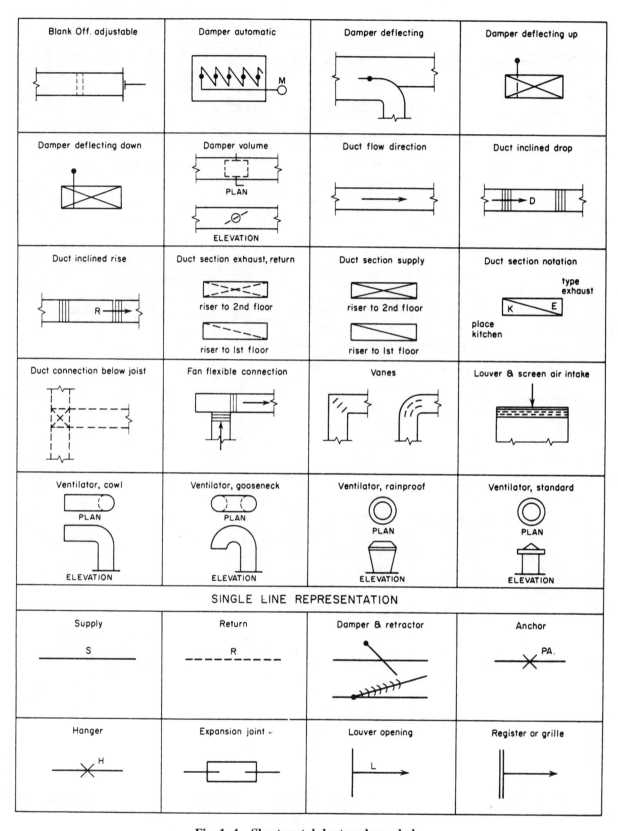

Fig. 1–4 Sheet metal ductwork symbols.

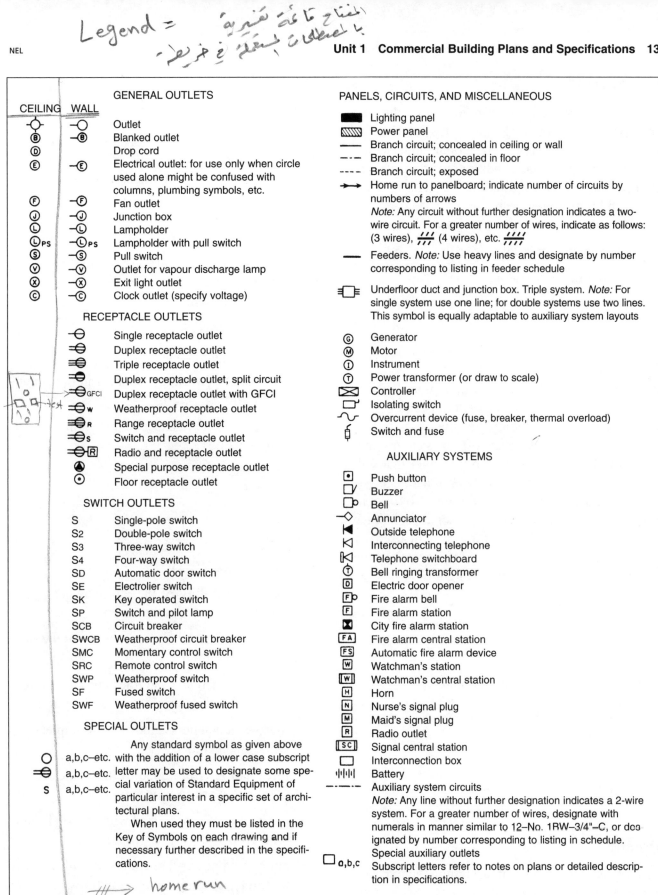

Fig. 1–5 Generic symbols for electrical plans.

ESTIMATING

Estimating is the process of determining the cost of a project by counting, measuring, and listing items of material; assigning labour units to the material based on the conditions of the installation; and determining direct job expenses, overhead, and return on investment.

The main steps in estimating are:

- Preparing a material take-off
- Tabulating and pricing
- Summarizing

Material Take-Off

A material take-off involves counting, measuring, and listing the quantities of materials and equipment necessary to complete the installation. Material is normally taken off by systems and listed on the take-off sheets in a manner that will permit labour units to be easily assigned.

The exact method and forms used by a contractor will be determined by personal preference or the computerized estimating system used by the contractor. One breakdown of systems in the order they would be taken off might be:

- Lighting fixtures (luminaires)
- Outlets
- Branch circuits
- Service, metering, and grounding
- Distribution equipment
- Feeders and busways
- Telephone
- Special systems

Pricing

Pricing is the process of assigning labour and prices to items of material that have been tabulated and listed on pricing sheets. Labour is assigned using standard labour units and adjusted to account for the conditions of the installation.

A standard labour unit is defined as the time required to handle and install an item under specified working conditions. Labour units are found in estimating manuals developed by such organizations as the Canadian Electrical Contractors Association and their provincial counterparts.

Summarizing

A summary bid sheet lists all of the projected costs of the project. The summary sheet includes such information as:

- Material costs
- Labour hours and costs
- Job factors (working height, multiple identical buildings, etc.)
- Direct job expenses (equipment, travel, warehousing)
- Overhead
- Profit

JOB MANAGEMENT

It is the responsibility of the job manager (whoever is running the job) to ensure that:

- tools and material arrive at the job at the time they are required.
- there are adequate facilities for storage and protection of material, tools, and equipment.
- there is adequate manpower to complete the work as scheduled.
- information is provided to the workforce to allow the work to proceed.
- the paperwork associated with the job is handled properly.
- company policies are followed with respect to safety, employment regulations, and job requirements.

The following steps will be helpful during the initial preparation for a project.

- Familiarize yourself with the project by studying the drawings and specifications.

- Prepare a list of materials required for the job. The estimator may provide you with the estimate sheets or you may decide to do your own material take-off and prepare a bill of materials yourself. Material should arrive on the site according to the activity that is being undertaken at the time. You will know how well you are doing by keeping records of all the material that you have ordered, what had to be returned and any delays in construction that result from lack of material. Material arriving on site should be inspected, inventoried, and stored in a clean, dry, and secure location. Packing slips should be carefully checked and back-ordered items recorded.

- Using electrical estimate sheets and the general contractor's construction schedule, determine when electrical activities should occur. The following is a list of the main construction events and corresponding electrical activities:

Construction Event	Electrical Activity
Site preparation	Trenching and underground work
Footings and foundations	Embedded work
Superstructure and floors	Embedded work and hangers
Building enclosure	Feeders, equipment, and branch circuits
Interior walls, partitions, and ceilings	Panelboards, branch circuits, and pulling wire
Trimming and decorating	Lighting
Floor finishing	Finishing

- Prepare a schedule of electrical construction activities that corresponds to the general contractors construction schedule. The activity schedule will be the basis for determining man-hour requirements, when material should be delivered to the site, and what tooling will be required. Fig. 1–6 is an example of a construction activity schedule.

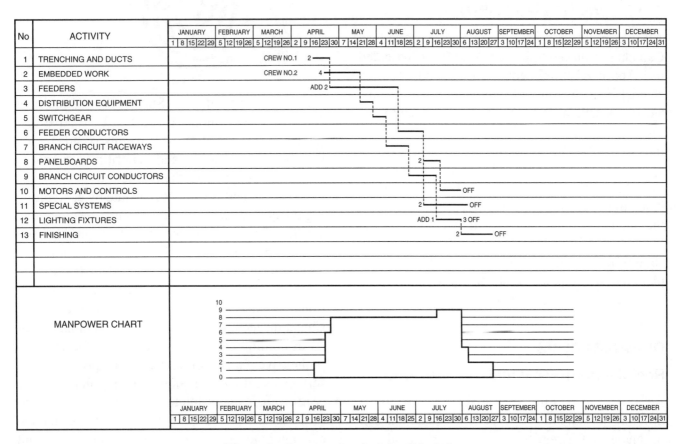

Fig. 1–6 Construction Activity Schedule.

CHANGE ORDERS

A Change Order is a change in the drawings or specifications that is made after the contract has been awarded. Change orders result from such things as:

- Requests for additional work from the owner
- Errors or omissions in the drawings and specifications
- Delays in construction

When an owner requests a change in the contract that will affect the contract price, an estimate of the costs involved in making the changes is prepared and a Request for Change Order is submitted to the owner. (This is a document which outlines the changes to be made, and authorizes the contractor to make the changes. It must be signed by the owner or owner's representative.) The most important thing about change orders is that they should be handled promptly and fairly.

APPROVAL OF EQUIPMENT

All electrical equipment sold or installed in Canada is required to be approved for its intended use. Approved equipment has met specified safety standards of federal and provincial governments. Currently several companies are accredited by the Standards Council of Canada to approve electrical equipment. Equipment that has been approved by an approval agency will have an identifying label such as CSA or ULC. For equipment that has not been approved, arrangements may be made for field approval by contacting the inspection authority in your province. They should also be able to provide you with a complete list of companies that can approve electrical equipment in your province.

ORGANIZATIONS

Registered Professional Engineer (PEng)

Although the requirements may vary slightly in different jurisdictions, the general statement can be made that a registered professional engineer has demonstrated his or her competence by graduating from college or university and passing a difficult licensing examination. Following the successful completion of the examination, the engineer is authorized to practise engineering under the laws of the province.

It is usually required that a registered professional engineer must supervise the design of any building to be used by the public. The engineer must indicate approval of the design by affixing a seal to the plans.

Information concerning the procedure for becoming a registered professional engineer and a definition of the duties of the professional engineer can be obtained by writing to the provincial government department that supervises licensing and registration.

CODES AND STANDARDS

Standards Council of Canada (SSC)

The objective of the Standards Council of Canada is to promote efficient and effective voluntary standardization as a means of advancing the national economy, and benefiting the health, safety, and welfare of the public.

The Council fulfills its mandate by accrediting organizations engaged in standard development, certification, testing and quality, and environmental registration. It is also responsible for approving National Standards of Canada (NSC).

The Council represents Canada on international standards bodies or agencies and accredits Canadian technical experts to international standards committees. The Council also offers a comprehensive family of information, products and services dealing with standards, regulations, and conformity assessment procedures. Through these activities the Council facilitates Canadian and international trade, protects consumers, and promotes international standards cooperation.

For more information please write to:

Standards Council of Canada
1200-45 O'Conner Street
Ottawa, ON K1P 6N7
Tel: (613) 238-3222
http://www.scc.ca

Canadian Standards Association (CSA)

The Canadian Standards Association (CSA) is a nonprofit organization that was formed in 1919 to develop standards for a variety of organizations and disciplines. These standards are the cornerstone for product certification, and are subsequently approved by provincial and federal government authorities to become recognized throughout the country.

Thousands of manufacturers work directly with the CSA on the certification process and the CSA mark appears on many new products every year. The *Canadian Electrical Code* is one of the many standards that are developed by the CSA. This code is then adopted by provincial authorities to become law in the various provinces.

For more information please write to:

Canadian Standards Association
Customer Service Department
178 Rexdale Boulevard
Etobicoke, ON M9W 1R3
Tel: (416) 747-4000
      1-800-463-6727
Fax: (416) 747-4149

Underwriters' Laboratories of Canada

Underwriters' Laboratories of Canada (ULC) is a nonprofit organization that operates laboratories and maintains a certification service for the testing and classification of devices, constructions, materials, and systems to determine their relation to life, fire, and property hazards.

ULC also publishes standards, specifications, and classification for products used in connection with on fire, accident, and property hazards.

For more information please write to:

Underwriters' Laboratories of Canada
7 Crouse Road
Scarborough, ON M1R 3A9
Tel: (416) 757-3611
      1-800-INFO-ULC
Fax: (416) 757-9540
http://www.ulc.ca
e-mail: ulcinfo@ulc.ca

Canadian Electrical Code

The original *Canadian Electrical Code* was developed in 1920 and was finalized in June 1927. At this time the revised draft was formally approved and printed as the *Canadian Electrical Code, Part I.* The purpose of the *C.E.C., Part I* is to prevent fire and shock, *Section 0.*

The *C.E.C., Part I* is the basic standard for the layout and construction of all electrical systems in Canada. However, some provincial and local codes contain specific amendments that must be adhered to in all electrical wiring installations under their jurisdiction. The authors encourage the reader to develop a detailed knowledge of the layout and content of the latest edition of the *C.E.C., Part I.* Also, refer to Table 1-2.

The *C.E.C., Part I* is divided into numbered sections, with each covering a major topic. These sections are further divided into numbered rules, subrules, paragraphs, and subparagraphs; related rules are often grouped together under a subsection heading. All references in the text are to the *Section* or to the *Section and rule number.* For example, *Rule 8-200(1)(a)(i)* in the text, refers to *Section 8, rule 200, subrule 1, paragraph a, subparagraph i.* Occasionally, a Section reference will include a clause. See Table 1-2. It is hoped that this explanation will assist the student in locating *C.E.C., Part I* references in the text.

To help the user of this text, relevant *C.E.C., Part I* sections are paraphrased where appropriate. However, the *C.E.C., Part I* must be consulted before any decision related to electrical installation is made.

Enforcement of the *C.E.C., Part I* is generally under the jurisdiction of the utility company providing electrical power in the particular province. In Ontario, for instance, electrical inspection is conducted by field inspectors employed by Ontario Hydro.

The *C.E.C., Part I* is revised and updated every four years.

Code Terms. The following terms are used throughout the *C.E.C., Part I*. It is important to understand the meanings of these terms.

Approved (as applied to electrical equipment): equipment has been submitted for examination and testing to an accredited certification organization and that the equipment conforms to appropriate Canadian Standards Association (CSA) standards.

Shall: Indicates a mandatory requirement.

Notwithstanding: In spite of.

Practicable: Feasible, possible.

Citing Code References

Every time that an electrician makes a decision concerning the electrical wiring, the decision should be checked by reference to the *C.E.C., Part I*. Usually this is done from memory. If there is any doubt in the electrician's mind, then the *C.E.C., Part I* should be referenced directly—just to make sure. It is a good idea to record the *C.E.C., Part I* location of the information—this is referred to as citing the reference. There is a very exact way that the location of an item is to be cited. The various levels of *C.E.C., Part I* referencing are shown in Table 1–3. Starting at the top of the table, each step becomes a more specific reference. A reference to *Section 2* includes all the information and requirements that are set forth in several pages. Citing a specific rule or exception narrows the reference to a few sentences or perhaps even a few words.

Table 1–3   Citing the *C.E.C., Part I*.

Division	Designation	Example
Section	0–84	Section 8
Rule	Even numbers	8–212
Subrule	Numbered sequence	8–106(1)
Paragraph	a–z	8–108(1)(d)
Subparagraph	Lower case/Roman numerals	8–204(2)(a)(i)
Clause	Capital A–Z	8–204(2)(b)(ii)(B)

C.E.C., Part I Use of SI (Metric) Measurements

The 2002 *C.E.C., Part I* includes both imperial and metric measurements. The metric system is known as the International System of Units (SI). Fig. 1–7 provides a list of SI units and symbols, as well conversion factors for previously used units.

The 2002 edition of the *C.E.C. Part I* indicates conduit sizes and box capacity in metric units. Conduit sizes are specified by the metric trade size followed by the traditional imperial designator in parentheses. Usable space and capacity of boxes are indicated in millilitres or cubic centimetres followed by the cubic inch value in parentheses. It is important to note that one millilitre is equal to one cubic centimetre.

A metric (SI) measurement is not shown for box size, wire size, horsepower designation for motors, and other "trade sizes" that are nominal rather than exact measurements.

This system is international and many compromises were made to accommodate regional practice. In the U.S. the practice is to use the period as the decimal marker and the comma to separate a string of numbers into groups of three for easier reading. In many countries, the comma is used as the decimal marker and spaces are left to separate the string of numbers into groups of three. The SI system, taking something from both, uses the period as the decimal marker and the space to separate a string of numbers into groups of three starting from the decimal point and counting in either direction (e.g., 12 345.789 99). The only exception to this occurs when there are four numbers on either side of the decimal point in the string. In this case, the third and fourth numbers from the decimal point are not separated (e.g., 2015.1415).

In the SI system, the units increase or decrease in multiples of 10, 100, 1000, and so on. For instance, one megawatt (1 000 000 watts) is 1000 times greater than one kilowatt (1000 watts).

The metric prefixes indicate the factor by which the basic unit is multiplied. For example, adding the prefix kilo to watt forms the new name kilowatt, meaning 1000 watts. Refer to Fig. 1–8 for prefixes used in the SI system.

Certain of the prefixes shown in Fig. 1–8 have a preference in usage. These prefixes are mega, kilo, centi, milli, micro, and nano. Thus, commonly used multiples of a metre are a kilometre (1000 metres), a centimetre (0.01 metre), and a millimetre (0.001 metre).

The advantage of the SI system is that the prefixes allow measurements to be written with few zeros, which lessens the possibility of confusion. For example, a four-foot lamp is approximately 1200 millimetres, or 1.2 metres.

Some common measurements of length in the imperial system are shown with their SI equivalents in Fig. 1–9.

ELECTRICAL INSPECTION

The Canadian Electrical Code requires that a permit be obtained for the inspection of all work with respect to the installation, alteration, or repair of electrical equipment. It is the responsibility of the electrical contractor or person doing the work to obtain the permit. When the installation has passed inspection, the inspection authority will issue a connection permit (sometimes referred to as authorization for connection or current permit) to the supply authority. The supply authority will then make connection to the installation.

Symbol	SI Unit	Multiplying Factor for Conversion to Previously Used Unit	Previously Used Unit
A	ampere	1	ampere
cm^3	cubic centimetre	0.061	cubic inch
°	degree (angle)	1	degree
°C	degree Celsius (temperature)	1.8 plus 32	degree Fahrenheit
h	hour(s)	1	hour(s) (time)
min	minute (time)	1	minute
Hz	hertz	1	cycle per second
MHz	megahertz	1	megacycles per second
J	joule	0.7376	foot-pound
kg	kilogram	2.205	pound
kJ	kilojoule	737.6	foot-pound
km	kilometre	0.621	mile
kPa	kilopascal	0.295	inch of mercury
		0.334	foot of water
		0.145	pound per square inch (psi)
lx	lux	0.093	foot-candle
L	litre	0.220	gallon
m	metre	3.281	feet
m^2	square metre	10.764	square feet
m^3	cubic metre	35.315	cubic feet
mm	millimetre	0.039 37	inch
mm^2	square millimetre	0.001 55	square inch
Pa	pascal	0.000 295	inch of mercury
		0.000 334	foot of water
		0.000 145	pound per square inch (psi)
Ω	ohm	1	ohm
V	volt	1	volt
W	watt	1	watt
μF	microfarad	1	microfarad

Fig. 1–7 Symbols for SI units.

mega	1 000 000	(one million)
kilo	1 000	(one thousand)
hecto	100	(one hundred)
deka	10	(ten)
	1	(one)
deci	0.1	(one-tenth) (1/10)
centi	0.01	(one-hundredth) (1/100)
milli	0.001	(one-thousandth) (1/1 000)
micro	0.000 001	(one-millionth) (1/1 000 000)
nano	0.000 000 001	(one-billionth) (1/1 000 000 000)

Fig. 1–8 SI prefixes and their values.

one inch	=	2.54	centimetres
	=	25.4	millimetres
	=	0.025 4	metre
one foot	=	12	inches
	=	0.304 8	metre
	=	30.48	centimetres
	=	304.8	millimetres
one yard	=	3	feet
	=	36	inches
	=	0.914 4	metre
	=	914.4	millimetres
one metre	=	100	centimetres
	=	1 000	millimetres
	=	1.093	yards
	=	3.281	feet
	=	39.370	inches

Fig. 1–9 Conversion factors for some common measurements of length.

REVIEW

Note: Refer to the *Canadian Electrical Code, Part I* or the plans where necessary.

1. What section of the specification contains a list of contract documents?

2. The requirement for temporary light and power at the job site will be found in what portion of the specification?

3. The electrician uses the Schedule of Drawings for what purpose?

Complete the following items by indicating the letter(s) designating the correct source(s) of information for:

4. Room width _____
5. Grading elevations _____
6. Ceiling height _____
7. Panelboard schedules _____
8. Exterior wall finishes _____
9. View of interior wall _____
10. Electrical outlet location _____
11. Electrical receptacle style _____
12. Swing of door _____

a. Site plan
b. Architectural floor plan
c. Elevations
d. Details
e. Electrical layout drawings
f. Specification
g. Sections
h. Electrical symbol schedule

Match the acronyms on the left with the phrase or word that best relates to that organization, document, or person.

13. SCC _____
14. *C.E.C.* _____
15. ULC _____
16. PE _____
17. CSA _____

a. Accrediting organizations
b. Seal
c. Manufacturers' standards
d. Listing service
e. Electrical code

Write the appropriate letters (*a, b,* or *c*) to indicate the proper interpretation of the *C.E.C., Part I.*

18. Must be done _____
19. May be done _____
20. Up to the electrician _____
21. Is required _____
22. Is allowed _____

a. shall
b. special permission
c. not allowed
d. allowed

23. List the drawings that are normally included in an electrical drawing set.

_____ _____

_____ _____

_____ _____

_____ _____

24. List the steps to be followed when working with a set of drawings.

_____ _____

_____ _____

_____ _____

_____ _____

25. Measure the length of each line using the scale indicated.   Answer

a) $\frac{1}{8}" = 1$ ft   _____   _____

b) $\frac{1}{4}" = 1$ ft   _____   _____

c) $\frac{1}{2}" = 1$ ft   _____   _____

d) $1\frac{1}{2}" = 1$ ft   _____   _____

e) $\frac{3}{8}" = 1$ ft   _____   _____

f) $\frac{3}{4}" = 1$ ft   _____   _____

g) $\frac{1}{4}" = 1$ ft   _____   _____

h) $\frac{1}{8}" = 1$ ft   _____   _____

i) $\frac{1}{4}" = 1$ ft   _____   _____

j) $\frac{1}{2}" = 1$ ft   _____   _____

k) 1 : 100   _____   _____

l) 1 : 50   _____   _____

m) 1 : 25   _____   _____

n) 1 : 75   _____   _____

o) 1 : 50   _____   _____

p) 1 : 125   _____   _____

q) 1 : 100   _____   _____

r) 1 : 25   _____   _____

s) 1 : 50   _____   _____

t) 1 : 25   _____   _____

UNIT 2

The Electric Service

OBJECTIVES

After completing the study of this unit, the student will be able to

- install power transformers to meet the *C.E.C., Part I* requirements
- draw the basic transformer connection diagrams
- recognize different service types
- connect metering equipment
- apply ground-fault requirements to an installation
- install a grounding system

The installation of the electric service to a building requires the cooperation of the electrician, the local utility company, and, in some cases, the electrical inspector. The availability of high voltage and the power company requirements determine the type of service to be installed. This unit will investigate several common variations in electrical service installations. The consumer's service for this building begins with a 300KVA, 120/208 V 3PH 4 Wire pad mounted transformer. (See Plot Plan.)

Pad-Mounted Transformers

Liquid-insulated transformers as well as dry-type transformers are used for this type of installation. These transformers may be fed by an underground or overhead service, Fig. 2–1. The secondary normally enters the building through a bus duct or large cable.

Underground Vault

This type of service is used when available space is an important factor and an attractive site is desired. The metering may be at the utility pole or in the building, Fig. 2–5.

Unit Substation

For unit substation installations, the primary runs directly to the unit substation where all of the necessary equipment is located, Fig. 2–7. The utility company requires the building owner to buy the equipment for a unit substation installation.

LIQUID-FILLED TRANSFORMERS

Modern liquid-filled transformers are filled with mineral oil or synthetic cooling liquids. The liquid performs several important functions: (1) it

24 Unit 2 The Electric Service

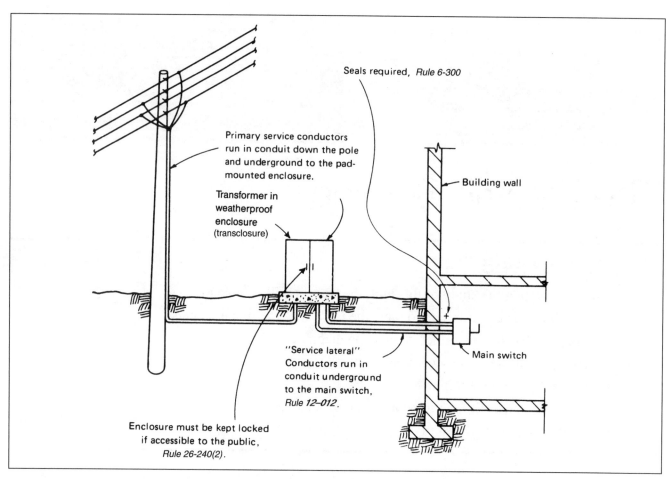

Fig. 2–1 Pad-mounted enclosure supplying underground service entrance.

Fig. 2–2 300 KVA, 120/208 V 3P 4 W pad mounted transformer.

is part of the required insulation dielectric, and (2) it acts as a coolant by conducting heat from the core and the winding of the transformer to the surface of the enclosing tank, which then is

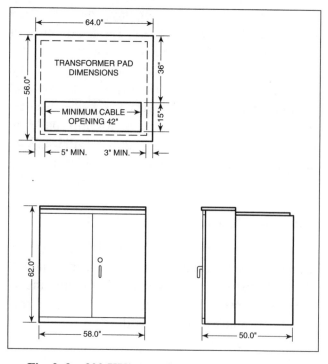

Fig. 2–3 300 KVA transformer pad dimensions.

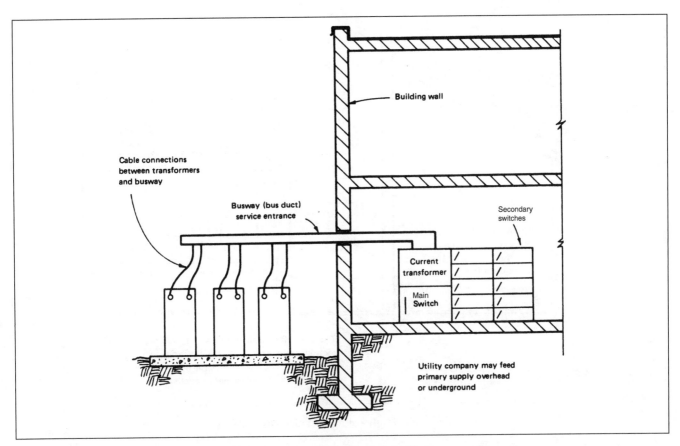

Fig. 2-4 Distribution transformers supplying bus duct service entrance.

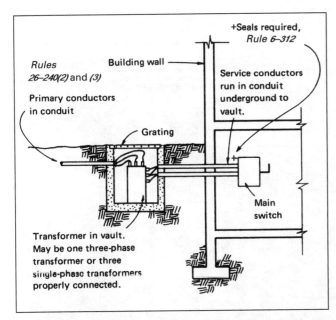

Fig. 2-5 Underground vault supplying underground service entrance.

cooled by radiation or fans. Many older transformers used askarel as a coolant since it is not flammable. Askarel is a trade name for oil containing PCBs. This type of oil is no longer permitted in applications such as transformers or ballasts. If it is found in an existing installation, approved steps regulated by the Ministry of the Environment must be taken to ensure containment of the area or the safe disposal of the oil.

DRY-TYPE TRANSFORMERS

Dry-type transformers are widely used because they are lighter than comparable liquid-filled transformers. Installation is simpler because there is no need to take precautions against liquid leaks.

Dry-type transformers are constructed so that the core and coil are open to allow for cooling by the free movement of air. Fans may be installed to increase the cooling. In this case, the transformer can be used at a greater load level. A typical dry-type transformer installation is shown in Fig. 2-7. An installation of this type is known as

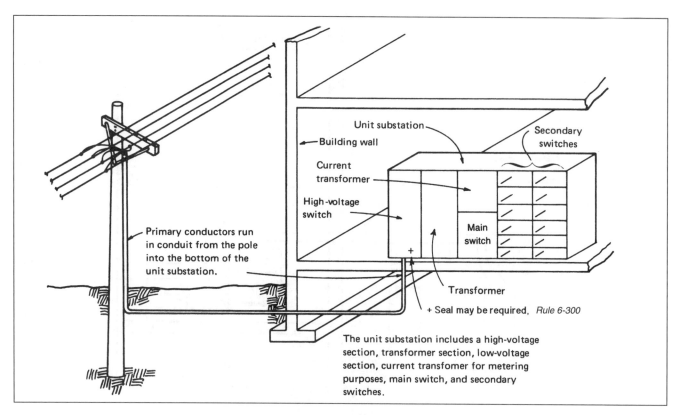

Fig. 2–6 High-voltage service entrance.

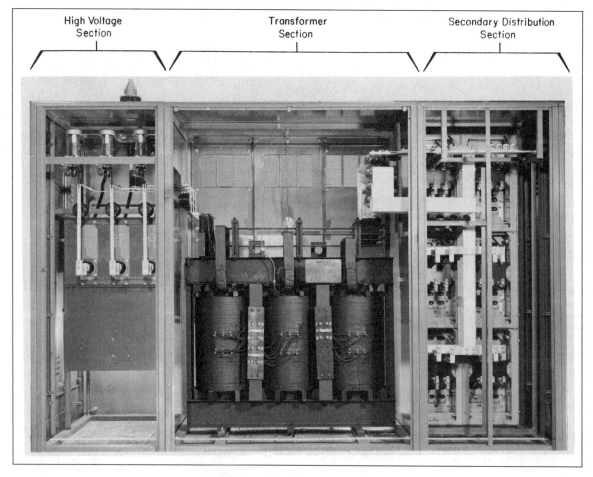

Fig. 2–7 A unit substation.

a *unit substation* and consists of three main components: (1) the high-voltage switch, (2) the dry-type transformer, and (3) the secondary distribution section.

TRANSFORMER OVERCURRENT PROTECTION

Rule 26–240 covers transformer installations in general; *Rules 26–252* and *26–254* group transformers into two voltage levels.

Requirements for the overcurrent protection of transformers rated over 750 volts are covered in *Rule 26–252* and *Table 50*.

Fig. 2–8 illustrates four of the more common situations found in commercial building transformer installations. The overcurrent devices as specified in *Rule 26–252* protect the transformer only. The conductors supplying or leaving the transformer may require additional overcurrent protection. (See Unit 17 of this text for information concerning fuses and circuit breakers.)

Fig. 2–9 shows one method of installing a dry-type transformer in a commercial or industrial building.

Rule 26–254 details the overcurrent protection for power and distribution transformers rated at 750 volts or less.

TRANSFORMER CONNECTIONS

A transformer is used in a commercial building primarily to change the transmission-line high voltage to the value specified for the building, such as 347/600Y or 120/208Y volts. A number of connection methods can be used to accomplish the changing of the voltage. The connection used depends upon the requirements of the building. The following paragraphs describe several of the more commonly used secondary connection methods.

Single-Phase System

Single-phase systems usually provide 120 and/or 240 volts with a two- or three-wire connection, Fig. 2–10. The centre tap of the transformer secondary shall be grounded in accordance with *Section 10*, as will be discussed later. Grounding is a vital safety measure and should be installed with great care.

Open Delta System

This connection scheme has the advantage of being able to provide either three-phase or three-phase and single-phase power using only two transformers. It is usually installed where there is a strong probability that the power requirement will increase, at which time a third transformer can be added. The open delta connection is illustrated in Fig. 2–11.

In an open delta transformer bank, 86.6% of the capacity of the transformers is available. For example, if each transformer in Fig. 2–11 has a 100-kVA rating, then the capacity of the bank is:

$$(100 \text{ kVA} + 100 \text{ kVA}) \times 86.6\% = 173 \text{ kVA}$$

Another way of determining the capacity of an open delta bank is to use 57.7% of the capacity of a full delta bank. The capacity of three 100-kVA transformers connected in full delta is 300 kVA. Thus, two 100-kVA transformers connected in open delta have a capacity of

$$300 \text{ kVA} \times 57.7\% = 173 \text{ kVA}$$

When an open delta transformer bank is to serve three-phase power loads only, the centre tap is not connected.

Four-Wire Delta System

This connection, also illustrated in Fig. 2–11, has the advantage of providing both three-phase and single-phase power from either an open (two transformers) or a closed (three transformers) delta system. A centre tap is brought out of one of the transformers, which is grounded and becomes the neutral conductor to a single-phase, three-wire power system. The voltage to ground from phases B and C, which are connected to the transformer that has been tapped, will be equal. These voltages are additive when measured phase to phase, for example, 120 volts to ground and 240 volts

Unit 2 The Electric Service

	Transformer rated impedance	Primary over 750 volts**	Secondary	Secondary
Transformers over 750 volts with primary and secondary protection. *Rule 26–252(4), Table 50.* **A**	not over 7.5%	• Max. fuse 300% • Max. breaker 600%	• Max. fuse 150% • Max. breaker 300%	• Max. fuse 250% • Max. beaker 250%
	over 7.5% but not over 10%	• Max. fuse 200% • Max. breaker 400%	• Max. fuse 125% • Max. breaker 250%	• Max. fuse 250% • Max. breaker 250%

	Primary over 750 volts**	Secondary
Transformers over 750 volts with primary protection only. **B**	• Maximum fuse—150%* • Maximum beaker—300%	• No overcurrent protection required
	* Where this percentage does not correspond to a standard fuse rating, the next higher standard rating or setting may be used.	

	Primary 750 volts or less	Secondary 750 volts or less
Transformers 750 volts and less with primary protection only. *Rules 26–254(1) and 26–256(1).* **C**	**Other than dry type** Maximum fuse size not to exceed 150% of the transformer's rated primary current, *Rule 26–254(I).* If the rated primary current is 9 amperes or more and 150% of this current does not correspond to a standard fuse or breaker size, then the next larger standard size may be used. For transformers having rated primary current of less than 9 amperes a fuse or nonadjustable-trip circuit breaker not to exceed 167% may be used, *Rule 26–254(2).* **Dry type** An individual overcurrent device is required. Maximum fuse or breaker size is not to exceed 125% of the transformer's rated primary current. Where 125% of the rated primary current does not correspond to a standard overcurrent device rating, the next higher standard rating may be used.	**Other than dry type** Transformer secondary protection is not required when the primary fuse does not exceed 150% of the rated primary current, *Rule 26–254(1).* **Dry type** Transformer secondary protection is not required when the primary fuse does not exceed 125% of the rated primary current, *Rule 26–256(1).* Transformer secondary protection is not required when the primary fuse does not exceed 125% of the rated primary current and the secondary conductors are rated at not less than 125% of the secondary current, *Rule 26-256(I).*

	Primary 750 volts or less	Secondary 750 volts or less
Transformers 750 volts and less with primary and secondary protection. *Rule 26–254(4) and 26–256(2).* **D**	**Other than dry type** Individual overcurrent protection is not required if the feeder overcurrent protection is not greater than 300% of the rated primary current and the secondary overcurrent protection is not greater than 125% of the rated secondary current. If the transformer is equipped with coordinated overload protection furnished by the manufacturer, individual primary overcurrent protection is not required if the primary feeder has overcurrent protection: • set not over six times primary current for transformers with not more than 7.5% impedance. • set not over four times primary current for transformers with more than 7.5% but not over 10% impedance. This may allow more than one transformer to be connected to one feeder. **Dry type** Individual overcuffent protection is not required if the feeder overcuffent protection is not greater than 300% of the rated primary current and the secondary overcurrent protection is not greater than 125% of the rated secondary current.	**Other than dry type** Maximum fuse or breaker is not to exceed 125% of the transformer's rated secondary current when the primary does not have individual overcurrent protection, *Rule 26-254(4).* For transformers having rated secondary current of 9 amperes or more, if the 125% sizing does not correspond to a standard size, then the next higher standard-size fuse or nonadjustable breaker may be used. For transformers having rated secondary current of less than 9 amperes, if the 125% sizing does not correspond to a standard size, then a fuse or breaker not to exceed 167% may be used. **Dry type** Maximum fuse or breaker is not to exceed 125% of the transformer's rated secondary current when the primary does not have individual overcurrent protection, *Rule 26-256(2).*
	**Individual primary overcurrent protection is not required if the primary feeder overcuffent device provides the required protection, *Rules 26-252(3) and 26-254(3).* This may allow more than one transformer to be connected to the same feeder.	

Fig. 2–8 Typical transformer overcurrent protection requirements.

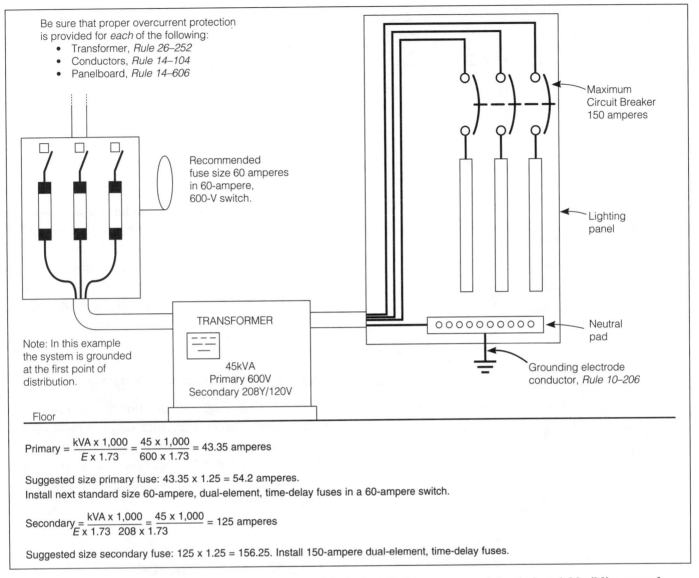

Fig. 2–9 Diagram of how a dry-type transformer might be installed in a commercial or industrial building served by 600 volts, three phase, where it is necessary to step down the voltage to 120/208 volts, three phase, four wire to supply lighting panels.

between phases. The voltage measured between the grounded tap and the A phase will be higher than 120 volts and lower than 240. This phase is called the *high leg* and cannot be used for lighting purposes. This high leg must be identified by using a red coloured conductor or by clearly marking it as the A phase conductor whenever it is present in a box or cabinet with the neutral of the system. See *Rule 4–036(4,5)*. This A phase is to be connected to the centre bus bar in panelboards and switchboards.

Three-Wire Delta System

This connection, illustrated in Fig. 2–12, provides only three-phase power. One phase of the system may be grounded, in which case it is often referred to as corner grounded delta. The power delivered and the voltages measured between phases remain unchanged. Overcurrent devices are not to be installed in the grounded phase. See *Rule 14–010*.

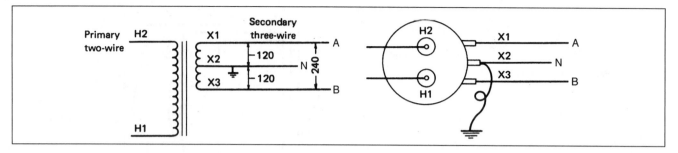

Fig. 2–10 Single-phase transformer connection.

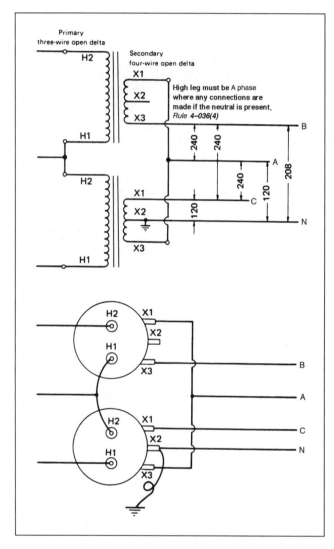

Fig. 2–11 Three-phase open delta connection for light and power.

wye (Y), Fig. 2–13. This system has the advantage of being able to provide three-phase power, and also permits lighting to be connected between any of the three phases and the neutral. Typical voltages available with this type of system are 120/208, 277/480, and 347/600. In each case, the transformer connections are the same.

Notes to All Connection Diagrams

All connection diagrams are shown with additive polarity and standard angular displacement. All three-phase connections are shown with the primary connected in delta. For other connections, it is recommended that qualified engineering assistance be obtained.

THE SERVICE ENTRANCE

The regulations governing the method of bringing the electric power into a building are established by the local utility company. These regulations vary considerably among utility companies. Several of the more common methods of installing the service entrance are shown in Fig. 2–1 and Figs. 2–4 through 2–6.

METERING

The electrician working on commercial installations seldom makes metering connections. However, the electrician should be familiar with the following two basic methods of metering.

Three-Phase, Four-Wire Wye System

The most commonly used system for modern commercial buildings is the three-phase, four-wire

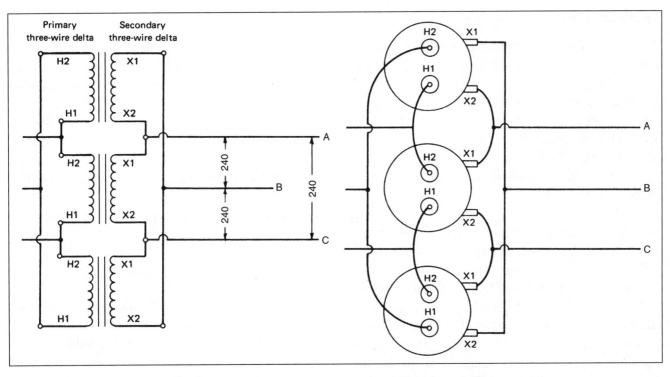

Fig. 2–12 Three-phase, delta-delta transformer connection.

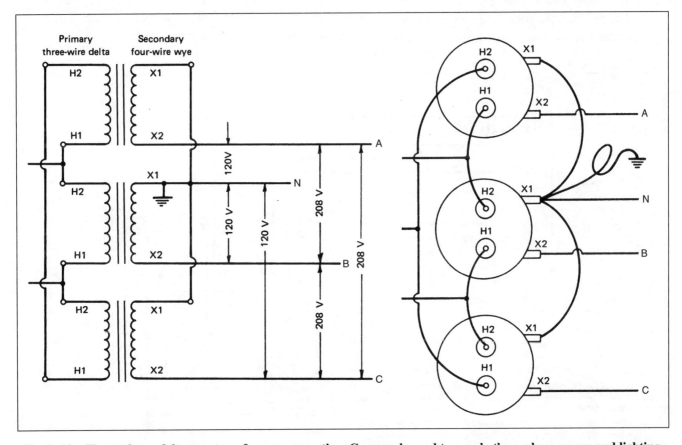

Fig. 2–13 Three-phase, delta-wye transformer connection. Commonly used to supply three-phase power and lighting.

High-Voltage Metering

When a commercial building is occupied by a single tenant, the utility may elect to meter the high-voltage side of the transformer. To accomplish this, a potential transformer and two current transformers are installed on the high-voltage lines and the leads are brought to the meter as shown in Fig. 2–14.

The left-hand meter socket in the illustration is connected to receive a standard socket-type watt-hour meter; the right-hand meter socket will receive in varhour meter (volt-ampere reactive meter). The two meters are provided with 15-minute demand attachments that register kilowatt (kW) and kilovolt-ampere reactive (kVAR) values respectively. These demand attachments will indicate the maximum usage of electrical energy for a 15-minute period during the interval between the readings made by the utility company. The rates charged by the utility company for electrical energy are based on the maximum demand and the power factor as determined from the two meters. A high demand or a low power factor will result in higher rates.

Low-Voltage Metering

Low-voltage metering of loads greater than 200 amperes is accomplished in the same manner as high-voltage metering. In other words, potential and current transformers are used. For loads of 200 amperes or less, the feed wires from the primary supply are run directly to the meter socket, Fig. 2–15. For multiple occupancy buildings, such as the commercial building discussed in this text, the meters are usually installed as a part of service-entrance equipment called a *switchboard*.

SERVICE-ENTRANCE EQUIPMENT

When the transformer is installed at a location far from the building, the service-entrance equipment consists of the service-entrance conductors, the main switch or switches, the metering equipment, and the secondary distribution switches, Fig. 2–16. The commercial building shown in the plans is equipped in this manner and will be used as an example for the following paragraphs. Table 2–1 provides the main service loading schedule.

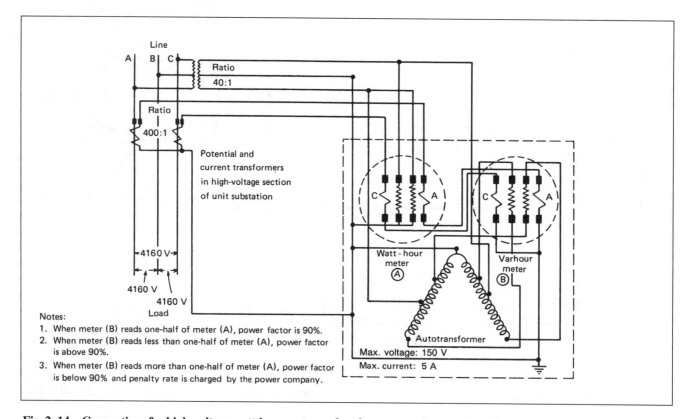

Fig. 2–14 Connections for high-voltage watt-hour meter and varhour meter. Demand attachments and timers are not shown.

Fig. 2–15   Meter socket.

Table 2–1   Main Service Loading Schedule

Tenant Area	Demand on Service Conductor
Owner's area	11 000 W
Bakery	43 435 W
Drug store	20 529 W
Insurance office	32 755 W
Beauty salon	21 825 W
Doctor's office	14 234 W
Subtotal	143 778 W
Boiler	150 000 W
Total	293 778 W

$$I = \frac{P}{E \times 1.73} = \frac{293\,778}{208 \times 1.73} = 825 \text{ A.}$$

Using standard-rated equipment and conductor ampacities based on *Table 2, C.E.C.* the minimum ampacity of the service conductors would be: $825 \times 1.25 = 1031$ A.

Using three conductors per phase: $\frac{1031}{3} = 344$ A.

The minimum conductor size from *Table 2* would be 400 kcmil.

Note: The conductors supplying this building will be based on the requirements for the secondary of the transformer supplying the building and will therefore be larger than required by this service calculation.

The Service

The service for the commercial building is similar to the service shown in Fig. 2–1 (page 24). A pad-mounted, three-phase transformer is located outside the building and rigid conduit running underground serves as the service raceway.

It will also be helpful to examine Fig. 2–16, Fig. 2–17, and the riser diagram on sheet E1 of the plans for the commercial building.

SERVICE DISCONNECTING MEANS

The *C.E.C., Part I* permits as many as four disconnect switches for services, *Rule 6–104*, unless special permission is given for more, *Rule 2–030*. The sum of the overcurrent protection installed in the switches may exceed the ampacity of the service-entrance conductors.

The service lateral is tapped into the main enclosure to feed one 800-ampere and one 600-ampere fusible main switch. These switches are shown in Figs. 2–17, 2–18A, and 2–18B. The 800-ampere main switch serves the boiler. The 600-ampere switch serves the secondary distribution switches consisting of two 200-ampere switches, four 100-ampere switches, and one 60-ampere switch.

The metering for the service entrance of this commercial installation is located in the third section of the main service-entrance equipment. The meters for six of the occupancies in the building are installed directly in the line. The seventh meter and current transformers are installed to measure the load to the boiler. The service equipment can be purchased in a single enclosure, such as the enclosure illustrated for the commercial building in this text. The contractor and electrician should consult the *C.E.C., Part I* and local regulations, and obtain the supply authority's approval of location and installation requirements before installing any service, *Rule 6–206*.

GROUNDING

Why is grounding necessary? Electrical systems and their associated conductors are grounded to keep voltage spikes to a minimum should lightning strike or other line disturbances occur. Grounding stabilizes normal voltage to ground. By bonding equipment grounding conductors to the grounding conductor of the system, the low impedance path for fault current will facilitate the

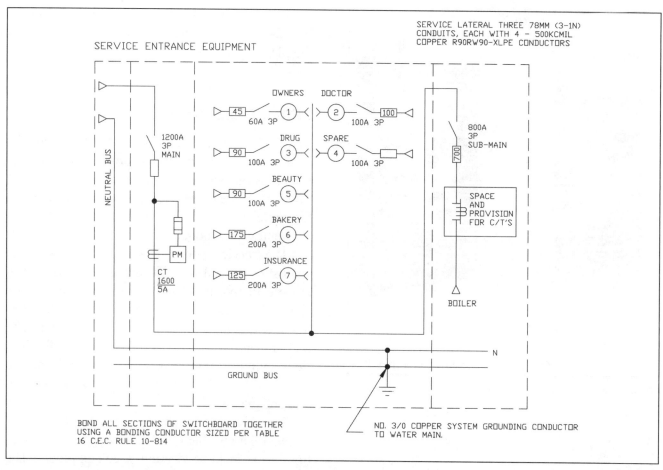

Fig. 2–16 Commercial building service-entrance equipment, one line diagram. (Courtesy of Erickson Electrical Equipment Co.)

operation of overcurrent devices under ground-fault conditions. To "facilitate" means to "make easier."

Electrical metallic conduits and other electrical equipment are grounded to keep the voltage to ground to a minimum, thus reducing the electrical shock hazard.

It is very important that the ground connections and the grounding system be properly installed. To achieve the best possible grounding system, the electrician must use the recommended procedures when installing the proper equipment. Fig. 2–19 illustrates some of the terminology used in *C.E.C., Section 10, Grounding and Bonding.*

Properly grounded electrical systems and enclosures provide a good path for ground currents if a ground fault occurs—for example, if insulation fails. The reason for this is that the lower the impedance (ac resistance to the current) of the grounding path, the greater the ground-fault current. This increased ground-fault current causes the overcurrent device protecting the circuit to respond faster. This is called *inverse time*, which means that the higher the current, the less time it will take to operate the overcurrent device.

For example, with 208 volts at 0.1 Ω impedance:

$I = E/Z = 208/0.1 = 2080$ amperes

but with 208 volts at 0.01 Ω impedance:

$I = E/Z = 208/0.01 = 20\ 800$ amperes

When an overcurrent device opens the circuit faster, less equipment damage results.

The amount of ground-fault damage to electrical equipment is related to (1) the response time of the overcurrent device and (2) the amount of current. One common term used to relate the time and current to the ground-fault damage is *ampere-squared seconds (I^2t):*

I^2t = amperes × amperes × time in seconds

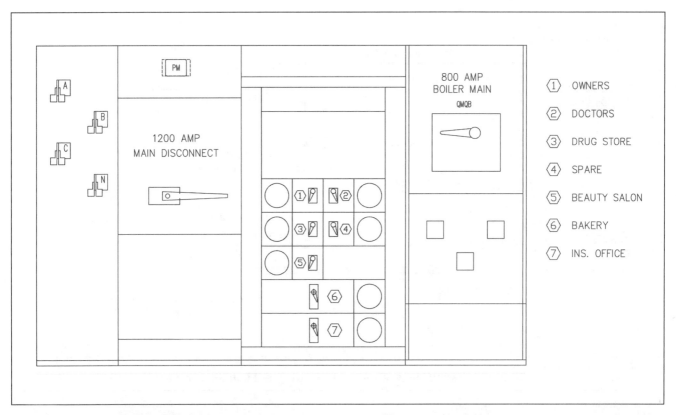

Fig. 2–17 Commercial building service-entrance equipment, pictorial view. (Courtesy of Erickson Electrical Equipment Co.)

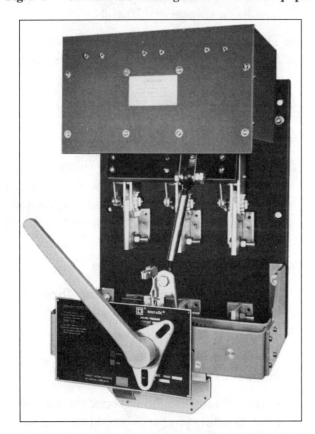

Fig. 2–18A Bolted pressure contact switch for use with Class J, Class L, or Class T high-capacity fuses. The service for the commercial building in the plans has two of these switches as the main disconnecting means.

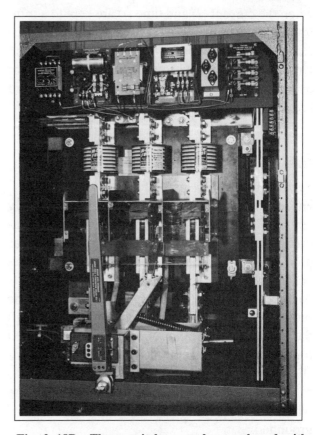

Fig. 2–18B These switches can be purchased with GFP, shunt tripping, auxiliary contacts, phase failure relays, and other optional attachments. Switch illustrated has GFP, phase failure relay, shunt tripping, and antisingle-phasing blown fuse indicator.

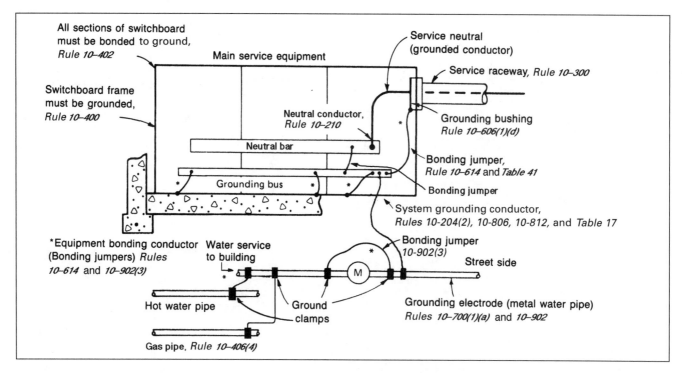

Fig. 2–19 Terminology of service grounding and bonding, *Section 10*.

It can be seen in the expression for I^2t that when the current (I) and time (t) are kept to a minimum, a low value of I^2t results. Lower values of I^2t mean that less ground-fault damage will occur. Units 17, 18, and 19 provide detailed coverage of overcurrent protective devices, fuses, and circuit breakers.

System Grounding

System grounding calls for grounding of the entire system rather than a single item. By "system" is meant, for example, the service neutral conductor, hot and cold water pipes, gas pipes, service-entrance equipment, and jumpers around water meters. If any of these system parts become disconnected or open in any way, the integrity of the system is still maintained through the other paths. In other words, everything is tied together. The concept of system grounding is shown in Fig. 2–20.

Fig. 2–20 and the following steps illustrate what can happen if an entire system is *not* grounded.

1. Bonding jumper A is not installed originally.

2. A live wire contacts the gas pipe. The gas pipe now has a voltage of 120 volts.

3. The insulating joint in the gas pipe results in a poor path to ground; assume the resistance is 8 ohms.

4. The 20-ampere fuse does not blow because:

   $I = E/R = 120/8 = 15$ amperes

5. If a person touches the "hot" (live) gas pipe and the water pipe at the same time, current flows through the person's body. If the body resistance is 12 000 ohms, then the current is:

   $I = E/R = 120/12\,000 = 0.01$ ampere

   This value of current passing through a human body can cause death.

6. The fuse is now "seeing" 15 amperes + 0.01 ampere = 15.01 amperes, so it still does not blow.

7. If the system grounding concept had been used, bonding jumper A would have kept the voltage difference between the water pipe and the gas pipe at zero, *Rule 10–406(4)*.

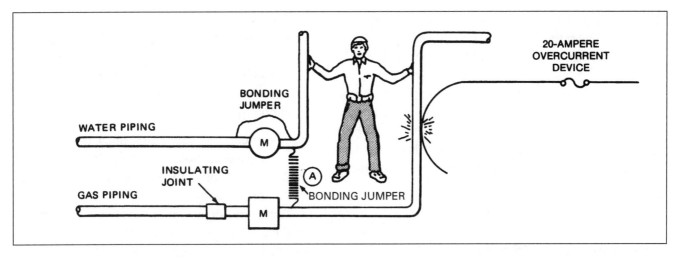

Fig. 2–20   System grounding.

Thus, the fuse would blow. If 3.05 m (10 ft) of No. 4 copper wire were used as the jumper, then the resistance of the jumper would be 0.003 08 ohms. The current is:

$I = E/R = 120/0.003\,08 = 38\,961$ amperes

(In an actual system, the impedance of all of the parts of the circuit would be much higher. Thus, a much lower current would result, but it would be enough to cause the fuse to blow.)

Advantages of System Grounding

The advantages of system grounding are set out in *Rule 10–002,* which gives the objectives of grounding and bonding.

These objectives can be briefly summarized as follows:

- to protect life from the danger of electric shock;
- to limit the voltage upon a circuit;
- to facilitate the operation of electrical apparatus and systems;
- to limit voltage on a circuit when exposed to lightning
- to limit ac circuit voltages to ground to 150 volts or less on circuits supplying interior wiring systems;
- to keep the potential voltage differentials between the different parts of the system at a minimum, thereby reducing shock hazard;
- to keep impedance of a ground path at a minimum, which results in higher current flow in the event of a ground fault. The lower the impedance, the greater is the current flow, and the faster the overcurrent device opens.

In Figs. 2–16 and 2–19, the main service equipment, the service raceways, the neutral bus, the grounding bus, and the hot and cold water pipes have been bonded together to form a *grounding electrode system.*

In the commercial building, at least 6.1 m (20 ft) of No. 3/0 bare copper conductor is installed in the footing to serve as an additional supplemental electrode. This conductor, buried in the concrete footing, is supplemental to the water pipe ground. Many interpretations of the *C.E.C., Part I* are possible concerning the grounding electrode systems concept. Therefore, the local inspection authority should be consulted. For instance, some electrical inspectors may not require the bonding jumper between the hot and cold water pipes shown in Fig. 2–19. They may determine that an adequate bond is provided through either type of water heater (electric or gas). Other electrical inspectors may require that the hot and cold water pipes be bonded together because some water heaters contain insulating fittings that are intended to reduce corrosion inside the tank caused by electrolysis. According to their reasoning, even though the water heater originally installed contains no fittings of this type, such fittings may be included in a future replacement heater, thereby necessitating a bond between the cold and hot water pipes. The local electrical inspector should be consulted for the proper requirements. However, bonding the pipes together, as shown in Fig. 2–19, is the recommended procedure.

Grounding Terms (*Section 0*)

Section 0 provides a definition for the various terms associated with grounding and bonding electrical systems.

Bonding: means a low-impedance path obtained by permanently joining all noncurrent carrying metal parts to assure electrical continuity. Any current likely to be imposed must be conducted safely.

Grounding: means a permanent path to earth with sufficient ampacity to carry any fault current liable to be imposed on it.

Bonding Conductor: means a conductor that connects the noncurrent-carrying parts of electrical equipment to the service equipment or grounding conductor.

Grounding Conductor: means the conductor used to connect the service equipment to the grounding electrode.

Grounding Electrode: means a buried metal water piping system, or metal rod, plate, or other means buried or driven into the ground so that it is in contact electrically with the ground.

Grounding Rules (*Section 10*)

When grounding service-entrance equipment, the following *C.E.C., Part I* rules must be observed:

- The system must be grounded when maximum voltage to ground does not exceed 150 volts, *Rule 10–106(1)(a)*.

- The system must be grounded if a neutral is used as a circuit conductor (for example, a 347/600-volt, wye-connected, three-phase, four-wire system), *Rule 10–106(1)(b)*. This is similar to the connection of the system shown in Fig. 2–13.

- The identified conductor must be grounded where the midpoint of one transformer is used to establish the grounded neutral circuit conductor, as on a 120/240-volt, three-phase, four-wire delta system, or a single-phase three-wire system, *Rule 10–210(1)*. Only one phase shall be grounded, *Rule 10–210(2)*. Refer to Fig. 2–11.

- *Rule 10–500* demands that an effective grounding path be established. The path to ground from all circuits, equipment, and metal enclosures (1) must be permanent and continuous, (2) must have adequate capacity to safely conduct any fault current that it might be called upon to carry, and (3) must have impedance low enough to limit the voltage to ground and to facilitate operation of the overcurrent device. Details of this subject are discussed in Units 17, 18, and 19.

- All grounding schemes shall be installed so that no objectionable currents will flow over the grounding conductors and other grounding paths, *Rule 10–200(1)*.

- The grounding electrode conductor must be connected to the supply side of the service disconnecting means, *Rule 10–204(1)(b)*. It must not be connected to any grounded circuit conductor on the load side of the service disconnect, *Rule 10–204(1)(d)*.

- The identified neutral conductor is the conductor that must be grounded, *Rule 10–210(1)*.

- Tie (bond) exposed noncurrent-carrying parts of the system together, *Rules 10–400, 10–402,* and *10–406*.

- The grounding conductor used to connect the grounded neutral conductor to the grounding electrode must not be spliced, *Rule 10–806(1)*. There are exceptions for installations where the grounding electrode conductor may be spliced by either exothermic welding or irreversible compression-type connectors that are listed for that purpose.

- The grounding conductor is to be sized according to *Table 17, Rule 10–812(a)*.
- The minimum conductor size for concrete-encased electrodes is found in *Table 43, Rule 10–702(2)*.
- The metal water supply system shall be bonded to the grounding conductor of the electrical system. *Rule 10–406(2), App. B.*
- The grounding conductor may be connected to a metal water pipe that is run underground for at least 3 metres beyond the building, or to artificial electrode, *Rule 10–700(1)*.
- The grounding conductor must be a minimum of No. 6 copper if the run is exposed and without metal covering or protection. This wire must be free from exposure to mechanical injury, *Rule 10–806(2)*.
- The grounding conductor may be uninsulated in most installations. See *Rule 10–806*.
- Bonding shall be provided around all insulating joints or sections of the metal piping system that may be disconnected, *Rules 10–616(1)(d), 10–902(3), and 10–908*.
- The connection to the grounding electrode must be accessible, *Rule 10–902(2)*.
- The grounding conductor must be connected tightly using the proper lugs, connectors, clamps, or other approved means, *Rule 10–908*.

In Fig. 2–21 the three most common types of artificial ground electrodes are illustrated. These are connected to the neutral in the same manner as a system grounding conductor from the public water main.

The first 1.5 metres of metal water piping entering the building is **not** considered to be "interior." The proper location to connect the grounding electrode conductor, the bonding conductors associated with the metal framing members, concrete-encased electrodes, and ground ring is anywhere on the first 1.5 metres of metal water piping after it enters the building. The first 1.5 metres may include a water meter.

- *Rule 10–700* does not permit using a metal underground gas piping system as the grounding electrode. However, where metal gas piping comes into a building, it must be bonded to all of the other metal piping that is required to be bonded together, *Rule 10–406(4)*.

BONDING

Rule 10–604 lists the parts of the service-entrance equipment that must be bonded. *Rule 10–606* lists the methods approved for bonding this service equipment. Grounding bushings, Fig. 2–22, and bonding jumpers are installed on service-entrance equipment to ensure a low-impedance path to ground if a fault occurs on any of the service-entrance conductors, *Rule 10–606(1)*. Service-entrance conductors are not fused at the service head. Thus, the short-circuit current on these conductors is limited only by the following: (1) the capacity of the transformer or transformers supplying the service equipment, and (2) the distance between the service equipment, and the transformers. The short-circuit current can easily reach 40 000 to 50 000 amperes or more in commercial buildings. These installations are usually served by a large-capacity transformer located close to the service-entrance equipment and the metering. **This extremely high fault current produces severe arcing, which is a fire hazard.** The use of proper bonding reduces this hazard to some extent.

Rule 10–614(1)(b) states that the main bonding jumpers must have an ampacity not less than the grounding electrode conductor. The grounding lugs on grounding bushings are sized by the trade size of the bushing. The lugs become larger as the size of the bushing increases.

The conductor may be sized according to *Table 41* if used in conjunction with two locknuts and a grounding bushing. *Rule 12–906(2)* states that if No. 8 AWG or larger conductors are installed in a raceway, an insulating bushing or equivalent must be used, Fig. 2–23. This bushing protects the wire from shorting to ground as it passes through the metal bushing. Combination bushings can be used. These bushings are metallic (for mechanical strength) and have plastic insulation. When conductors are installed in electrical metallic tubing, they can be protected at the fittings by the use of connectors with insulated throats.

If the conduit bushing is made of insulating material only, as in Fig. 2–23, then two locknuts must be used, Fig. 2–24. *Rule 10–610(b)*.

Various types of ground clamps are shown in Figs. 2–25 through 2–28. These clamps and their connection to the grounding electrode must conform to *Rule 10–908*.

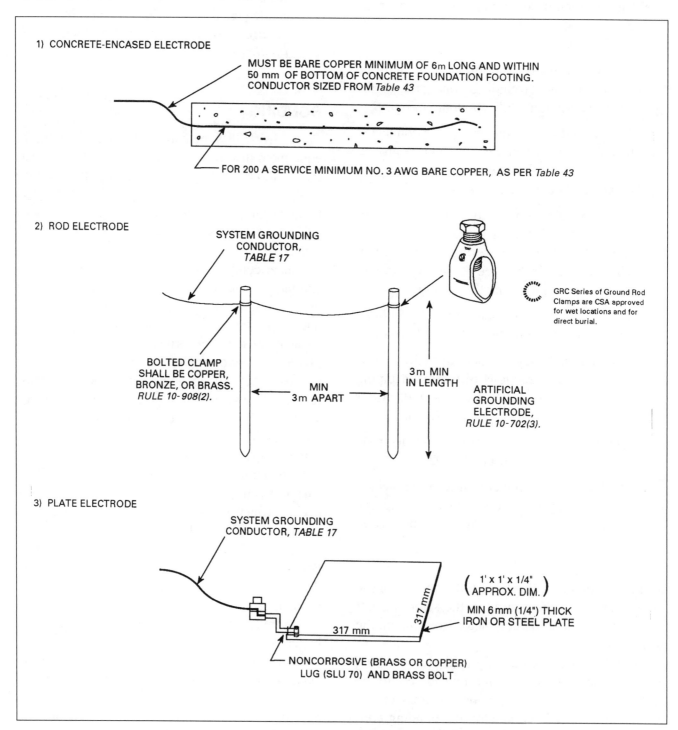

Fig. 2–21 Three common types of artificial grounding electrodes.

Fig. 2–22 Insulated ground bushing with bonding lug.

Fig. 2–23 Insulated plastic bushings.

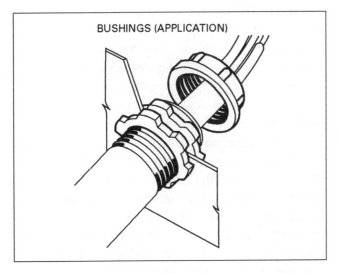

Fig. 2–24 The use of locknuts. See *Rule 10–610(b)*.

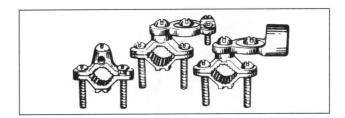

Fig. 2–25 Typical ground clamps.

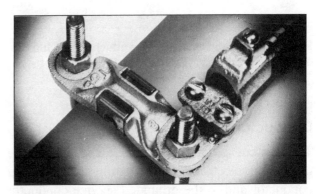

Fig. 2–26 Grounding conductor connection to water pipe.

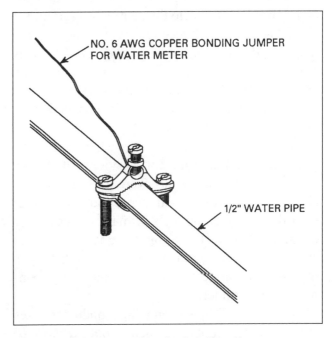

Fig. 2–27 Ground clamp for bonding jumper connection at water meter.

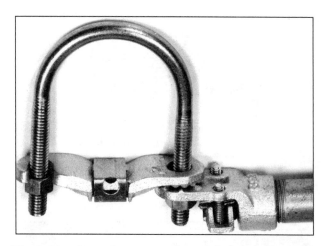

Fig. 2–28 Ground clamp used to attach grounding conductor to well casing. (Courtesy of Thomas & Betts.)

Sizing the System Grounding Conductors

The system grounding conductor connects the grounding electrode to the equipment bonding conductor and to the grounded service conductor (neutral) on the supply side of the service disconnecting means. In the commercial building, this means that the system grounding conductor connects the main water pipe (grounding electrode) to the ground bus (equipment bonding conductor connection point) and to the neutral bus (grounded circuit conductor connection point) **only** on the line side of the main service. *Rule 10–516* does not permit the grounded circuit conductor to be used for bonding equipment or raceways to ground on the load side of the connection to ground. See Fig. 2–19 (page 36).

Table 17 is referred to when selecting grounding conductors for services of an ac system, where there is no overcurrent protection ahead of the service-entrance conductors other than the utility company's primary or secondary overcurrent protection. This table is also used to size the ground conductor when establishing a new service on the secondary side of a wye-connected transformer.

In *Table 17,* the grounding conductor sizes are given for copper conductors only. *Tables 16* and *17* are based upon the fact that bare copper conductors and their bolted connections can withstand

- one ampere
- for five seconds
- for every 30 circular mils (0.0152 mm^2).

Insulated (75°C) copper conductors can withstand

- one ampere
- for five seconds
- for every 42.25 circular mils (0.0214 mm^2).

When bare conductors are in the same raceway or cable as insulated conductors, always apply the insulated conductor withstand limitations.

To size a grounding electrode conductor for a service with three parallel 500-kcmil service conductors, it is necessary to total the current carrying capacity of the three conductors and select a grounding electrode conductor for an equivalent-sized service conductor. The size of the grounding conductor is based on the total ampacity of the paralleled conductors, taking into consideration the number of conductors and the type of insulation. In this case, the grounding electrode conductor is No. 3/0 copper (see Fig. 2–16). Fig. 2–29 shows another example.

The main bonding jumper, see Figs. 2–30 and 2–31, connects the ground bushings of the service conduit to the grounding bus bar in the service disconnect. The jumper is sized according to *Table 41*. The service to the commercial building consists of three 500-kcmil Type RW90 XLPE conductors in parallel. They have a total current carrying capacity of 1185 amperes. In *Table 41*, this requires a No. 3/0 bonding jumper.

According to *Rule 4–022(3)*, the minimum size of a neutral conductor is based on the maximum unbalanced current or the size of the grounding conductor required for the system. This will insure that the neutral will be able to carry any fault current that might be imposed on it.

When parallel conductors are used, *Rule 10–204(3)* requires a neutral to be run in each raceway that contains the conductors. Based on a computed neutral load of 333 amps, the required neutral conductor for three runs in parallel would be 333 / 3 = 111 amps. If a RW75 conductor is

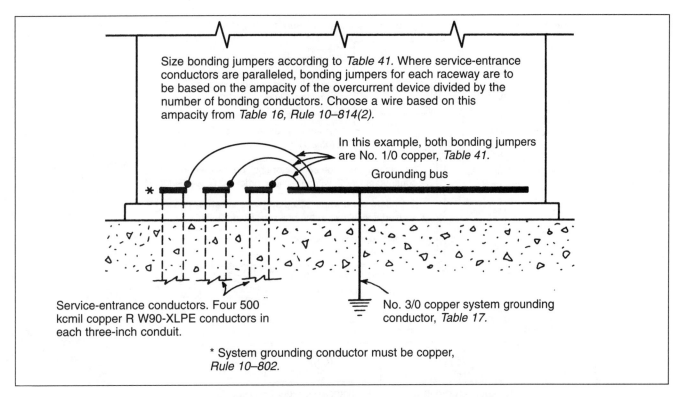

Fig. 2–29 Typical service entrance (grounding and bonding).

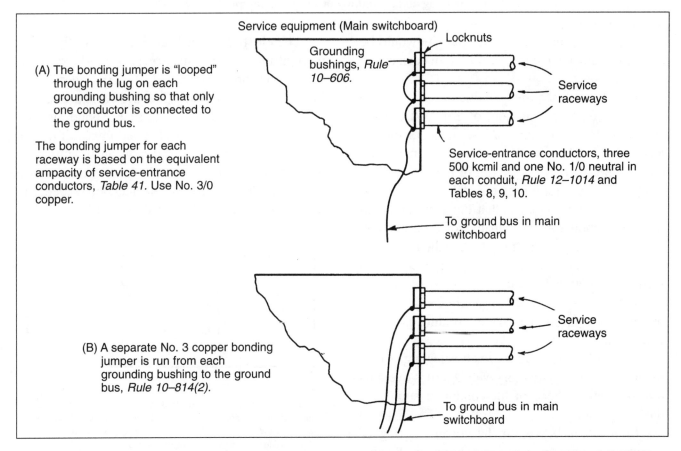

Fig. 2–30 Sizing of main bonding jumper for the service on the commercial building, *Rule 10–616* and *Table 41*.

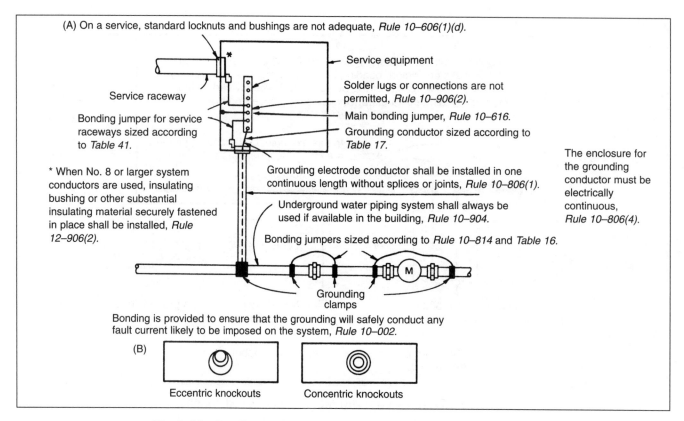

Fig. 2–31 Bonding and grounding of service equipment, *Section 10*.

used, *Table 2* indicates that No. 2 AWG would be the minimum size neutral permitted in each of the three raceways.

Rule 12–108 requires conductors that are run in parallel to be a minimum of 1/0 AWG. However, for service conductors in metal raceways, *Rule 10–204* permits neutral (grounded) conductors smaller than 1/0 to be run in parallel as long as they meet the requirements of *Rule 4–022(3)*.

Rule 10–500 demands that the grounding and bonding conductors be capable of carrying any fault current that they might be called upon to carry under fault conditions. This subject is covered in much greater detail in Units 17, 18, and 19.

Refer to Figs. 2–29 through 2–32 for the proper procedures to be used in bonding electrical equipment.

Table 16 is referred to when selecting equipment grounding conductors when there is overcurrent protection ahead of the conductor supplying the equipment.

Table 16 is based on the current setting or rating, in amperes, of the overcurrent device installed ahead of the equipment (other than service-entrance equipment) being supplied.

The electric boiler branch circuit in the commercial building consists of two 500-kcmil conductors per phase protected by 700-ampere fuses in the main switchboard. The use of flexible metal conduit at the boiler means that a bonding wire must be run through each conduit as shown in Fig. 2–33. A detailed discussion on flexible connections is given in Unit 11. An illustration of a typical application for the electric boiler feeder in the commercial building is shown in Fig. 2–31.

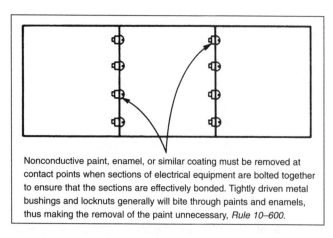

Fig. 2–32 Ensuring bonding when electrical equipment is bolted together.

GROUND-FAULT PROTECTION (*RULE 14–102*)

The *C.E.C., Part I* requires the use of ground-fault protection (GFP) devices on services that meet the conditions outlined in *Rule 14–102*. Thus, GFP devices are installed

- on solidly grounded wye services above 150 volts to ground, not over 750 volts between phases, that are rated at 1000 amperes or more (for example, on 347/600-volt systems)
- on solidly grounded wye services below 150 volts to ground rated at 2000 amperes or more (for example, on 120/208-volt systems)
- to operate at 1200 amperes or less
- so that the maximum time of opening the service switch or circuit breaker does not exceed one (1) second for ground-fault currents of 3000 amperes or more
- to limit damage to equipment and conductors on the *load side* of the service disconnecting means (GFP will *not* protect against damage caused by faults occurring on the *line side* of the service disconnect.)

When a fuse–switch combination serves as the service disconnect, the fuses must have adequate interrupting capacity to interrupt the available fault current (*Rule 14–012*), and must be capable of opening any fault currents exceeding the interrupting rating of the switch during any time that the GFP system will not cause the switch to open.

GFP is not required on

- delta-connected three-phase systems
- ungrounded, wye-connected three-phase systems
- 120/240-volt single-phase systems rated at less than 2000 amperes
- 120/208-volt three-phase four-wire systems rated at less than 2000 amperes
- systems over 750 volts, for example, 2400/4160 volts
- service disconnecting means rated at less than 1000 amperes

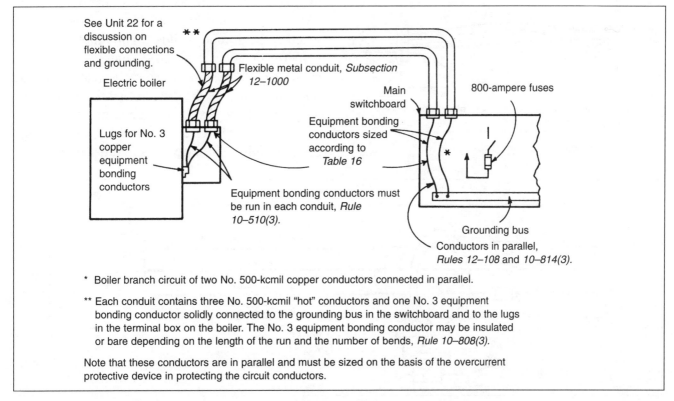

Fig. 2–33 Sizing bonding conductors for the electric boiler feeder. Note that these conductors are in parallel and must be sized on the basis of the overcurrent protective device protecting the circuit conductors.

The time of operation as well as the ampere setting of the GFP device must be considered carefully to ensure that the continuity of the electrical service is maintained. The time of operation of the device includes: (1) the sensing of a ground fault by the GFP monitor, (2) the monitor signalling the disconnect switch to open, and (3) the actual opening of the contacts of the disconnect device (either a switch or a circuit breaker). The total time of operation may result in a time lapse of several cycles or more. (See Units 17 and 18).

GFP circuit devices were developed to overcome a major problem in circuit protection: the low-value phase-to-ground arcing fault, Fig. 2–34. The amount of current that flows in an arcing phase-to-ground fault can be low when compared to the rating or setting of the overcurrent device. For example, an arcing fault can generate a current flow of 600 amperes. Yet, a main breaker rated at 1600 amperes will allow this current to flow without tripping, since the 600-ampere current appears to be just another *load* current. The operation of the GFP device assumes that under normal conditions the total instantaneous current in all of the conductors of a circuit will exactly balance, Fig. 2–35. Thus, if a current coil is installed so that all of the circuit conductors run through it, the normal current measured by the coil will be zero. If a ground fault occurs, some current will return through the grounding system and an unbalance in the conductors will result. This unbalance is then detected by the GFP device, Fig. 2–36.

The purpose of GFP devices is to sense and protect equipment against *low-level ground faults*. GFP monitors do not sense phase-to-phase faults, three-phase faults, or phase-to-neutral faults. These monitors are designed to sense phase-to-ground faults only.

Large (high magnitude) ground-fault currents can cause destructive damage even though a GFP device is installed. The amount of arcing damage depends upon (1) how much current flows and (2) the length of time that the current exists. For example, if a GFP device is set for a ground fault of 500 amperes and the time setting is six cycles, then the device will need six cycles to signal the switch or circuit breaker to open the circuit, whether the ground fault is slightly more than 500 amperes or as large as 20 000 amperes. The six cycles needed to signal the circuit breaker plus the operation time of the switch or breaker may be long enough to permit damage to the switchgear.

The damaging effects of high-magnitude ground faults, phase-to-phase faults, three-phase

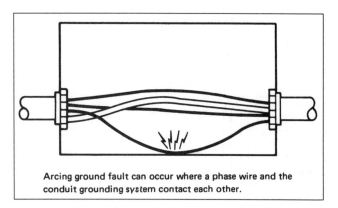

Arcing ground fault can occur where a phase wire and the conduit grounding system contact each other.

Fig. 2–34 **Ground faults.**

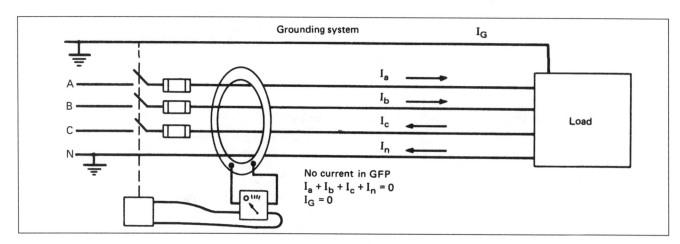

Fig. 2–35 **Normal condition.**

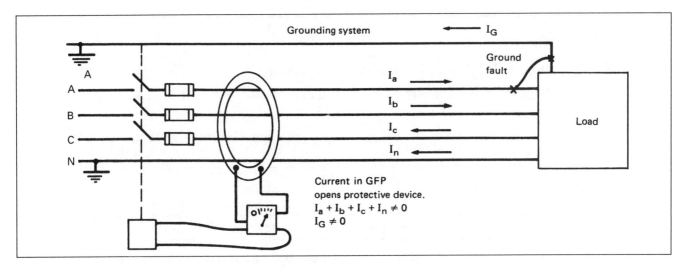

Fig. 2-36 Abnormal condition.

faults, and phase-to-neutral faults can be reduced substantially by the use of current-limiting overcurrent devices. These devices reduce both the peak let-through current and the time of opening once the current is sensed. For example, a ground fault of 20 000 amperes will open a current-limiting fuse in less than one-half cycle. In addition, the peak let-through current is reduced to a value much less than 20 000 amperes (see Units 17 and 18).

GFP for equipment, *Rule 14-102*, is not to be confused with personal Ground Fault Protection (GFCI or GFI). (GFCI protection is covered in detail in *Electrical Wiring: Residential*.)

The GFP is connected to the normal fused switch or circuit breaker that serves as the circuit protective device. The GFP is adjusted so that it will signal the protective device to open under abnormal ground-fault conditions. The maximum setting of the GFP is 1200 amperes. In the commercial building in the plans, the service voltage is 120/208 volts three phase. The voltage to ground on this system is 120 volts. If the ampacity of the service is 2000 amperes or more, then GFP is required, *Rule 14-102(1)(b)*. The electrician can follow a number of procedures to minimize the possibility of an arcing fault. Examples of these procedures follow:

- Ensure that conductor insulation is not damaged when the conductors are pulled into the raceway.
- Ensure that the electrical installation is properly grounded and bonded.
- Ensure that locknuts and bushings are tight.
- Ensure that all electrical connections are tight.
- Tightly connect bonding jumpers around concentric and/or eccentric knockouts.
- Ensure that conduit couplings and other fittings are installed properly.
- Check insulators for minute cracks.
- Install insulating bushings on all raceways.
- Insulate all bare bus bars in switchboards when possible.
- Ensure that conductors do not rest on bolts or other sharp metal edges.
- Do not allow electrical equipment to become damp or wet either during or after construction.
- Ensure that all overcurrent devices have an adequate interrupting capacity.
- Do not work on live panels.
- Be careful when using *fish tape*, since the loose end can become tangled with the electrical panelboard.
- Be careful when working with live parts; do not drop tools or other metal objects on top of such parts.

REVIEW

Note: Refer to the *C.E.C., Part I* or the plans as necessary.

1. A 300-kVA, dry-type transformer bank has a three-phase, 600-volt delta primary and a three-phase, 208-volt wye secondary.

   a. What is the proper size of fuse, in amperes, that must be installed in the secondary?

   _____
   _____
   _____
   _____

   b. What is kVA output if one transformer goes bad and it is necessary to use an open delta bank?

   _____
   _____
   _____
   _____

2. Draw the secondary connections for a three-phase, four-wire, wye-connected transformer bank. Label the locations where the voltage across a transformer is 120 volts.

3. A panelboard is added to an existing panelboard installation. Two knockouts, one near the top and the other near the bottom, are cut in the adjoining sides of the boxes. Indicate on the drawing the proper way of extending the phase and neutral conductors to the new panelboard. Line side lugs are suitable for two conductors.

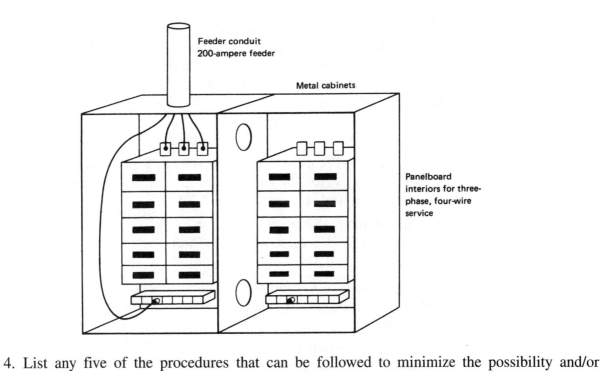

4. List any five of the procedures that can be followed to minimize the possibility and/or severity of an arcing fault.

5. A 5-horsepower, three-phase, 230-volt motor is installed. The motor requires the installation of an equipment grounding conductor in the same conduit as the motor branch circuit. The full-load rating of the motor is 15.2 amperes. The circuit consists of No. 12 Type TWN75 wire with an ampacity of 20 amperes. However, the circuit is provided with 30-ampere short-circuit protection. What size copper conductor is used for the equipment grounding conductor?

6. The service of the commercial building consists of three _____-mm (_____-inch) conduits, each containing three _____-kcmil phase conductors plus one No. _____ neutral conductor.

7. What is the purpose of grounding systems and enclosures?

8. What is the kVA capacity of each of the following? (Circle the correct answer.)
   a. Three 75-kVA transformers connected closed delta.
      (1) 225 kVA   (2) 150 kVA   (3) 130 kVA
   b. Two 75-kVA transformers connected open delta.
      (1) 225 kVA   (2) 150 kVA   (3) 130 kVA
   c. Three 150-kVA transformers wye-connected.
      (1) 300 kVA   (2) 450 kVA   (3) 225 kVA

9. Of the following, which is the proper size copper grounding electrode conductor for a 200-ampere service that consists of No. 3/0 phase conductors? (Circle the correct answer.)
   a. No. 4 AWG   b. No. 3 AWG   c. No. 2 AWG

10. A service is supplied by three 350-kcmil conductors per phase. What is the minimum size copper grounding electrode conductor to be used? (Circle the correct answer.)
    a. No. 1/0 AWG   b. No. 2/0 AWG   c. No. 3/0 AWG

11. Ground-fault sensing equipment is required on certain types of services. For the following systems, indicate where this equipment is used by inserting in the blanks either R (Required) or N (Not Required).

    a. 120/208volt three-phase system _____

    b. 347/600-volt three-phase system _____

    c. delta systems _____

    d. single-phase systems _____

    e. 1200-ampere service, three-phase/four-wire wye, 277/480 volts _____

    f. 800-ampere service _____

12. The *C.E.C., Part I* requires that if a metal water piping system is available on the premises, the electrical system must be grounded to this water piping system. What type of wire must be used? Cite the pertinent rule. _____

13. Equipment is grounded and bonded so that in the event of a fault, the ground path will be able to
    _____

14. *Rule 10–806(1)* states that a grounding electrode conductor shall not be spliced. However, there are some circumstances where it may be impossible to install a grounding electrode conductor in one piece, without splicing. In such situations, *Rule* _____ does allow the grounding electrode conductor to be spliced by means of _____.

15. Does the *C.E.C., Part I* permit the use of a metal underground gas piping system as the grounding electrode for a service? _____

16. The engineering calculations for an 800-ampere service entrance call for two 500-kcmil copper conductors per phase, connected in parallel. The neutral calculations show that the neutral conductors need only be No. 3 copper. Yet, the specifications and riser diagram show that the neutral conductor is No. 1/0 copper, run in parallel. The riser diagram and specifications call for two conduits, each containing three 500-kcmil phase conductors, plus one No. 1/0 neutral conductor. Explain why the neutral conductor is sized No. 1/0.

UNIT 3

Branch Circuit and Feeders

OBJECTIVES

After completing the study of this unit, the student will be able to

- calculate the minimum allowable loading for a given area
- calculate the current in each line of a three-phase, four-wire system
- determine the correct size for the branch-circuit protection
- determine correct size feeders
- derate conductors as required
- calculate the voltage drop in single-phase circuits

The electrician often must assume the responsibility for making decisions on the sizing, protection, and installation of branch circuits. Even on well-engineered projects, the electrician will be asked to make numerous choices based on economic considerations. In the commercial building plans, the selection of the wire type to be used and, therefore, the sizing of the conduit, the derating of the conductor, and the allowance for voltage drop are decisions to be made by the electrician.

BRANCH-CIRCUIT CALCULATION

According to *Section 0* of the *C.E.C., Part I*, a branch circuit is that portion of the wiring system between the final overcurrent device protecting the circuit and the outlet(s). Thus, all outlets are supplied by branch circuits. An outlet is defined in *Section 0* as a point on the wiring system at which current is taken to supply equipment for a specific use. It can be seen that once the outlet requirements are tabulated, the number and sizes of the branch circuits can be determined.

Rule 8–210 governs the required calculations for the expected feeder loads and the number of branch circuits necessary.

It is important to note that *Rule 8–106(2)* requires that the minimum feeder capacity must be based on the load calculated using *Rule 8–210* and *Table 14* or the actual connected load requirements, whichever is larger.

To illustrate the procedure for determining the branch circuits, the branch-circuit loads in the drugstore are divided into three categories: lighting loads, motor loads, and other loads. The *C.E.C., Part I* requirements for load calculations are summarized as follows:

- Use *Table 14* (Table 3–1 in text) to find the watts per square metre for general lighting and receptacle loads or use the connected load if it is larger.

Table 3–1
(See Rule 8–210)
Watts Per Square Metre and Demand Factors for Services and Feeders for Various Types of Occupancy

Type of Occupancy	Watts Per Square Metre	Demand Factor Percent	
		Service Conductors	Feeders
Stores, Restaurants	30	100	100
Offices			
First 930 m^2	50	90	100
All in excess of 930 m^2	50	70	90
Industrial and Commercial	25	100	100
Churches	10	100	100
Garages	10	100	100
Storage Warehouses	5	70	90
Theatres	30	75	95
Armouries and Auditoriums	10	80	100
Banks	50	100	100
Barber Shops and Beauty			
Parlours	30	90	100
Clubs	20	80	100
Court Houses	20	100	100
Lodges	15	80	100

Table 3–1 *Table 14 of the C.E.C., Part I.*

- For show window lighting use the rating of the equipment installed, *Rule 8–210(b)*.
- The sum of the noncontinuous loading and 125% of the continuous loading shall not exceed the rating of the circuit, *Rule 8–104(1–3)*.
- Allow actual rating for specific loads, *Rule 8–210(b)*.
- For motors, see *Section 28*.
- For transformers, see *Section 26*.
- For electric heating, see *Section 62*.
- For appliances, see *Section 26*.
- For air-conditioning and refrigeration, see *Section 28*.

GENERAL LIGHTING LOADS

Lighting for Display Area of Store

The minimum lighting load to be included in the calculations for a given type of occupancy is determined from *Table 14* or the actual connected load, whichever is greater. For a "store" occupancy, the table indicates that the unit load per square metre is 30 watts. Therefore, the minimum lighting load allowance for the first floor of the drugstore is:

18.2 m × 7.04 m × 30 W/m^2 = 3844 W

Since this value is a minimum, it is also necessary to determine the connected load. The greater of the two values will become the computed lighting load, according to *Rule 8–106(2)*.

The connected load for the first floor of the drugstore is:

27 Style I luminaires × 87 volt-amperes	= 2349	volt-amperes
4 Style E luminaires × 144 volt-amperes	= 576	volt-amperes
15 Style D luminaires × 74 volt-amperes	= 1110	volt-amperes
2 Style N luminaires × 60 volt-amperes	= 120	volt-amperes
Total connected load	= 4155	volt-amperes

As presented in detail in a later unit, the illumination (lighting) in the main area of the drugstore is provided by fluorescent luminaires (fixtures) equipped with two lamps and a ballast required to operate the lamps. According to the luminaire schedule (E4), this lamp–ballast combination will use 75 watts of power and has a load rating of 87 volt-amperes. (Watts are what we pay for, volt-amperes are what we design for.) It will be the convention in this text to use the word "load" when referring to the watts or the volt-ampere rating of the load. The total connected load, 4155 volt-amperes, is used to determine the computed loading according to *Rule 8–106(2)*. In past practice, it has been common to round luminaire volt-ampere values upward when tabulating the load. Now, with increasing utilities costs, clients often want an accurate estimate of anticipated energy costs. This requires an accurate presentation of the loads.

Storage Area Lighting

The minimum load allowance for this storage space is:

96.86 m^2 × 30 W/m^2 = 2906 W

The connected load for the storage area (basement) is:

9 Style L luminaires × 87 volt-amperes = 783 volt-amperes

Show Window Lighting

Rule 8–210(b) requires the use of the actual connected load for the purposes of determining branch-circuit and feeder sizing. The connected load for the show window is:

3 receptacle outlets × 500 volt-amperes	= 1500 volt-amperes
1 lighting track × 1650 volt-amperes	= <u>1650</u> volt-amperes
Total connected load	= 3150 volt-amperes

Determining the Lighting Loading

All of the lighting in the drugstore area has now been tabulated and the next step is to determine the wattage values that comply with *Rule 8–210*. The minimum demand watts according to *Rule 8–210* and *Table 14* is:

General store area	3844 watts
Storage area	<u>2906</u> watts
Total minimum allowance	6750 watts

The total connected lighting load for the drugstore is:

General lighting	4155 watts
Storage area lighting	783 watts
Receptacles	17 @ 120 VA = 2040
Total connected load	6978 watts

Because the connected load is greater than the minimum allowance, the connected load, which is 6978 watts, will be used in the calculations for the branch circuits and feeders.

MOTOR LOADS (APPLIANCES AND AIR-CONDITIONING AND REFRIGERATION EQUIPMENT)

Rule 28–700 should be referred to for air-conditioning and refrigeration equipment that incorporates hermetic refrigerant motor-compressors.

Units 9 and 11 of this text contain much additional information relative to the *C.E.C., Part I* requirements for appliance and motor applications.

The drugstore is provided with air-conditioning equipment, Fig. 3–1. The data on the nameplate are as follows:

Voltage: 208 volts, three-phase, three-wire, 60 hertz

Hermetic refrigerant compressor-motor: rated load current 20.2 amperes, 208 volts, three-phase

Evaporator motor: full load current 3.2 amperes, 208 volts, single-phase

Condenser motor: full load current 3.2 amperes, 208 volts, single-phase

Minimum circuit ampacity: 31.65 amperes

Overcurrent protection: 50 amperes, time-delay fuse

Locked rotor current: 140 amperes

It is important to note that when the nameplate specifies a maximum size fuse the equipment must be protected by fuses; using circuit breakers is not approved and would be a violation of *Rule 2–100(4)*. When the nameplate specifies the maximum size fuse or circuit breaker, then either may be used as the branch-circuit protection.

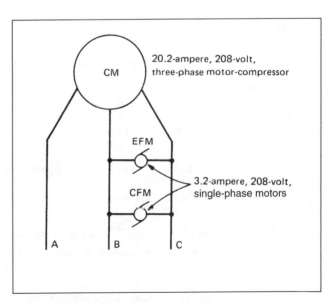

Fig. 3–1 Internal load connections of a typical air-conditioner.

BRANCH CIRCUITS

The actual number of branch circuits to be installed in the drugstore is determined by reference to the specifications, or to the panelboard schedule shown on sheet E1 of the plans. The number of branch circuits is based on information recorded in Table 3–2, the drugstore feeder loading schedule. The following discussion will illustrate the process employed to arrive at the final number of circuits.

The specifications (see Appendix) for this commercial building direct the electrician to install copper conductors of a size not smaller than No. 12 AWG (American Wire Gauge). The maximum load on a No. 12 AWG copper conductor is limited by the maximum allowable overcurrent protective device, which is 20 amperes.

Continuous loads limit the branch-circuit loads to 80% of the rating of the overcurrent protection device. For the 120-volt circuit this would limit the load to $(120 \times 20 \times 0.8) = 1920$ volt-amperes. All of the lighting load in Table 3–2 is considered continuous.

7788 W/1920 W per branch circuit	= 5 branch circuits
Add:	
Receptacle circuits (5–15 R receptacles) 1440 W per branch circuit	= 3 branch circuits
Roof circuit	= 1 branch circuit
Sign circuit	= 1 branch circuit
Cooling circuit	= <u>1</u> branch circuit
Minimum required for drugstore	= 11 branch circuits

Other factors, such as the switching arrangement and convenience of installation, are also

Table 3–2 Drugstore Feeder Loading Schedule

Item	Count	VA/unit	Demand Load	Installed Load	Demand factor Feeder	Demand factor Service	Use Feeder	Use Service
MINIMUM BASIC LOAD								
(C.E.C. Rule 8-210)	225 m²	30 W/m²	6750		1	1	6750	6750
INSTALLED BASIC LOADS								
Style D luminaires	15	74		1110			1110	1110
Style E luminaires	4	144		576			576	576
Style I luminaires	27	87		2349			2349	2349
Style N luminaires	2	60		120			120	120
Style L luminaires	9	87		783			783	783
General receptacles	21	120		2520			2520	2520
Basic installed loads							7458	7458
BASIC LOAD		(Use the greater of the demand or installed loads, *Rule 8–106(2)*.)					7458	7458
SPECIAL LOADS								
Show window rec.	3	500		1500			1500	1500
Track lighting	15	110		1650			1650	1650
Sign	1	1200		1200			1200	1200
Roof Receptacle	1	120		120			120	120
Motors	**Volts**	**FLA**	**Phase**					
Cooling unit								
• compressor	208	20.2	3	7269	1.25		9086	7269
• evaporator motor	208	3.2	1	666			666	666
• condenser motor	208	3.2	1	666			666	666
Total load							22 346	20 529

$$I = \frac{P}{E \times 1.73} = \frac{22\,346}{208 \times 1.73} = 62.10 \text{ A}$$

Assuming continuous loads, standard-rated equipment, and TWN75 copper conductors (*Table 2*), the ampere rating of the circuit would be:

$$\frac{62.1}{0.8} = 77.6 \text{ A}$$

The feeder would be 4 No. 4 AWG T90 copper conductors in a 27-mm (1-in) raceway, supplied by a 100-ampere disconnect c/w 90-ampere fuses.

important considerations in determining the actual number of circuits used. For example, in the drugstore 17 circuits are scheduled.

CONSIDERATIONS WHEN SIZING WIRE

When sizing wire it is important to consider the following four points:

1. the type of installation and location
2. the type of equipment rating, 100% or 80%, *Rule 8–104*, *Appendix B*
3. the number of current-carrying conductors and the ambient temperature
4. the voltage drop on the feeder or branch circuit

Each of these considerations provides specific derating factors that must be compared in order to find the correct size wire to be installed. The factor that requires the **largest size** wire is the determining factor in deciding the wire size to be installed, *Rule 8–104(6)*.

1. *The type of installation and location.*
   The four types of installation are:
   - free air
   - direct burial, *Rule 8–104(7)*
   - conduit
   - cable

   The type used depends on the location—whether it is wet, damp, dry, corrosive, flammable, or hazardous, for example.

2. *The rating of equipment, Rule 8–104(4,5)* and *Appendix B*. This is divided into two main subdivisions (refer to Table 3–3):

   A. For 100%-rated equipment, *Rule 8–104(4)*, the maximum continuous load is:
   - 85% of the circuit rating for free air installations (The circuit rating is determined by the lesser of the overcurrent device or the wire ampacity, *Rule 8–104(1)*.)
   - 100% of the circuit rating for conduit or cable installations

Table 3–3

Type of Installation	Code Ampacity Table	Derating Factors for Listed Ampacity of Conductors	
		Percent of Continuous Operation of Equipment	
	Rule 8–104, App. B	100%	80%
Free Air	Cu *Table 1*	8–104(4)(b)	8–104(5)(b)
	Al *Table 2*	0.85	0.70
Conduit or Cable	Cu *Table 2*	8–104(4)(a)	8–104(5)(a)
	Al *Table 4*	1.00	0.80
Direct Burial (Free Air Rating)	Cu *Table 1*	8–104(7)	8–104(7)
	Al *Table 3*	0.85	0.70
Direct Burial (Cables in Contact or Inadequate Spacing)	Cu *Table 2*	8–104(7)	8–104(7)
	Al *Table 4*	0.85	0.70

Table 3–3 Derating factors based on equipment rating, *Rule 8–104(4,5)*.

   B. For 80%-rated equipment, *Rule 8–104(5)*, the maximum continuous load is:
   - 70% of the circuit rating for free air installations.
   - 80% of the circuit rating for conduit or cable installations

   Most equipment is 80%-rated equipment. This means that it is not marked for 100% operation and is capable of supplying a continuous load at 80% of its marked and/or rated ampacity.

   EXAMPLE: A 100-ampere switch that is 80% rated is capable of supplying a continuous load of 80 amperes.

3. *The number of current-carrying conductors and the ambient temperature.* See *Rule 4–004*. Two tables specify derating factors that apply when current-carrying conductors are in contact with each other.
   Table 5B is for free air installations with two to four conductors in contact.
   Table 5C is for conduit or cable installations where more than three current-carrying conductors are in contact with each other.
   It is important to note that the ampacity of conductors size 1/0 AWG and larger in

56 Unit 3 Branch Circuit and Feeders

underground installations is not solely determined from *Tables 1* to *4*. Conductor ampacity is determined by first referring to *Diagrams B4–1* to *B4–4* in *Appendix B* for an appropriate installation configuration. Once the installation configuration has been determined, refer to *Tables D8A* to *D11B* in *Appendix D* for the appropriate ampacity. Be sure to read the notes at the end of the tables to ensure that the proper table is being used.

The ampacity of the conductor shall be the lower of this value or the value calculated from *Tables 1 to 4* with application of *Rule 8–104(4,5)*.

Another consideration is the ambient temperature, especially the maximum probable air temperature surrounding the conduit or cable. This is important due to the positive temperature coefficient of resistance and the temperature rating of the insulation of the wires.

Table 5A provides the derating factors for various ambient temperatures.

4. *Voltage drop.* The *C.E.C., Part I* permits a total voltage drop on a feeder and branch circuit to be 5%. However, it permits up to 3% on either the feeder or branch circuit. If the feeder voltage drop is 3% the branch circuit voltage drop cannot be greater than 2%. If the branch circuit voltage drop is 3% the feeder voltage drop cannot be greater than 2%. *Rule 8–102* and *Table D3* provide the information necessary to ensure safe operating levels of voltage. The voltage drop is usually a maximum of 2% on the feeder and 3% on the branch circuit.

Table D3 may be used to size the wire where long runs will have a significant voltage drop, to ensure that the maximum percent is not exceeded.

To make a safe and proper installation, the electrician will select a wire that will satisfy each of the various derating factors.

EXAMPLE: A No. 12-2 AC90 copper cable is used to feed an electric space heating unit located 38 metres from the panel. The area is rated a dry location. The heater draws 16 amperes at 240 volts. It is fed from a 20-ampere, two-pole 80% rated (not marked 100%) circuit breaker. The maximum ambient temperature is 50°C (122°F).

The branch circuit must satisfy all four of the considerations discussed above.

1. *The type of installation and location.* The installation is in cable, therefore ampacity will be taken from *Table 2* of the *C.E.C., Part I* (Table 3–4). *Table 19* approves Type AC90 for dry locations.

2. *The type of equipment rating.* 100% or 80%, *Rule 8–104, Appendix B*. The breaker is not marked as 100% rated, therefore the maximum continuous current that it is rated to carry is 80% of its rating or 20 amperes × 0.8 = 16 amperes.

   Rule 8–104(5)(a) allows the maximum continuous load to be 80% of the circuit rating.

   The No. 12 wire from *Table 2*, column 4 has a 20-ampere allowable ampacity and the breaker is rated at 20 amperes, therefore the rating of the circuit is 20 amperes. *Rule 8–104(1)*.

   Maximum continuous load = 20 amperes × 0.8
   = 16 amperes

3. *The number of current-carrying conductors and the ambient temperature.* There are two current-carrying conductors. Referring to *Table 5C* (Table 3–5) the correction factor is 1.00. For an ambient temperature over 30°C *Table 5A* (Table 3–6) gives the correction factor. Column 4 of *Table 5A* at 50°C gives a correction factor of 0.80.

   Therefore, the maximum current permitted on the No. 12 wire is: 20 amperes × 1.00 × 0.80 = 16 amperes, *Rule 4–004(1)(b) and (8)*.

4. *The voltage drop on the feeder or branch circuit.* The maximum voltage drop permitted on a branch circuit is 3%. *Table D3* gives a method of calculating the maximum length of wire in a two-wire circuit with the voltage drop on the circuit permitted by the *C.E.C., Part I*. This may be up to 3%, *Rule 8–102(1)*.

   The calculation involves:

- AWG of the wire
- current on the wire (load)
- distance of the conductor run (e.g., 38 metres from panel to outlet)

Table 2
ALLOWABLE AMPACITIES FOR NOT MORE THAN 3 COPPER CONDUCTORS IN RACEWAY OR CABLE
Based on Ambient Temperature of 30°C*

	Allowable Ampacity					
	60°C†	75°C†	85–90°C†	110°C†	125°C†	200°C†
Size AWG kcmil	Type TW	Types RW75, TW75	Types R90, RW90 T90 Nylon / Paper / Mineral-insulated Cable**	See Note (1)	See Note (1)	See Note (1)
14	15	15	15	30	30	30
12	20	20	20	35	40	40
10	30	30	30	45	50	55
8	40	45	45	60	65	70
6	55††	65	65	80	85	95
4	70	85	85	105	115	120
3	80	100	105	120	130	145
2	100	115	120	135	145	165
1	110	130	140	160	170	190
0	125	150	155	190	200	225
00	145	175	185††	215	230	250
000	165	200	210	245	265	285
0000	195	230	235	275	310	340
250	215	255	265	315	335	—
300	240	285	295	345	380	—
350	260	310	325	390	420	—
400	280	335	345	420	450	—
500	320	380	395	470	500	—
600	355	420	455	525	545	—
700	385	460	490	560	600	—
750	400	475	500	580	620	—
800	410	490	515	600	640	—
900	435	520	555	—	—	—
1000	455	545	585	680	730	—
1250	495	590	645	—	—	—
1500	520	625	700	785	—	—
1750	545	650	735	—	—	—
2000	560	665	775	840	—	—
	Col. 1	Col. 2	Col. 3	Col. 4	Col. 5	Col. 6	Col. 7

* See *Table 5A* for the correction factors to be applied to the values in Columns 2 to 7 for ambient temperatures over 30°C.
The ampacity of aluminum-sheathed cable is based on the type of insulation used on the copper conductors.

† These are maximum allowable conductor temperatures for 1, 2, or 3 conductors run in a raceway, or 2 or 3 conductors run on a cable and may be used in determining the ampacity of other conductor types in *Table 19*, which are so run, as follows: From *Table 19* determine the maximum allowable conductor temperature for that particular type; then from *Table 2* determine the ampacity under the column of corresponding temperature rating.

** These ratings are based on the use of 90°C insulation on the emerging conductors and for sealing. By special permission, mineral-insulated cable may be used at higher temperatures without decrease in allowable ampacity, provided that insulation and sealing material approved for such higher temperature are used.

†† For 3-wire 120/240 and 120/208 residential services or subservices the allowable ampacity for sizes No. 6 and No. 210 AWG shall be 60 amperes and 200 amperes respectively. In this case the 5% adjustment per *Rule 8–106(1)* cannot be applied.
See *Table 5C* for the correction factors to be applied to the values in Columns 2 to 7 where there are more than three conductors in a run of raceway or cable.

NOTE: These ampacities are only applicable under special circumstances where the use of insulated conductors having this temperature rating is acceptable.

Table 3–4 *Table 2* of the *C.E.C., Part I.*

Unit 3 Branch Circuit and Feeders

- rated conductor temperature
- percent of allowable ampacity
- type of conductor, Cu or Al (see *Table D3, Note 5*)
- percent voltage drop on feeder or branch circuit (2% or 3%)
- voltage applied to circuit

The following formula may be used to find the maximum length of a run of wire for a given voltage drop.

L = T.D3 dist. × % V.D. × D.C.F. × volts/120

where:

L = Maximum length of two-wire copper conductor run in metres.

T.D3 dist. = The distance shown in *Table D3* for the size of wire (AWG heading on top of table) and the actual load in amperes (Current, Amperes column).

% V.D. = The maximum percent voltage drop allowed on the circuit. *Note:* This value should be shown as a whole number, not as a percentage.

D.C.F. = Distance correction factor. This is taken from the table in *Table D3, Note 3*. The correction factor depends on the rated conductor temperature and the percent-

Table 5C
AMPACITY CORRECTION FACTORS FOR *TABLES 2* AND *4*

Number of Conductors	Ampacity Correction Factor
1–3	1.00
4–6	0.80
7–24	0.70
25–42	0.60
43 and up	0.50

Table 3–5   *Table 5C* of the *C.E.C., Part I.*

Table 5A
(See *Rules 4–004(8)* and *12–2212* and *Tables 1, 2, 3, 4, 57, 58,* and *D3*)

Correction Factors Applying to *Tables 1, 2, 3,* and *4*

Ampacity Correction Factors for Ambient Temperatures Above 30°C
(These correction factors apply, column for column, to *Tables 1, 2, 3,* and *4*.
The correction factors in column 2 also apply to *Table 57*.)

Correction Factor

Ambient Temperature °C	60°C Type TW	75°C Types RW75, TW75	85–90°C Types R90, RW90, T90, NYLON	110°C See Note (2)	125°C See Note (2)	200°C See Note (2)
40	0.82	0.88	0.90	0.94	0.95	1.00
45	0.71	0.82	0.85	0.90	0.92	1.00
50	0.58	0.75	0.80	0.87	0.89	1.00
55	0.41	0.65	0.74	0.83	0.86	1.00
60	—	0.58	0.67	0.79	0.83	0.91
70	—	0.35	0.52	0.71	0.76	0.87
75	—	—	0.43	0.66	0.72	0.86
80	—	—	0.30	0.61	0.69	0.84
90	—	—	—	0.50	0.61	0.80
100	—	—	—	—	0.51	0.77
120	—	—	—	—	—	0.69
140	—	—	—	—	—	0.59
Col. 1	Col. 2	Col. 3	Col. 4	Col. 5	Col. 6	Col. 7

NOTES: (1) The ampacity of a given conductor type at these higher ambient temperatures is obtained by multiplying the appropriate value from *Table 1, 2, 3,* or *4* by the correction factor for that higher temperature.
(2) These ampicites are only applicable under special circumstances where the use of insulated conductors having this temperature rating are acceptable.

Table 3–6   *Table 5A* of the *C.E.C., Part I.*

age of allowable ampacity (% allowable ampacity = actual load amperes/Allowable conductor ampacity × 100).

Volts = The voltage that the circuit operates at (e.g., 120, 208, 240).

Using our example of No. 12 AC90 with a 3% voltage drop, the maximum length of run is:

L = T.D3 dist. × % V.D. × D.C.F. × volts/120

= 6.1 × 3 × 1.00 × 240/120

= 36.6 m

Since the outlet is 38 metres from the panel, the wire size must be increased to a No. 10 copper conductor, *Rule 8–104(6)*. The No. 10-2 AC90 copper cable is the smallest size that will satisfy *all* of the derating requirements.

DETERMINING CONDUCTOR SIZE AND TYPE

Just as an electric motor can burn out, so can the insulation on a conductor become damaged as a result of extremely high temperatures. The insulation can soften, melt, and finally break down, causing grounds, shorts, and possible equipment damage and personal injury.

The source of this damaging heat comes from the surrounding room temperature (ambient temperature) and the heat generated by the current in the wire. The heat generated in the wire can be calculated by:

Watts = I^2R

Thus, the more current-carrying conductors that are installed in a single raceway or cable, the more damaging is the cumulative effect of the heat generated by the conductors. This heat must be kept to a safe, acceptable level by either restricting the number of conductors in a raceway or cable, or limiting the amount of current that a given size conductor may carry. This means that various derating factors and correction factors must be applied to the ampacity of the conductor being installed. *Ampacity* is defined as "the current in amperes that a conductor can carry continuously under the condition of use without exceeding its temperature rating."

When called upon to connect electrical loads such as lighting, motors, heating, refrigeration, or air-conditioning equipment, the electrician must have a working knowledge of how to select the proper size and type of conductors to be installed. Installing the right size wire will assure that the voltage at the terminals of the equipment is not below the minimums set out in the *C.E.C., Part I*. It also assures that the insulation will be able to withstand the temperature conditions that exist. The conduit must also be the correct size to facilitate pulling in the conductors without damage.

There are a number of things that must be considered when selecting wires. The *C.E.C., Part I* must be followed carefully. It contains terms such as *total demand load, calculated load, intermittent load, noncontinuous load, connected load, load current rating, maximum allowable load current, rated load current, derating factors, correction factors, ambient temperature,* and *ampacity*. These terms have a specific meaning and are very important in the selection of conductors. The electrician should become familiar with each of them.

Table 19 lists the types of wire and cable that are available for use. A person selecting a conductor should always refer to this table to determine if the conductor insulation is appropriate for the conditions of installation.

Tables 1, 2, 3, and *4* are referred to regularly by electricians, engineers, and electrical inspectors when information about wire sizing is needed. These tables show the allowable ampacities of insulated conductors. The temperature limitations, the types of insulation, the material the conductor is made of (copper, copper-clad aluminum, aluminum), the conductor size in AWG or in kcmil (thousand circular mils), and the ampacity of the conductor are indicated in the appropriate table. Correction factors for high ambient temperatures are shown in *Table 5A*. See Table 3–6 (page 58) in the text.

Table 2 (Table 3–4, page 57) is used most since it is referred to when the copper conductors are installed in a raceway or cable. All of these tables can be found in the middle section of the *C.E.C., Part I*.

CORRECTION FACTORS FOR AMBIENT TEMPERATURE

The ampacities as shown in *Table 1, 2, 3,* or *4* must be multiplied by the appropriate correction factor from *Table 5A*.

For example, consider a No. 3 TW75 conductor, with an initial ampacity of 100 amperes, being installed in a boiler room where the ambient temperature is expected to reach 42°C. The maximum permitted load current for this conductor would be:

100 amperes $\times$ 0.82 = 82 amperes

DERATING FACTORS FOR MORE THAN THREE CURRENT-CARRYING CONDUCTORS IN ONE RACEWAY

Conductors must be derated according to *Table 5C*

- if more than three current-carrying conductors are installed in a raceway or cable, Fig. 3–2(A).
- when single conductors or cable assemblies are stacked or bundled, as in a cable-tray for lengths over 600 mm (2 ft) without spacing, Fig. 3–2(B).
- if the number of cable assemblies (e.g., non-metallic sheathed cable) are run together for distances over 600 mm (2 ft), Fig. 3–2(C).

It is important to note that the *C.E.C., Part I* refers to current-carrying conductors for the purpose of derating when more than three conductors are installed in a raceway or cable. Following are the basic rules:

- **Do** count all current-carrying wires.
- **Do** count neutrals of a three-wire circuit or feeder when the system is four-wire, three-phase, wye connected, Fig. 3–3.
- **Do** count neutrals of a four-wire, three-phase, wye-connected circuit or feeder when the major portion of the load is electric discharge lighting (fluorescent, mercury vapour, high-pressure sodium, etc.), data processing, and other loads where third harmonic (180 Hz) currents flow in the neutral, Fig. 3–4.
- **Do not** count neutrals of a three-wire, single-phase circuit or feeder where the neutral carries only the unbalanced current of the "hot" phase conductors, Fig. 3–5.
- **Do not** count equipment bonding conductors that are run in the same conduit with the circuit conductors, Fig. 3–6. However, the bonding conductors must be included when calculating conduit fill.

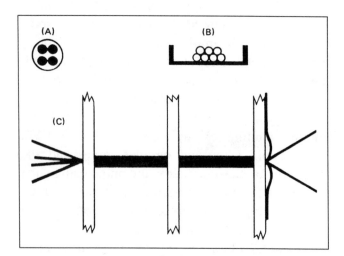

Fig. 3–2 Bundled and stacked conductors.

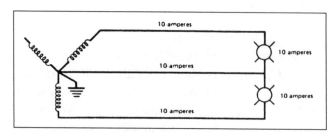

Fig. 3–3 The neutral of a three-wire circuit supplied by a four-wire, three-phase, wye-connected system will carry the vector sum of the "hot" conductors of the circuit. All three of these conductors are considered to be current-carrying conductors. See *Rule 4–004(4)*.

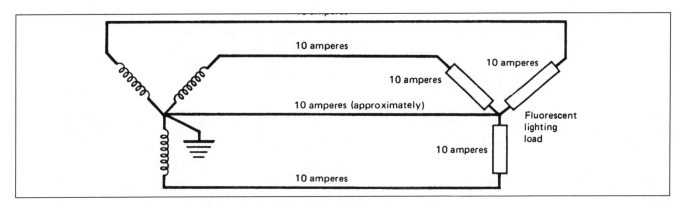

Fig. 3–4 The neutral of a four-wire, three-phase circuit supplying electric discharge lighting will carry approximately the same current as the "hot" conductors of the circuit. All four of these conductors are considered to be current-carrying conductors. See *Rule 4–022(2)(a)*.

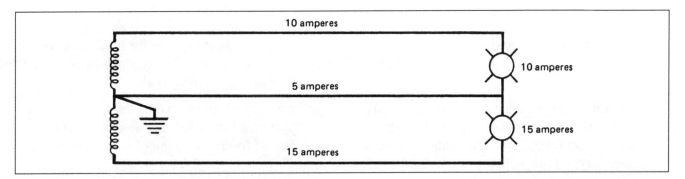

Fig. 3–5 The neutral in the diagram is not to be counted as a current-carrying conductor. See *Rule 4–004(3)*.

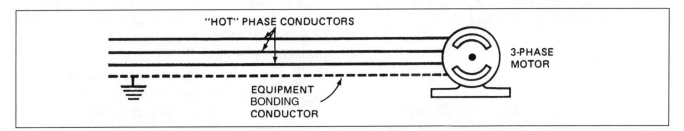

Fig. 3–6 In this diagram, there are three current-carrying conductors. The equipment grounding conductor must be included when determining conduit fill. The equipment grounding conductor does not have to be counted when determining conductor ampacity.

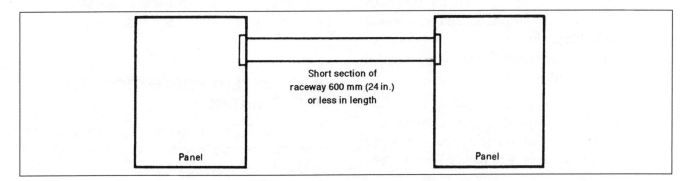

Fig. 3–7 It is not necessary to derate the conductors in short sections of raceway as illustrated in the diagram, but all of the conductors must be included when calculating conduit fill.

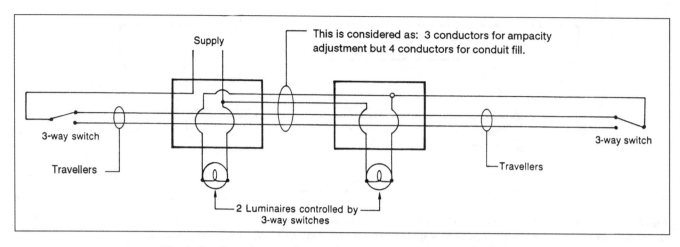

Fig. 3–8 Counting conductors in a three-way switch installation.

- **Do not** derate conductors in short sections of raceway 600 mm (2 ft) or less in length, Fig. 3–7.

- **Do not** count the noncurrent-carrying "travellers" on a three-way or four-way switching arrangement, Fig. 3–8. The actual conductor count is used for purposes of conduit fill calculations. Only one of the travellers is carrying current at any one time, so only one conductor need be counted when determining the ampacity adjustment.

FEEDERS

The discussion so far has concentrated on the branch circuit. This is a necessary first step in determining the size of the feeder that will carry the combined load of all the branch circuits in the designated area. The specific requirements for feeders are given in *Rule 8–210* for buildings other than schools, hospitals, and residential buildings (including hotels, motels, and dormitories). Much of the information has been presented in the discussion of branch circuits.

A feeder

- must have an ampacity no less than the sum of the loads of the branch circuits supplied by the feeder, except as reduced for demand factors.

- may be reduced in size for loads that are not operated simultaneously, such as air-conditioning and heating loads.

- neutral may be sized only for the unbalanced portion of the load. See *Rule 4–022(1)*. See Unit 8 for a discussion of neutral currents.

- does not need to be larger than the conductors for the service supplying the feeder.

The demand factors listed in *Table 14* are allowed because in these specific cases it is highly unlikely that all the loads would be energized at the same time, *Rule 8–210(a)*.

The same logic applies to noncoincident loads as to the demand factors, *Rule 8–106(3)*.

The neutral of a system carries only the unbalanced current in many load conditions. In such cases, it need not be the same size as the ungrounded conductors, *Rule 4–022*, but in no case may a neutral service conductor be smaller than the size set forth in *Table 17* for the grounding electrode conductor. See *Rule 4–022(3)(b)*.

OVERCURRENT PROTECTION AND CIRCUIT RATING

The details for the selection of overcurrent protective devices are covered in Units 17, 18, and 19, and throughout the text.

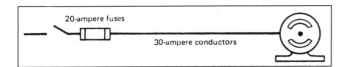

Fig. 3–9 Where a larger conductor is installed for any reason, such as to minimize voltage drop, the ampere rating of the overcurrent device determines the branch-circuit classification. In this example, 30-ampere conductors have been installed and connected to the 20-ampere overcurrent device. This circuit is classified as a 20-ampere circuit. *Rule 8–104(1)*.

The rating of a circuit depends upon the rating of the overcurrent protection device or the ampacity of the wire, **whichever is less.** See *Rule 8–104(1)* and Fig. 3–9.

Electrical equipment is rated to carry continuous loads at either 100% of its rating or 80% of its rating. A standard 15-ampere circuit breaker is permitted to carry 80% of its rating (12 amps) when the circuit conductors are rated to carry 15 amps, are run in a raceway or cable, and have an insulation temperature rating of 60 to 90°C.

A load is considered to be continuous, if in normal operation, it will be energized for more than one hour in any two-hour period. See *Rule 8–104(3)*. Larger loads are considered continuous if they are energized for more than three hours in any six-hour period.

The minimum rating of a branch circuit when standard-rated equipment is used is 125% of the full-load current if the loads are continuous. If the loads are not continuous, then the rating of the branch circuit is 100% of the non-continuous loads. For combined loads the minimum rating of the circuit is the non-continuous loads +125% of the continuous loads.

Conductor overcurrent protection is covered in *Rule 14–104*. The basic requirement is that conductors shall be protected at their ampacity, but several conditions exist when the overcurrent protection can exceed the ampacity. Commonly used maximum overcurrent protection is shown in Table 3–7.

(Refer to *Tables 2, 4,* and *13* of the *C.E.C., Part I*)

Wire Size	Copper	Aluminum
No. 14 AWG	15 amperes	(*not* permitted)
No. 12 AWG	20 amperes	15 amperes
No. 10 AWG	30 amperes	25 amperes

Table 3–7 Maximum overcurrent protection.

When the ampacity of a conductor does not match that of a standard-size fuse or circuit breaker, the conductor is considered adequately protected by installing the next higher standard-size overcurrent protection device up to 600 amperes. If the rating is above 600 amperes, then the next lower standard size must be selected.

VOLTAGE DROP

Rule 8–102 requires that the voltage drop in a branch circuit, or in a feeder, not exceed 3% of the system voltage. Also, the total voltage drop, including both the branch circuit and the feeder, should not exceed a total of 5% of the system voltage, Fig. 3–10.

To determine the voltage drop, two factors must be known: (1) the resistance of the conductor, and (2) the current in the conductor. For example, the drugstore window display receptacle outlets are to be loaded to 1500 volt-amperes or

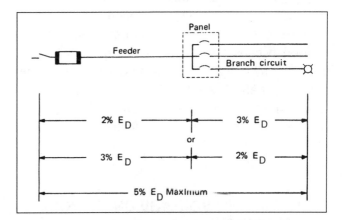

Fig. 3–10 For light, power, and lighting loads it is required that the combined voltage drop shall not exceed 5% on the feeder and branch-circuit wiring. As in the above diagram, if the feeder has a 2% voltage drop, then the branch circuit shall have not over 3% voltage drop. See *Rules 8–102(1)*.

12.5 amperes. The distance from the panelboard to the centre receptacle is 25.9 m (85 ft). The resistance of a No. 12 uncoated copper conductor is 6.33 ohms/km (1.93 ohms per 1000 feet). Therefore the total resistance is:

(6.33 ohms/1000 m) × 2 × 25.9m = 0.328 ohm

(1.93 ohms/1000 ft) × 2 × 85 ft = 0.328 ohm

The factor of 2 in the preceding equation is included since both the phase conductor and the neutral conductor carry current; thus, the resistance of each must be included. The voltage drop is now defined by Ohm's Law, $E = IR$.

E = 12.5 amperes × 0.3281 ohm = 4.1 volts

The allowable voltage drop, according to *Rule 8–102*, is:

120 volts × 0.03 = 3.6 volts (0.03 = 3%)

Accordingly, it is necessary to use No. 10, which would have a voltage drop of 2.57 volts, for this circuit.

This method of determining the voltage drop neglects any inductance in the circuit. However, inductance usually is insignificant at the short distances that are encountered in commercial wiring. When it is desired to calculate the voltage drop, taking into consideration both the resistance and the reactance, available tables and charts can be used to simplify the calculations. Tables 3–8 and 3–9 show how to determine conductor sizes and voltage loss. Voltage drop is an important consideration because the higher temperature conductors now installed in electrical systems permit the use of a smaller size of conductor for a given load. These smaller conductors often result in excessive voltage loss in the circuit conductors.

DETERMINING VOLTAGE DROP (VOLT LOSS) USING TABLES

With higher current ratings on new insulations, it is extremely important to keep *volt loss* in mind.

How to Figure Volt Loss

The calculation for volt loss is as follows:

$$\text{Volt loss} = \frac{L \times I \times \text{T.V.}}{304\,800}$$

where:

L = length of conduit in metres

I = current in amperes

T.V. = value shown in table for kind of conduit and size of wire to be used.

EXAMPLE: Calculate the volt loss for No. 6 copper wire in 55 metres (180.4 ft) of steel conduit—three-phase, 40-ampere load at 80% power factor.

$$\text{Volt loss} = \frac{55 \times 40 \times 745}{304\,800}$$

$$= 5.38 \text{ volts}$$

(For a 240-volt circuit the percentage voltage drop is 5.38 / 240 × 100 or 2.24%.)

These tables take into consideration *reactance on AC circuits* as well as resistance of the wire.

Remember on short runs to check to see that the size and type of wire indicated has sufficient ampere capacity.

How to Select Size of Wire

The calculation to arrive at the correct size of wire is:

$$\frac{\text{V.L.}}{L \times I} \times 304\,800$$

where:

V.L. = Volt loss

L = Length in metres of one wire

I = Current in amperes

Look under the column applying to the type of current and power factor for the figure nearest, but not above, your result and you have the size of wire needed.

Table 3–8 Copper Conductors — Ratings and Volt Loss†

Conduit	Wire Size	Ampacity 60°C Wire	Ampacity 75°C Wire	Ampacity 90°C Wire	Direct Current	Volt Loss Three Phase (60 Cycle, Lagging Power Factor) 100%	90%	80%	70%	60%	Single Phase (60 Cycle, Lagging Power Factor) 100%	90%	80%	70%	60%
Steel Conduit	14	20*	20*	25*	6140	5369	4887	4371	3848	3322	6200	5643	5047	4444	3836
	12	25*	25*	30*	3860	3464	3169	2841	2508	2172	4000	3659	3281	2897	2508
	10	30	35*	40*	2420	2078	1918	1728	1532	1334	2400	2214	1995	1769	1540
	8	40	50	55	1528	1350	1264	1148	1026	900	1560	1460	1326	1184	1040
	6	55	65	75	982	848	812	745	673	597	980	937	860	777	690
	4	70	85	95	616	536	528	491	450	405	620	610	568	519	468
	3	85	100	110	490	433	434	407	376	341	500	501	470	434	394
	2	95	115	130	388	346	354	336	312	286	400	409	388	361	331
	1	110	130	150	308	277	292	280	264	245	320	337	324	305	283
	0	125	150	170	244	207	228	223	213	200	240	263	258	246	232
	00	145	175	195	193	173	196	194	188	178	200	227	224	217	206
	000	165	200	225	153	136	162	163	160	154	158	187	188	184	178
	0000	195	230	260	122	109	136	140	139	136	126	157	162	161	157
	250	215	255	290	103	93	123	128	129	128	108	142	148	149	148
	300	240	285	320	86	77	108	115	117	117	90	125	133	135	135
	350	260	310	350	73	67	98	106	109	109	78	113	122	126	126
	400	280	335	380	64	60	91	99	103	104	70	105	114	118	120
	500	320	380	430	52	50	81	90	94	96	58	94	104	109	111
	600	355	420	475	43	43	75	84	89	92	50	86	97	103	106
	750	400	475	535	34	36	68	78	84	88	42	79	91	97	102
	1000	455	545	615	26	31	62	72	78	82	36	72	84	90	95
Nonmagnetic Conduit (Lead-covered Cables or Installation in Fibre or Other Nonmagnetic Conduit, etc.)	14	20*	20*	25*	6140	5369	4876	4355	3830	3301	6200	5630	5029	4422	3812
	12	25*	25*	30*	3860	3464	3158	2827	2491	2153	4000	3647	3264	2877	2486
	10	30	35*	40*	2420	2078	1908	1714	1516	1316	2400	2203	1980	1751	1520
	8	40	50	55	1528	1350	1255	1134	1010	882	1560	1449	1310	1166	1019
	6	55	65	75	982	848	802	731	657	579	980	926	845	758	669
	4	70	85	95	616	536	519	479	435	388	620	599	553	502	448
	3	85	100	110	470	433	425	395	361	324	500	490	456	417	375
	2	95	115	130	388	329	330	310	286	259	380	381	358	330	300
	1	110	130	150	308	259	268	255	238	219	300	310	295	275	253
	0	125	150	170	244	207	220	212	199	185	240	254	244	230	214
	00	145	175	195	193	173	188	183	174	163	200	217	211	201	188
	000	165	200	225	153	133	151	150	145	138	154	175	173	167	159
	0000	195	230	260	122	107	127	128	125	121	124	147	148	145	140
	250	215	255	290	103	90	112	114	113	110	104	129	132	131	128
	300	240	285	320	86	76	99	103	104	102	88	114	119	120	118
	350	260	310	350	73	65	89	94	95	94	76	103	108	110	109
	400	280	335	380	64	57	81	87	89	89	66	94	100	103	103
	500	320	380	430	52	46	71	77	80	82	54	82	90	93	94
	600	355	420	475	43	39	65	72	76	77	46	75	83	87	90
	750	400	475	535	34	32	58	65	70	72	38	67	76	80	83
	1000	455	545	615	26	25	51	59	63	66	30	59	68	73	77

*The overcurrent protection for conductor types marked with an (*) shall not exceed 15 amperes for No. 14 AWG, 20 amperes for No. 12 AWG, and 30 amperes for No. 10 AWG copper; or 15 amperes for No. 12 AWG and 25 amperes for No. 10 AWG aluminum and copper-clad aluminum after any correction factors for ambient temperature and number of conductors have been applied.
†Figures are L-L for both single phase and three phase. Three-phase figures are average for the three phases.

(Courtesy of Copper Industries, Bussmann Division.)

EXAMPLE: Calculate the correct size of copper wire in 55 metres of steel conduit for a three-phase, 40-ampere load at 80% power factor. Maximum volt loss from local code not to exceed 5.5 volts.

$$\frac{5.5}{55 \times 40} \times 304\,800 = 762$$

Select the number from the table, three phase at 80% factor power that is nearest but not greater than 762. This number is 745, which indicates the size of wire needed: No. 6 AWG.

Table 3-9 Aluminum Conductors — Ratings and Volt Loss[†]

Conduit	Wire Size	Ampacity 60°C Wire	Ampacity 75°C Wire	Ampacity 90°C Wire	Direct Current	Volt Loss Three Phase (60 Cycle, Lagging Power Factor) 100%	90%	80%	70%	60%	Single Phase (60 Cycle, Lagging Power Factor) 100%	90%	80%	70%	60%
Steel Conduit	12	20*	20*	25*	6360	5542	5039	4504	3963	3419	6400	5819	5201	4577	3948
	10	25	30*	35*	4000	3464	3165	2836	2502	2165	4000	3654	3275	2889	2500
	8	30	40	45	2520	2251	2075	1868	1656	1441	2600	2396	2158	1912	1663
	6	40	50	60	1616	1402	1310	1188	1061	930	1620	1513	1372	1225	1074
	4	55	65	75	1016	883	840	769	692	613	1020	970	888	799	708
	3	65	75	85	796	692	668	615	557	497	800	771	710	644	574
	2	75	90	110	638	554	541	502	458	411	640	625	580	529	475
	1	85	100	115	506	433	432	405	373	338	500	499	468	431	391
	0	100	120	135	402	346	353	334	310	284	400	407	386	358	328
	00	115	135	150	318	277	290	277	260	241	320	335	320	301	278
	000	130	155	175	252	225	241	234	221	207	260	279	270	256	239
	0000	155	180	205	200	173	194	191	184	174	200	224	221	212	201
	250	170	205	230	169	148	173	173	168	161	172	200	200	194	186
	300	190	230	255	141	124	150	152	150	145	144	174	176	173	168
	350	210	250	280	121	109	135	139	138	134	126	156	160	159	155
	400	225	270	305	106	95	122	127	127	125	110	141	146	146	144
	500	260	310	350	85	77	106	112	113	113	90	122	129	131	130
	600	285	340	385	71	65	95	102	105	106	76	110	118	121	122
	750	320	385	435	56	53	84	92	96	98	62	97	107	111	114
	1000	375	445	500	42	43	73	82	87	89	50	85	95	100	103
Nonmagnetic Conduit (Lead-covered Cables or Installation in Fibre or Other Nonmagnetic Conduit, etc.)	12	20*	20*	25*	6360	5542	5029	4490	3946	3400	6400	5807	5184	4557	3926
	10	25	30*	35*	4000	3464	3155	2823	2486	2147	4000	3643	3260	2871	2480
	8	30	40	45	2520	2251	2065	1855	1640	1423	2600	2385	2142	1894	1643
	6	40	50	60	1616	1402	1301	1175	1045	912	1620	1502	1357	1206	1053
	4	55	65	75	1016	883	831	756	677	596	1020	959	873	782	688
	3	65	75	85	796	692	659	603	543	480	800	760	696	627	555
	2	75	90	100	638	554	532	490	443	394	640	615	566	512	456
	1	85	100	115	506	433	424	394	360	323	500	490	455	415	373
	0	100	120	135	402	346	344	322	296	268	400	398	372	342	310
	00	115	135	150	318	277	281	266	247	225	320	325	307	285	260
	000	130	155	175	252	225	234	223	209	193	260	270	258	241	223
	0000	155	180	205	200	173	186	181	171	160	200	215	209	198	185
	250	170	205	230	169	147	163	160	153	145	170	188	185	177	167
	300	190	230	255	141	122	141	140	136	130	142	163	162	157	150
	350	210	250	280	121	105	125	125	123	118	122	144	145	142	137
	400	225	270	305	106	93	114	116	114	111	108	132	134	132	128
	500	260	310	350	85	74	96	100	100	98	86	111	115	115	114
	600	285	340	385	71	62	85	90	91	91	72	98	104	106	105
	750	320	385	435	56	50	73	79	82	82	58	85	92	94	95
	1000	375	445	500	42	39	63	70	73	75	46	73	81	85	86

*The overcurrent protection for conductor types marked with an (*) shall not exceed 15 amperes for No. 14 AWG, 20 amperes for No. 12 AWG, and 30 amperes for No. 10 AWG copper; or 15 amperes for No. 12 AWG and 25 amperes for No. 10 AWG aluminum and copper-clad aluminum after any correction factors for ambient temperature and number of conductors have been applied.
[†]Figures are L-L for both single phase and three phase. Three-phase figures are average for the three phases.

(Courtesy of Copper Industries, Bussmann Division.)

ENERGY SAVING CONSIDERATIONS

The rapidly escalating cost of energy encourages the consideration of installing one-size-larger conductor than is required by the *C.E.C., Part I*. This is particularly true for the smaller size conductors, i.e., Nos. 14, 12, 10, and 8.

The watt loss (heat) in the copper conductors can be calculated as follows:

Watts $= I^2R$

Consider the energy cost comparisons for a typical circuit with a 16-ampere load at a distance of 30.5 m (100 ft) from the panel.

The resistance of No. 12 copper is 6.33 milliohms per metre (0.193 ohm per 100 feet) and the resistance of No. 10 copper is 3.97 milliohms per metre (0.121 ohm per 100 feet).

EXAMPLE 1: Using the No. 12 copper wire, the total watt loss in the conductor for 10 hours per day for 250 days is:

$$\text{Watts} = I^2R$$
$$= 16 \times 16 \times 0.006\,33 \times 2 \times 30.5$$
$$= 98.8$$

$$\text{kWh per year} = \frac{98.8 \times 10 \text{ hours per day} \times 250 \text{ days per year}}{1000}$$
$$= 247$$

Cost of watt loss @ $0.10 per kWh equals $24.70 per year.

EXAMPLE 2: Using the No. 10 copper wire, the total watt loss in the conductor is:

$$\text{Watts} = I^2R$$
$$= 16 \times 16 \times 0.003\,97 \times 0.121 \times 2 \times 30.5$$
$$= 62.0$$

$$\text{kWh per year} = \frac{62.0 \times 10 \text{ hours per day} \times 250 \text{ days per year}}{1000}$$
$$= 155$$

Cost of watt loss @ $0.10 per kWh equals $15.50 per year.

COMPARISON: Suppose that the list price of No. 12 copper Type T90 conductor is $45.00 per 300 metres. Therefore the cost of 61 m (200 ft) is $9.15.

Suppose that the list price of No. 10 copper Type T90 conductor is $75.00 per 300 metres. Therefore the cost of 61 m (200 ft) is $15.25.

The difference in wire cost is $6.10.

The estimated saving is $24.70 − $15.50 = $9.20 per year.

The investment in No. 10 copper Type T90 conductors would be returned in less than eight months.

If the electrical contractor installs 3000 metres of No. 10 copper instead of 3000 metres of No. 12 copper, a considerable energy saving will be experienced. However, the contractor must consider whether or not the *C.E.C., Part I* requires the use of larger-size conduit. (See *Table 6*.)

The larger-size conductors also keep voltage drop to a minimum, which permits the connected electrical equipment to operate more efficiently.

Connections for the wires on the wiring devices terminal could be more time consuming. The electrician might prefer to attach No. 12 pigtails to the terminals of wiring devices and then splice these to the No. 10 circuit conductors.

Reduction of heat produced by the wiring conductors will decrease air-conditioning costs for the building, but also increase heating costs.

The savings in energy consumption through the use of larger conductors is worthy of consideration.

ALUMINUM CONDUCTORS

The conductivity of aluminum is not as great as that of copper for a given wire size. For example, checking *Table 1*, the ampacity for copper No. 12 Type R90 is 25 amperes, whereas the ampacity for aluminum taken from *Table 3* for No. 12 Type R90 is 20 amperes. As another example, a No. 8 Type T90 Nylon copper wire (*Table 2*) has an ampacity of 45 amperes, whereas a No. 8 Type T90 Nylon aluminum (*Table 4*) or copper-clad aluminum wire has an ampacity of only 30 amperes. It is important to check the ampacities listed in *Table 1, 2, 3, or 4* and the footnotes to these tables.

Resistance is an important consideration when installing aluminum conductors. An aluminum conductor has a higher resistance compared to a copper conductor for a given wire size, which therefore causes a greater voltage drop.

Common Connection Problems

Some common problems associated with aluminum conductors when not properly connected may be summarized as follows:

- A corrosive action is set up when dissimilar wires come in contact with one another when moisture is present.

- The surface of aluminum oxidizes as soon as it is exposed to air. If this oxidized surface is not broken through, a poor connection results. When installing aluminum conductors, particularly in large sizes, an inhibitor is brushed onto the aluminum conductor, then the conductor is scraped with a stiff brush where the connection is to be made. The process of scraping the conductor breaks through the oxidation, and the inhibitor keeps the air from coming into contact with the conductor. Thus, further oxidation is prevented. Aluminum connectors of the compression type usually have an inhibitor paste factory-installed inside the connector.

- Aluminum wire expands and contracts to a greater degree than does copper wire for an equal load. This is another possible cause of a poor connection. Crimp connectors for aluminum conductors are usually longer than those for comparable copper conductors to provide greater contact area for the connection.

PROPER INSTALLATION PROCEDURES

Proper, trouble-free connections for aluminum conductors require terminals, lugs, and/or connectors that are suitable for the type of conductor being installed. See Table 3–10.

Records indicate that the majority of failures in the use of aluminum wire have occurred because of poor installation. The electrician should take special care to make sure that all connections are thoroughly tight. The type of screwdriver where the blade is rotated by pushing on the handle should not be used for installation of terminals, since it often gives a false impression as to the tightness of the connection.

Terminals on receptacles and switches must be suitable for the conductors being attached. Some equipment, such as the connections to circuit breakers, are required by CSA and the manufacturer to be tightened with a torque wrench to a specified torque. Manufacturer's literature should be consulted and strictly followed when torquing requirements are set forth.

**Table 3–10
Conductors Permitted to Connect to Terminals**

Type of Device	Marking on Terminal or Connector	Conductor Permitted
15- or 20-ampere receptacles and switches	CO/ALR	aluminum, copper, copper-clad aluminum
15- and 20-ampere receptacles and switches	NONE	copper, copper-clad aluminum
30-ampere and greater receptacles and switches	AL/CU	aluminum, copper, copper-clad aluminum
30-ampere and greater receptacles and switches	NONE	copper only
Screwless pressure terminal connectors of the push-in type	NONE	copper or copper-clad aluminum
Wire connectors	AL/CU	aluminum, copper, copper-clad aluminum
Wire connectors	NONE	copper only
Wire connectors	AL	aluminum only
Any of above devices	COPPER OR CU ONLY	copper only

Wire Connections

When splicing wires or connecting a wire to a switch, fixture, circuit breaker, panelboard, meter socket, or other electrical equipment, some type of wire connector is required.

Wire connectors are known in the trade by such names as *screw terminal, pressure terminal connector, wire connector, wing nut, wire nut, Scotchlok™, split-bolt connector, pressure cable connector, solderless lug, soldering lug, solder lug,* and others. Soldering-type lugs are not often used today. Solderless connectors, designed to establish connections by means of mechanical pressure, are quite common. Examples of some types of wire connectors, and their uses, are shown in Fig. 3–11.

As with the terminals on wiring devices (switches and receptacles), wire connectors must be marked *AL* when they are to be used with aluminum conductors only. This marking is found on the connector itself or on or in the shipping carton.

CONNECTORS USED TO CONNECT WIRES TOGETHER IN COMBINATIONS OF NO. 18 AWG THROUGH NO. 6 AWG. THEY ARE TWIST-ON, SOLDERLESS, AND TAPELESS.	 WIRE CONNECTORS VARIOUSLY KNOWN AS WIRE NUT™, WING NUT™, AND SCOTCHLOK™
CONNECTORS USED TO CONNECT WIRES TOGETHER IN COMBINATIONS OF NO. 16, NO. 14, AND NO. 12 AWG. THEY ARE CRIMPED ON WITH A SPECIAL TOOL, THEN COVERED WITH A SNAP-ON INSULATING CAP.	 CRIMP-TYPE WIRE CONNECTOR AND INSULATING CAP
SOLDERLESS CONNECTORS ARE AVAILABLE IN SIZES NO. 14 AWG THROUGH 500 KCMIL. THEY ARE USED FOR ONE SOLID OR ONE STRANDED CONDUCTOR ONLY, UNLESS OTHERWISE NOTED ON THE CONNECTOR OR ON ITS SHIPPING CARTON. THE SCREW MAY HAVE A SCREWDRIVER SLOT, OR SOCKET.	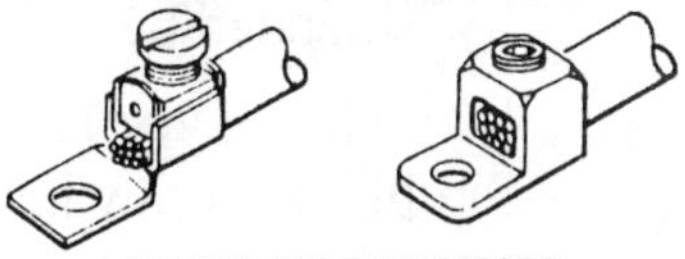 SOLDERLESS CONNECTORS
COMPRESSION CONNECTORS ARE USED FOR NO. 8 AWG THROUGH 1000 KCMIL. THE WIRE IS INSERTED INTO THE END OF THE CONNECTOR, THEN CRIMPED ON WITH A SPECIAL COMPRESSION TOOL.	 COMPRESSION CONNECTOR
SPLIT-BOLT CONNECTORS ARE USED FOR CONNECTING TWO CONDUCTORS TOGETHER, OR FOR TAPPING ONE CONDUCTOR TO ANOTHER. THEY ARE AVAILABLE IN SIZES NO. 10 AWG THROUGH 1000 KCMIL. THEY ARE USED FOR TWO SOLID AND/OR TWO STRANDED CONDUCTORS ONLY, UNLESS OTHERWISE NOTED ON THE CONNECTOR OR ON ITS SHIPPING CARTON.	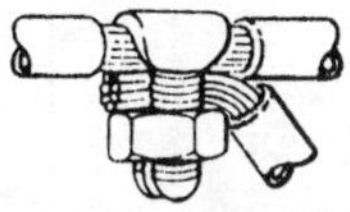 SPLIT-BOLT CONNECTOR

Fig. 3–11 Types of wire connectors.

Connectors marked *AL/CU* are suitable for use with aluminum, copper, or copper-clad aluminum conductors. This marking is found on the connector itself or on or in the shipping carton.

Connectors not marked *AL* or *AL/CU* are for use with copper conductors only.

Unless specifically stated on or in the shipping carton, or on the connector itself, conductors made of copper, aluminum, or copper-clad aluminum may not be used in combination in the same connector. Combinations, when permitted, are usually limited to dry locations only.

REVIEW

Note: Refer to the *C.E.C., Part I* or the plans as necessary.

1. For each of the following, indicate the unit load that is to be included in the branch-circuit calculations.

    a. A restaurant _____

    b. A school room _____

    c. A corridor in a school _____

    d. A corridor in a dwelling _____

2. A 208-volt, single-phase load rated at 7280-volt-amperes is located 30 m (100 ft) from the panelboard. The branch circuit is installed in a conduit to serve this load.

    a. The current in each conductor is _____ amperes.

    b. The minimum Type T90 Nylon conductor size for each conductor is _____ AWG.

    c. The voltage drop will be _____ %.

3. When Type R90 conductors are installed in an ambient temperature of 60°C the ampacity as shown in *Table 2* must be corrected by a _____ multiplier.

4. When four Type RW75 current-carrying conductors are installed in one raceway, the ampacity is derated by a factor of _____.

5. Four No. 1 Type RW75 copper current-carrying conductors are installed in a conduit running through a room where the ambient temperature is 45°C. The maximum load that can be connected to each of these conductors is _____ amperes.

6. Complete the following chart to select conductors in compliance with the *C.E.C., Part I*.

Conductor Type	RW75	RW90 XLPE	T90 Nylon
Ambient Temperature	27°C	33°C	27°C
Number of Current Carrying Conductors	2	4	3
Noncontinuous Loading	0	19.5A	0
Continuous Loading	13A	0	100A
Derating Factor			
Conductor Size			
Conductor Ampacity			
Terminations			
Circuit Ampacity			
Minimum Overcurrent Protection			
Maximum Overcurrent Protection			

7. The ampacity for aluminum conductors, for a given size and type, is less than that for copper conductors. Refer to *Table 3* or *4* and complete the following chart.

Conductor Size & Type	Copper Ampacity	Aluminum Ampacity
No. 12 AWG RW75		
No. 10 AWG RW75		
No. 3 AWG RW75		
500 kcmil RW75		

8. Is it permissible for an electrician to connect aluminum, copper, or copper-clad aluminum conductors together in the same connector? Yes _____ No _____

9. Terminals of switches and receptacles marked CO/ALR are suitable for use with _____, _____, and _____ conductors.

10. Wire connectors marked AL/CU are suitable for use with _____, _____, and _____ conductors.

11. A wire connector bearing no marking or reference to AL, CU, or ALR is suitable for use with (answer yes or no)

    a. copper conductors only. _____

    b. aluminum conductors only. _____

    c. copper and aluminum conductors. _____

12. Draw a line to match the proper *C.E.C., Part I* section to the specific subject covered.

    a. Section 28   1. Grounding

    b. Section 26   2. Fuses

    c. Section 26   3. Appliances

    d. Section 10   4. Motors

    e. Section 14   5. Air-conditioning equipment

13. When terminating aluminum wire to a device with a wire binding terminal screw, describe the correct termination procedure and provide the rule number. _____

    _____

    _____

    _____

    _____

14. The branch-circuit rating for an appliance shall not be less than the _____ rating of the appliance.

15. The branch-circuit overcurrent protective device for a typical electric water heater must not exceed 150% of the appliance's ampere rating. The appliance is marked "7.5 kW, 240 volts, single phase." What is the maximum overcurrent protective device permitted according to *Section 14*?

    _____

16. A disconnect switch is 20 metres from the motor and the motor controller. The motor is visible from the disconnect switch.

    a. Does the *C.E.C., Part I* consider this to be "within sight"? _____

    b. Give the reference that supports your answer to part (a). _____

    c. How should the installation be changed to meet code? _____

    _____

17. Appliances that have motor plus other loads (e.g., resistance heaters) are required to be marked with their (minimum) (maximum) overcurrent protection and (minimum, maximum) supply circuit conductor ampacity. Circle the correct answers.

18. An electric bake oven is rated at 7.5 kW, 208 volts, three phase. The terminals in the appliance, the disconnect switch, and the panel are suitable for 75°C conductors. Determine the following. Show your work.

    a. Current draw

Unit 3 Branch Circuit and Feeders

b. Minimum supply conductor ampacity

c. Size of Type R90 conductors

d. Ampere rating of fuses for branch-circuit protection

e. Size (ampere rating) of disconnect switch

f. Conduit size

UNIT 4

House Circuits (Owner's Circuits)

OBJECTIVES

After completing the study of this unit, the student will be able to

- diagram the proper connections for photocells and timer-controlled lights
- list and describe the main parts of an electric boiler control
- understand how to connect a sump pump circuit

Each occupant of the commercial building is responsible for electric power used within that occupant's area. It is the owner's responsibility to provide the power to light the public areas and to provide heat for the entire building. The circuits supplying the power to those areas and the devices for which the building owner is responsible are referred to as house circuits or owner's circuits.

LOADING SCHEDULE

The loading schedule for the house circuits indicates that they use the greatest proportion of the electric power metered in the commercial building (see Table 4–1). These circuits are fed from the house panel. The loading schedule can be divided into three parts:

- the boiler feed, which has a separate 800-ampere main switch in the service equipment.

- the emergency power system, including the equipment and devices that must continue to operate if the utility power fails.
- the remaining circuits that do not require connection to the emergency system.

LIGHTING CIRCUITS

Those lighting circuits considered to be the owner's responsibility can be divided into five groups, according to the method used to control the circuit:

- continuous operation
- manual control
- automatic control
- timed control
- photocell control

Continuous Operation

The luminaires installed at the top of each of the stairways to the second floor of the commercial building are in continuous operation. Since the stairways have no windows, it is necessary to provide artificial light even in the daytime. The power to these luminaires is supplied directly from the panelboard.

Manual Control

A conventional lighting control system supplies the remaining second-floor corridor lights as well as several other miscellaneous lights. All of the lights on manual control can be turned on as needed.

Automatic Control

The sump pump in the Motors section of Table 4–1, is an example of equipment that is operated by an automatic control system. No human intervention is necessary—the control system starts and stops the pump so as to keep the liquid in the sump between the preset maximum and minimum levels.

Timed Control

The lights at the entrance to the commercial building are controlled by a time clock located near the house panel in the boiler room. The circuit is connected as shown in Fig. 4–1. When the time clock is first installed, it must be adjusted to

Table 4–1   Owner's Area Calculation

Item	Count	VA/unit	Demand Load	Installed Load	Demand factor Feeder	Demand factor Service	Use Feeder	Use Service
MINIMUM BASIC LOAD								
(C.E.C. Rule 8-210)	66 m²	25 W/m²	1650 W		1	1	1650	1650
INSTALLED BASIC LOADS								
Style E luminaires	4	144		576			576	576
Style H luminaires	4	87		348			348	348
Style J luminaires	3	150		450			450	450
Style K luminaires	3	192		576			576	576
Style L luminaires	6	87		522			522	522
Style N luminaires	4	60		240			240	240
General receptacles	8	120		960			960	960
Installed basic loads							3672	3672
BASIC LOAD	(Use the greater of the demand or installed loads, *Rule 8–106(2)*.)						3672	3672
SPECIAL LOADS								
Telephone outlets	2	1250					2500	2500
Outdoor sign	1	1060					1060	1060
Motors	**Volts**	**FLA**	**Phase**					
Sump pump	120	9.4	1	1128	1.25		1410	1128
Circulating pump	120	4.4	1	528			528	528
Circulating pump	120	4.4	1	528			528	528
Circulating pump	120	4.4	1	528			528	528
Circulating pump	120	4.4	1	528			528	528
Circulating pump	120	4.4	1	528			528	528
Special loads subtotal							4050	3768
Total load							11 282	11 000

$$I = \frac{P}{E \times 1.73} = \frac{11\,282}{208 \times 1.73} = 31.35 \text{ A}$$

Assuming all loads are continuous, using standard-rated equipment, and T90/TWN75 copper conductors from *Table 2*, the ampere rating of the circuit would be:

$$\frac{31.35}{0.8} = 39.19 \text{ A}$$

The feeder would be 4 No. 8 T90/TWN75 copper conductors in a 21-mm ($\frac{3}{4}$-in) conduit, supplied by a 60-A disconnect c/w 45-A fuses.

the correct time. Thereafter, it automatically controls the lights. The clock motor is connected to the emergency panel so that the correct time is maintained in the event that the utility power fails. In the case of daylight saving time, a manual adjustment is required; forward in the spring and backward in autumn.

Photocell Control

The three lights located on the exterior of the building are controlled by individual photocells, Fig. 4–2. Control consists of a light-sensitive photocell and an amplifier that increases the photocell signal until it is sufficient to operate a relay that controls the light. The identified conductor (white) of the circuit must be connected to the photocell to provide power for the amplifier and relay.

SUMP PUMP CONTROL

The sump pump is used to remove water entering the building because of sewage line backups, water main breakage, minor flooding due to natural causes, or plumbing system damage within the building. Since the sump pump is a critical item, it is connected to the emergency panel. The pump motor is protected by a manual motor starter, Fig. 4–3, and is controlled by a float switch, Fig. 4–4. When the water in the sump

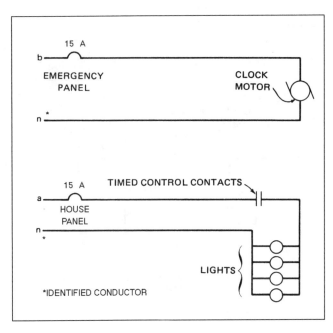

Fig. 4–1 Timed control of lighting.

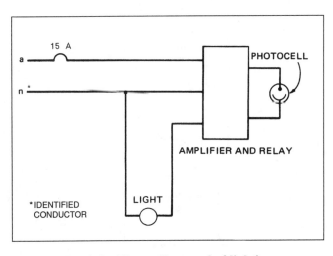

Fig. 4–2 Photocell control of lighting.

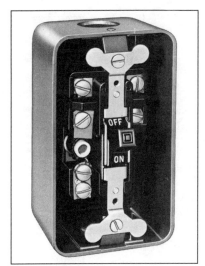

Fig. 4–3 Manual motor controller for sump pump.

Fig. 4–4 Float switch.

rises, a float is lifted that mechanically completes the circuit to the motor and starts the pump, Fig. 4–5. When the water level falls and the float lowers, the pump shuts off.

One type of sump pump available commercially is controlled by a water-pressure switch. The water pushes against a flexible neoprene gasket to operate a small integral switch within the submersible pump. This type of sump pump does not use a float.

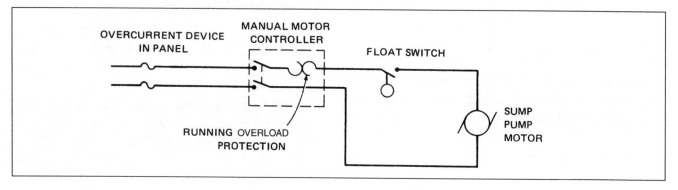

Fig. 4–5 Sump pump control.

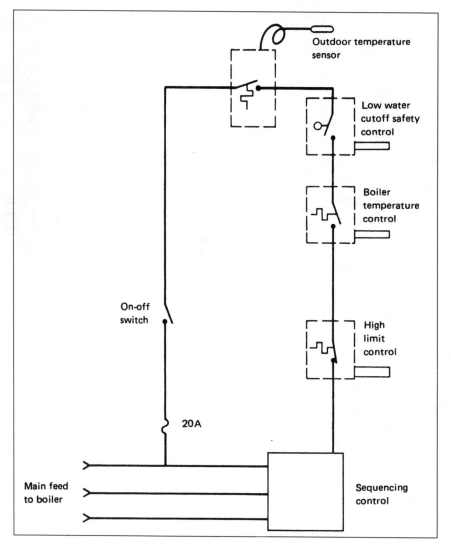

Fig. 4–6 Boiler control diagram.

BOILER CONTROL

An electrically heated boiler supplies heat to all areas of the commercial building. The boiler has a full-load rating of 200 kW. As purchased, the boiler is completely wired except for the external control wiring, Fig. 4–6. A heat sensor, plus a remote capillary tube and bulb, Fig. 4–7, is mounted so that the bulb is on the outside of the building. The bulb must be mounted where it will not receive direct sunlight and must be spaced at least 13 mm (1/2 in) from the brick wall. If these mounting instructions are not followed, the bulb will give inaccurate readings. The heat sensor is adjusted so that it closes when the outside temperature falls below a set point, usually 18°C (65°F). The sensor remains closed until the temperature increases to a value above the differential setting, approximately 21°C (70°F). Once the outside temperature causes the heat sensor to close, the boiler automatically maintains a constant water temperature and is always ready to deliver heat to any zone in the building that requires it.

A typical heating control circuit is shown in Fig. 4–8, while Fig. 4–9 is a schematic drawing of a typical connection scheme for the boiler heating elements.

The following points summarize the wiring requirements for a boiler:

- *Section 62* covers the rules for fixed electric space-heating equipment.
- The requirements for heating equipment that use fossil fuels are covered in *Subsection 26–800*.
- Heating equipment of this size will be supplied from its own branch circuit or feeder.
- A disconnect means must be installed to disconnect the ungrounded supply conductors.
- The disconnecting means for fossil-fueled heating equipment must be located so that a person operating the disconnect will not have to walk near or past the heating equipment in order to operate the disconnect, *Rule 28–806(5,6,7)*.

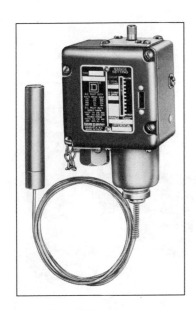

Fig. 4–7 Heat sensor with remote mounting bulb.

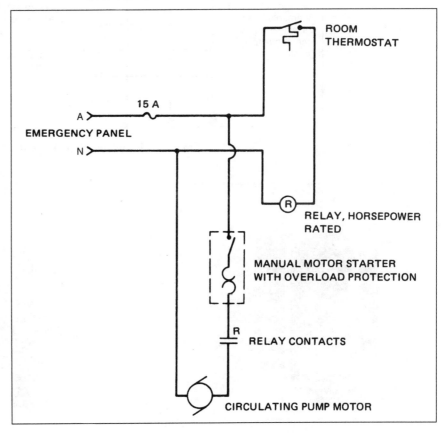

Fig. 4–8 Heating control circuit.

- The branch-circuit conductor sizes are to be based on 125% of the rated load.
- Overload protective devices are based on the service factor of the motor and are to be sized at 115% or 125% of the rated full load current, *Rule 28–306*.

It should be noted that the boiler for the commercial building is not sized for a particular heat loss; its selection is based on certain electrical requirements. Each of the building occupants has a heating thermostat located within their particular area. The thermostat operates a relay, Fig. 4–10, that controls a circulating pump, Fig. 4–11,

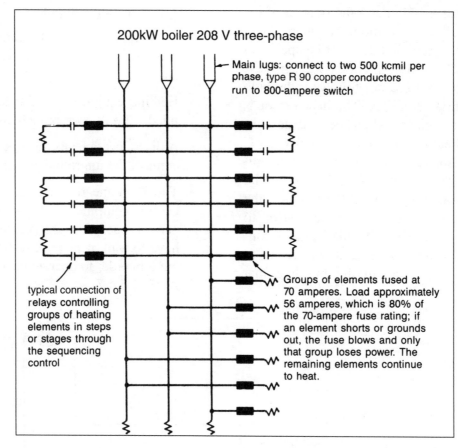

Fig. 4–9   Boiler heating elements.

Fig. 4–10   Relay in CSA Type 1 enclosure.

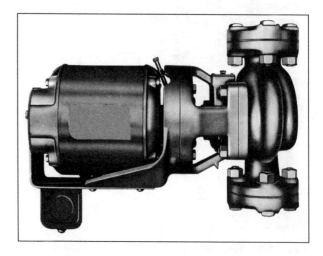

Fig. 4–11   Water circulating pump.

in the hot water piping system serving the area. The circuit to the circulating pump is supplied from the emergency panel so that the water will continue to circulate if the utility power fails, Fig. 4–12. In subfreezing weather, this continuous circulation prevents the freezing of the boiler water; such freezing can damage the boiler and the piping system. A thermostat controls a solenoid that opens and closes a zone valve to allow water to flow through the heater unit.

EMERGENCY AND EXIT LIGHTING

The requirement to install safety systems such as emergency and exit lights is based on Parts 3 and 9 of the National/Provincial Building Code and municipal bylaws. The installation of fire alarm systems is covered by *Section 32* of the *C.E.C., Part 1*. The installation of emergency and exit lights is covered by *Section 46* of the *C.E.C., Part 1*.

Appendix G of the C.E.C., Part 1 lists references to the electrical installation of fire safety systems required by the *National Building Code of Canada*.

Emergency Lighting

Emergency light fixtures will normally have incandescent or fluorescent lamps. They must have at least two lamps so that if one lamp fails, the area will not be left in total darkness. They

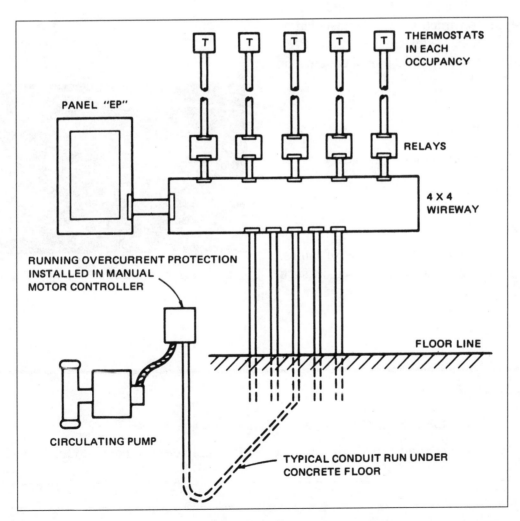

Fig. 4–12 Installation of Panel "EP" relays, thermostats, conduits, wireway, and motor controller to connect circulating pumps for each occupant.

may be supplied by a central emergency power system or from individual units.

Emergency lighting is required to provide a lighting level of 10 lux for a period of one-half to two hours (depending on the occupancy and the height of the building) from the time the normal power is lost. Many installations exceed these requirements.

Installation Requirements

Emergency lighting circuits must be kept independent of all other wiring and must not enter boxes, fittings, or raceways containing the conductors of any other system, except that they may enter transfer switches and light fixtures supplied from both the normal and emergency power supplies. *Rule 46-108*. (See Fig. 5-10.)

Emergency lighting circuits in buildings of noncombustible construction must be installed in a metallic raceway or cable. Rigid PVC and ENT may be used if embedded in 50 mm of concrete or masonry.

CENTRAL SUPPLY

See Unit 5 for a discussion of emergency power systems.

UNIT EQUIPMENT

An emergency lighting unit consisting of lamps, battery, and charger is commonly used in such applications as churches, schools, and offices where a central emergency supply is not available. *Subsection 46-300 C.E.C., Part 1* covers the installation requirements of unit equipment.

Some installation requirements of unit equipment are:

- Units must be mounted at a height of 2.0 m A.F.F. if possible.

- The receptacle supplying power to the unit must be mounted at a height of 2.5 m A.F.F. and not be farther than 1.5 m from the unit.

- If remote lamps are used, the voltage drop on the circuit supplying the lamps must not exceed 5% of the unit output voltage. See *Table D4, C.E.C., Part 1* for an example of determining voltage drop on unit equipment.

Figure 4–13   Emergency lighting unit.

REVIEW

Note: Refer to the *C.E.C., Part I* or the plans as necessary.

1. What size conduit(s) and wire are required to supply electric power to the boiler?

_____

_____

2. An overload protective device with a rating of _____ ampere(s) should be installed in the manual motor controller for the sump pump. (Show calculations.)

3. There are _____ circulating pumps.

4. The total load connected to the house panel is _____ volt-ampere(s).

5. The outside temperature sensor should not be installed where the _____ or the _____ can affect its reading.

6. Find the maximum trip rating for an inverse time circuit breaker used to provide short-circuit protection to a circulating pump motor branch circuit. (Show calculations.)

7. List four methods used to control lighting circuits originating in the house panel. Indicate a positive or negative feature for each type of control.

   a. _____
   b. _____
   c. _____
   d. _____

8. The disconnect means for the boiler must be located
   _____.

9. Conductors supplying an electrically heated boiler must have an ampacity of at least _____% of the rated load as required by *C.E.C., Part I, Rule* _____.

10. The sump pump operation is controlled by a(n) _____.

11. What is the actual current in each of the conductors serving the boiler when all elements are on? (Show calculations.)

UNIT 5

Emergency Power Systems

OBJECTIVES

After completing the study of this unit, the student will be able to
- select and size an emergency power system
- install an emergency power system

Many provincial and local codes require that equipment be installed in public buildings to ensure that electric power is provided automatically if the normal power source fails. The electrician should be aware of the special installation requirements for these emergency systems. *Section 46* of the *C.E.C., Part I* covers emergency system requirements.

SOURCES OF POWER (*RULES 46–200 TO 46–210*)

Battery-Powered Lighting (*RULE 46–202(1)(a)*)

When the need for emergency power is confined to a definite area, such as a stairway, and the power demand in this area is low, self-contained battery-powered units are a convenient and efficient means of providing power. In general, these units are wall mounted and are connected to the normal source by permanent wiring methods. Under normal conditions, this regular source powers a regulated battery charger to keep the battery at full power. When the normal power fails, the battery is automatically connected to one or more lamps that provide enough light to the area to permit its use, such as lighting a stairway sufficiently to allow people to leave the building. Battery-powered units are commonly used in stairwells, hallways, shopping centres, supermarkets, and other public structures.

Central Battery Power

If the power demand is high (excluding the operation of large motors), central battery-powered systems are available. These systems usually operate at 32 volts and can service a large number of lamps.

SPECIAL SERVICE ARRANGEMENTS

If separate utility power sources are available, *Rule 6–102(1)(a)*, Fig. 5–1, emergency power can be provided at a minimum cost. If protection is required to cover power failures within the building only, then a connection may be made ahead of the main switch, Fig. 5–2. However all of the equipment must have an adequate interrupting rating for the available short-circuit current (see Unit 18).

Unit 5 Emergency Power Systems 83

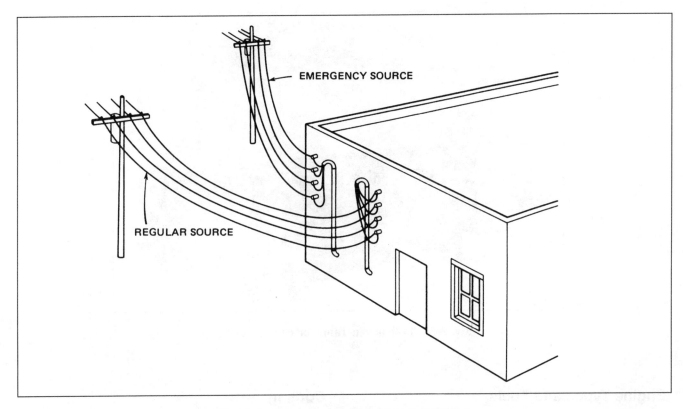

Fig. 5–1 Separate services, *Rule 6–102(1)(a)*.

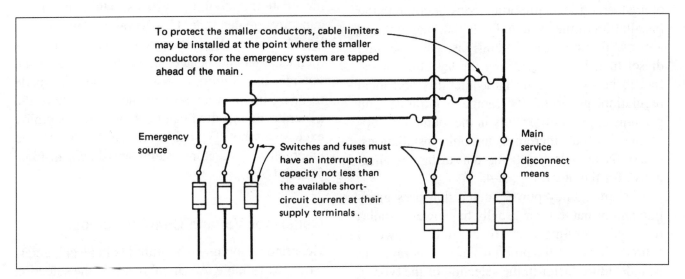

Fig. 5–2 Possible connection ahead of service disconnect means, Rule 32–204.

EMERGENCY GENERATOR SOURCE (RULE 46–202(3))

Generator sources may be used to supply emergency power. The selection of such a source for a specific installation involves consideration of the following factors:

- the engine type
- the generator capacity
- the load transfer controls

A typical generator for emergency use is shown in Fig. 5–3.

Fig. 5–3 Typical generator for emergency use.

Engine Types and Fuels

The type of fuel to be used in the driving engine of a generator is an important consideration in the installation of the system. Fuels that may be used are low pressure gas, natural gas, gasoline, and diesel fuel. Factors affecting the selection of the fuel to be used include its availability and local regulations governing its storage. Natural gas and gasoline engines differ only in the method of supplying the fuel; therefore, in some installations, one of these fuels may be used as a standby alternative for the other, Fig. 5–4.

An emergency power source that uses gasoline and/or natural gas usually has lower installation and operating costs than a diesel-powered source. However, the problems of fuel storage can be a deciding factor in the selection of the type of emergency generator. Gasoline is not only dangerous, but becomes stale after a relatively short period of time and thus cannot be stored for long periods. If natural gas is used, the Btu content must be greater than 1100 Btu per cubic foot. Diesel-powered generators require less maintenance and have a longer life. A diesel system is usually selected for installations having large power requirements, since the initial costs closely approach the costs of systems using other fuel types.

Cooling

Smaller generator units are available with either air or liquid cooling systems. Units having a capacity greater than 15 kilowatts generally use liquid cooling. For air-cooled units, it is recommended that the heated air be exhausted to the outside. In addition, a provision should be made to bring in fresh air so that the room where the generator is installed can be kept from becoming excessively hot. Typical installations of air-cooled emergency generator systems are shown in Figs. 5–5 and 5–6.

Generator Voltage Characteristics

Generators having any required voltage characteristic are available. A critical factor in the selection of a generator for a particular application is that it must have the same voltage output as the normal building supply system. For the commercial building covered in this text, the generator selected provides 120/208-volt, wye-connected, three-phase, 60-hertz power.

Capacity (*Rule 46–100*)

It is an involved, but extremely important, task to determine the correct size of the engine-driven

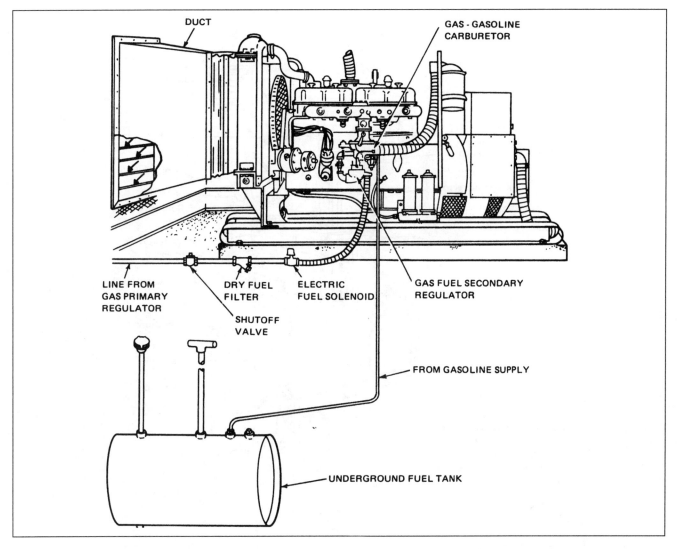

Fig. 5–4 Typical installation of a natural gas/gasoline engine-driven generator.

power system so that it has the minimum capacity necessary to supply the selected equipment. If the system is oversized, additional costs are involved in the installation, operation, and maintenance of the system. However, an undersized system may fail at the critical time when it is being relied upon to provide emergency power. To size the emergency system, it is necessary to determine all of the equipment to be supplied with emergency power. If motors are to be included in the emergency system, then it must be determined if all of the motors can be started at the same time. This information is essential to ensure that the system has the capacity to provide the total starting inrush kVA required.

The starting kVA for a motor is equivalent to the locked rotor kVA of the motor. The locked rotor kVA value is independent of the voltage

Table 430–7(b) Locked-Rotor Indicating Code Letters

Code Letter	Kilovolt-Amperes per Horsepower with Locked Rotor		
A	0	—	3.14
B	3.15	—	3.54
C	3.55	—	3.99
D	4.0	—	4.49
E	4.5	—	4.99
F	5.0	—	5.59
G	5.6	—	6.29
H	6.3	—	7.09
J	7.1	—	7.99
K	8.0	—	8.99
L	9.0	—	9.99
M	10.0	—	11.19
N	11.2	—	12.49
P	12.5	—	13.99
R	14.0	—	15.99
S	16.0	—	17.99
T	18.0	—	19.99
U	20.0	—	22.39
V	22.4	—	and up

Reprinted with permission from NFPA 70–1993, *National Electrical Code*®, Copyright© 1992, National Fire Protection Association, Quincy, Massachusetts 02269. This reprinted material is not the complete and official position of the NFPA on the referenced subject, which is represented only by the standard in its entirety.

Table 5–1 Locked-rotor indicating code letters.

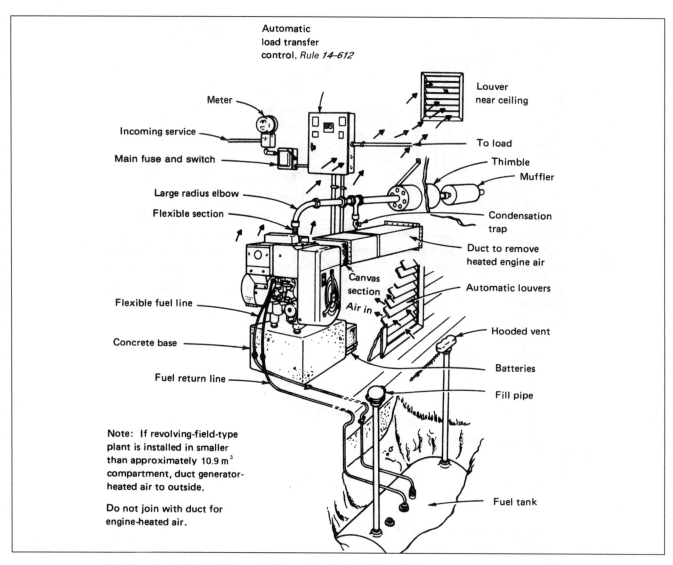

Fig. 5–5 Typical installation (pressure).

characteristics of the motor. Thus, a 5-horsepower, Code E motor requires a generator capacity of (see Table 5–1):

5 hp × 4.99 kVA/hp = 24.95 kVA

If two motors are to be started at the same time, the emergency power system must have the capacity to provide the sum of the starting kVA values for the two motors.

For a single-phase motor rated at less than $\frac{1}{2}$ horsepower, Table 5–2 lists the approximate kVA values that may be used if exact information is not available. The power system for the commercial building must supply the following maximum kVA:

Five $\frac{1}{6}$-horsepower C-type motors
5 × 1.85 kVA   = 9.25 kVA

One $\frac{1}{2}$-horsepower Code L motor
$\frac{1}{2}$ × 9.99 kVA   = 4.99 kVA

Lighting and receptacle load from
emergency panel schedule   = 4.51 kVA
Total   18.75 kVA

Thus, the generator unit selected must be able to supply this maximum kVA load as well as the continuous kVA requirement for the commercial building.

Continuous kVA requirement:
Lighting and receptacle load   4.510 kVA
Motor load   3.763 kVA
Total   8.273 kVA

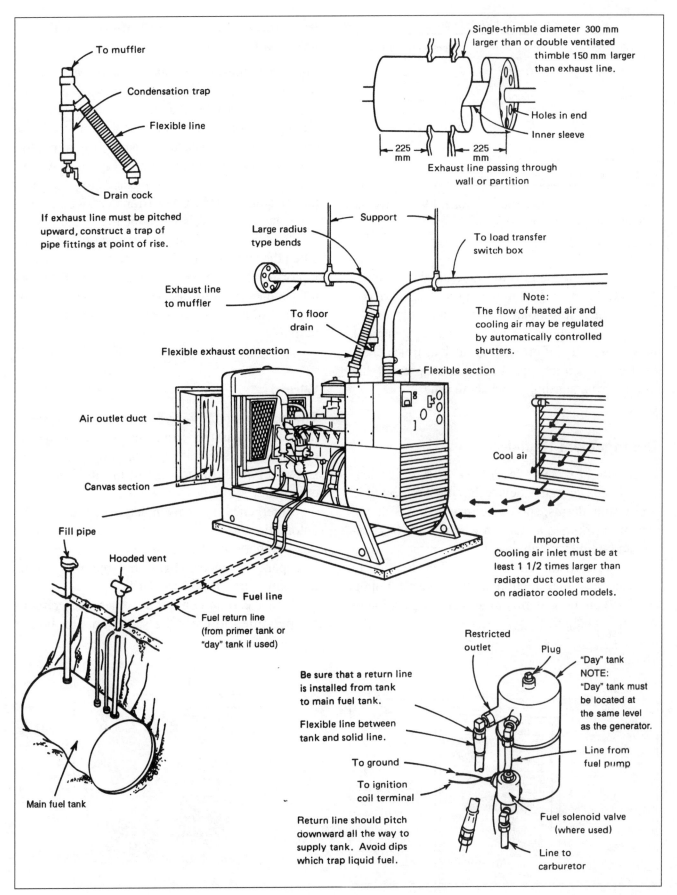

Fig. 5-6 Typical installation (gasoline).

88 Unit 5 Emergency Power Systems

Approximate kVA Values

HP	Type	Locked Rotor kVA
1/6	C	1.85
1/6	S	2.15
1/4	C	2.50
1/4	S	2.55
1/3	C	3.0
1/3	S	3.25

C = Capacitor-start motor
S = Split-phase motor

Table 5–2 Approximate locked-roter kVa values for fraction horsepower motors.

A check of manufacturers' data shows that a 12-kVA unit is available having a 20-kVA maximum rating for motor starting purposes and a 12-kVA continuous rating. A smaller generator may be installed if provisions, such as time delays on the contactors, can be made to prevent the motors from starting at the same time.

Derangement Signals

According to *Rule 46–208(1)*, a *derangement signal*, i.e., a signal device having both audible and visual alarms, should be installed outside the generator room in a location where it can be readily and regularly observed. The purposes of a device such as the one shown in Fig. 5–7 are to indicate any malfunction in the generator unit, any load on the system, or the correct operation of a battery charger.

Automatic Transfer Equipment *(Rule 46–204(1))*

Equipment must be provided to start the engine of the emergency generator if the main power source fails and to transfer the supply connection from the regular source to the emergency source. These operations can be accomplished by a control panel such as the one shown in Fig. 5–8. A voltage-sensitive relay is connected to the main power source. This relay (transfer switch) activates the control cycle when the main source voltage fails, Fig. 5–9. The generator motor is started when the control cycle is activated. As soon as the motor reaches the correct speed, a set of contactors is energized to disconnect the load from its normal source and connect it to the generator output.

Wiring

The branch-circuit wiring of emergency systems must be separated from the standard system except for the special conditions noted. Fig. 5–10 shows a typical branch-circuit installation for an emergency system.

Emergency systems must be controlled by automatic transfer equipment, *Rule 46–204(1)*. An automatic light-activated device may be used to separately control lights not required during daylight hours, *Rule 46–204(2)*.

Under certain conditions, emergency circuits are permitted to enter the same junction box as

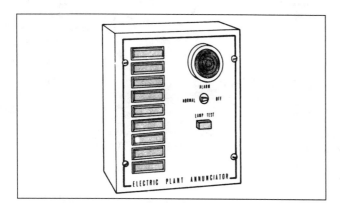

Fig. 5–7 Audible and visual signal alarm annunciator.

Fig. 5–8 Automatic transfer control panel.

Unit 5 Emergency Power Systems 89

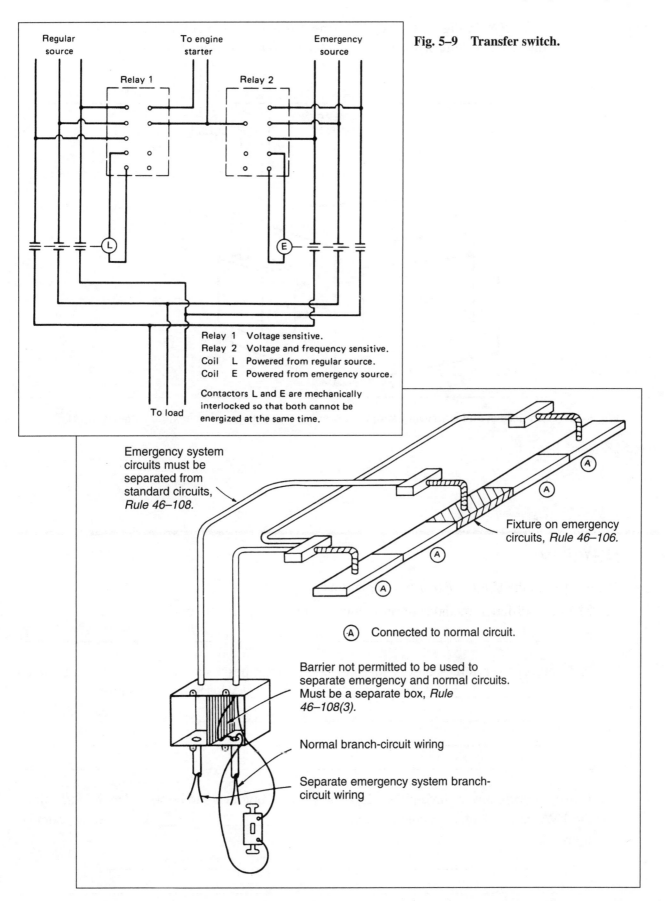

Fig. 5-9 Transfer switch.

Fig. 5-10 Branch-circuit wiring for emergency systems and normal systems.

normal circuits, *Rule 46–108(3)*. Fig. 5–11 shows an exit light that contains two lamps: one connected to the normal circuit, and the other connected to the emergency circuit.

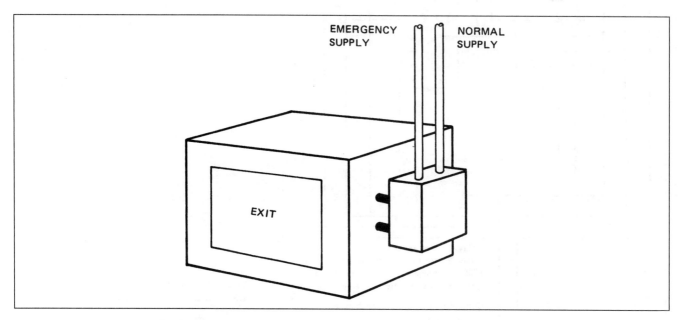

Fig. 5–11 Both normal and emergency supply may enter the same junction box, *Rule 46–108(3)(b)*.

REVIEW

Note: Refer to the *C.E.C., Part I* or the plans as necessary.

1. List three different methods of providing an emergency power source.

   a. _____, b. _____, c. _____

2. A $7\frac{1}{2}$-horsepower, three-phase, 230-volt Code F motor has a starting rating ranging between _____ kVA and _____ kVA. (Show calculations.)

   _____
   _____
   _____

3. If two $7\frac{1}{2}$-horsepower, three-phase, 230-volt Code F motors are to be started at the same time, the kVA capacity of the generator required is _____ kVA. (Show calculations.)

   _____
   _____
   _____

4. If two 7½-horsepower, three-phase, 230-volt Code F motors are to be operated by a generator, the minimum size generator, in kVA capacity, that can be used with selective control of the motors is _____ kVA. (Show calculations.)

_____

_____

_____

5. Name three conditions that can be indicated by a derangement signal device.

   a. _____, b. _____, c. _____

6. Emergency circuit wiring must be kept (together with) (separate from) normal circuit wiring. (Circle the correct answer.)

7. Three-way, four-way, and series-connected switches (are) (are not) permitted to control emergency lighting. (Circle the correct answer.)

UNIT 6

Fire Alarm and Safety Systems

OBJECTIVES

After completing the study of this unit, the student will be able to
- define the four stages of a fire
- list the four classes of fire
- list four fire signatures
- explain the purpose of a fire alarm system
- determine types of occupancy in which a fire alarm system is required
- list components of passive systems of fire prevention
- identify and define manual and automatic initiation devices
- list types of fire suppression systems
- determine minimum zoning requirements of commercial buildings for fire alarm systems
- list components of emergency standby systems for fire alarm systems
- identify the *Section 32* requirements for the installation of commercial fire alarm systems

SPECIAL TERMINOLOGY

The following special terminology applies to fire alarm systems.

Ancillary device — A device from another protection system that is actuated by the fire alarm system.

Annunciator — A device used to visually display signals received from a fire alarm system.

Class A circuit — A class A circuit (return loop circuit) is a circuit that begins at a pair of terminals in the control panel, makes a continuous loop through all the devices connected to the circuit, and terminates at a second pair of terminals in the control panel. A class A circuit does not have an end-of-line resistor.

Class B circuit — A class B circuit (terminated circuit) is a circuit which starts at a pair of terminals in the control panel, has one continuous path connecting the devices on the circuit, and terminates at an end-of-line resistor that may be mounted in the field beyond the last device or in the panel.

Electrical supervision — is the ability to detect a fault condition that would interfere with the normal operation of the circuit.

Fire detector — A device that can sense a fire condition. It includes heat and smoke detectors.

Trouble signal — An audible and visual signal resulting from failure of equipment or faults on circuits.

Alarm signal — An audible signal used to advise occupants of a building of an (emergency) fire condition.

Single-stage fire alarm system — is a system that will cause an alarm signal to sound on all the audible alarm devices of the system when any pull station or automatic detector is operated.

Two-stage fire alarm system — A system in which the operation of the first pull station or automatic detector initiates an alert signal. This is followed by an alarm signal if the alert signal is not acknowledged within five minutes.

FOUR STAGES OF A FIRE

In order to understand the purpose of a fire alarm system it is necessary to understand what fire is and how it works.

A fire consists of four main stages:

1. incipient stage
2. smouldering stage
3. flame stage
4. heat stage

In each of these stages there are identifying characteristics that may be detected by various means. Refer to Fig. 6–1. The earlier the detection of a fire, the less the damage and the quicker it may be extinguished.

- *Incipient Stage*: This is the first stage of a fire. There is no smoke and the heat being generated is almost undetectable. The fire is just beginning to appear.

- *Smouldering Stage*: This is the second stage of a fire. Smoke is now visible and various toxic gases are present. The fire is now readily detectable. It is important to note that life cannot be sustained where the smoke level in the air exceeds 6%. Most fire fatalities are caused by smoke inhalation and subsequent damage to lung tissue. Even at this stage some self-contained breathing apparatus (SCBA) is necessary to sustain life.

- *Flame Stage*: This is the third stage of a fire. Flame is now present but may be invisible, depending on the type of material being consumed.

- *Heat Stage*: This is the fourth and final stage of a fire. In this stage structural steel can buckle and the shattering of glass windows adds to the hazards. Uncontrolled heat and rapidly expanding air complete the development of the fire.

Fire is the third leading cause of accidental death in Canada.

Fire produces characteristic signatures, namely the presence of aerosols, the formation of smoke, visible light from flames, and the radiation of heat. See Fig. 6–2. Fires produce smoke and toxic gases that can overwhelm occupants if they are sleeping. Most fatalities are caused from the inhalation of smoke and toxic gases rather than from burns.

Heavy smoke reduces visibility and causes disorientation on the part of those trying to escape. Thus there is a great need for clear and easily understood signs to indicate the exits from the building and the location of fire fighting equipment.

FOUR CLASSES OF FIRE

The four classes into which fire can be divided are:

Class A—Wood, paper, rags

Class B—Vapour, air, gas, paint

Class C—Electrical

Class D—Combustible metals

A manual fire extinguisher will list the classes of fire that it is designed to extinguish on a label that is attached to the cylinder that holds the extinguishing agent.

Different extinguishing agents are required for the various classes of fire. Water under pressure is the most common. It deprives the fire of oxygen as it expands and turns into steam. Powder, foam, carbon dioxide, and halon are used in specialized situations.

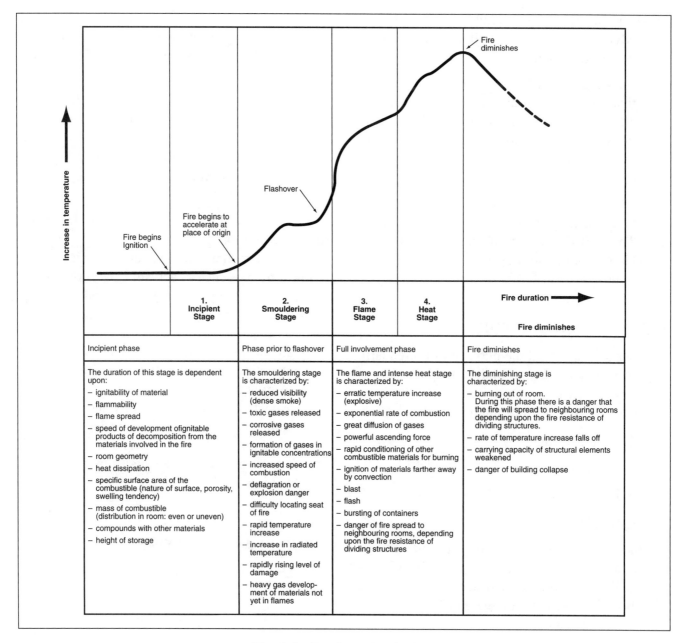

Fig. 6–1 Development of a fire.

FIRE TRIANGLE

In order to have successful combustion all of the components of the fire triangle must be met. The triangle consists of three main parts: heat, oxygen, and fuel. There is a fourth part, called the chain reaction component, which assists the fire. This will be discussed later.

Fire will not be able to start or grow if any of the following occurs:

1. Fuel is removed or controlled.

2. Oxygen is removed or controlled.

3. Heat is removed or controlled.

4. Chain reaction is removed or controlled.

The fire may be extinguished by smothering, because this will remove oxygen to below the critical level. It can also be extinguished by dilution of the combustible material or by cooling the material below the fire point.

Fire is a chemical reaction involving the three basic elements and the resultant continuing chain reaction. An unimpeded chain reaction is essential for its progress. The fire triangle of fuel,

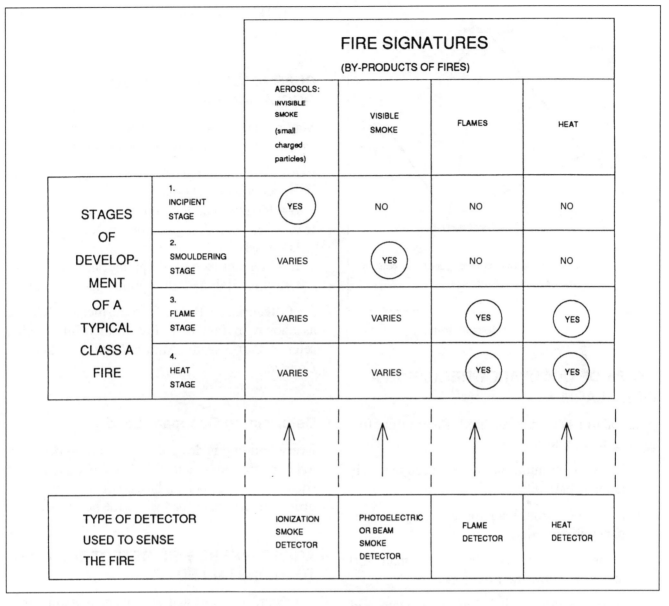

Fig. 6–2 Fire signatures.

oxygen, and heat thus evolves into the fire tetrahedron, Fig. 6–3. Any type of fire fighting or extinguishing revolves around removing one component of the chain reaction.

PURPOSE OF FIRE ALARM SYSTEMS

The primary purpose of fire alarm systems is to protect lives. The secondary purpose is to protect property and to initiate fire fighting in the event of a fire.

It should be noted that the best way to fight a fire is to prevent it from starting by avoiding circumstances that could initiate a fire. Modern early warning fire detection devices enable fires to be detected and acted upon before they can threaten life and property.

Once the alarm has been initiated, every occupant of the building can be automatically warned so that safe, controlled evacuation of the building can occur. In addition, the system can have the provision to automatically signal a monitoring company, which can then contact the local fire department.

Whenever a fire occurs with significant loss of life, the ensuing investigation will endeavour to find the cause and make recommendations to upgrade the fire code and building code. This will help prevent these situations from recurring.

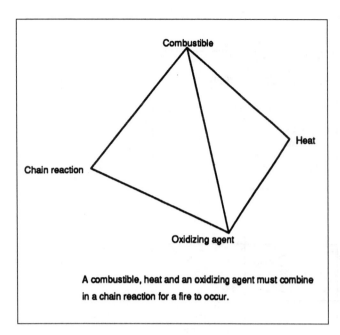

Fig. 6–3 Fire tetrahedron.

TYPES OF OCCUPANCY REQUIRING A FIRE ALARM

A fire alarm system is required if any of the following conditions are met:

- if the maximum permitted occupancy is greater that 300
- if the occupant load for any major occupancy is exceeded
- if the occupant load is greater than 150 above or below the first storey
- if the building, including basements and levels below grade, is more than three storeys in height
- if there is a contained use area, such as found in jails or mental health centres
- if there is an impeded egress zone
- if there is an interconnected floor space, such as open staircases between floors
- if the building contains a daycare facility with an occupancy greater than 40

These requirements are intended for guidance only and the authority having jurisdiction (AHJ) should be contacted to ensure that local building code and fire code regulations are complied with. The AHJ is usually the fire department for the city or region.

CLASSIFICATION OF MAJOR OCCUPANCIES

Major occupancies are divided into six main categories, with the following designations:

  A—Assembly, e.g., theatre, arena, restaurant, churches, schools
  B—Institutional, e.g., hospitals, jails
  C—Residential, e.g., homes, hotels
  D—Business
  E—Mercantile, e.g., shopping plazas
  F—Industrial, e.g., flour mills, garages, dairies

Categories A, B, and F are further subdivided, as shown in Fig. 6–4. For further information refer to Part 3 of the *National Building Code of Canada*.

Determining Occupant Load

Every building is designed for a certain occupant load by the architects. This is calculated on a square metre per person basis. For example, in an office the occupant load is 9.3 m^2 per person.

AREAS WHERE FIRE DETECTORS ARE TO BE INSTALLED

- Storage rooms that are outside dwelling units
- Service rooms that are outside dwelling units
- Janitor's rooms
- Elevator shafts
- Rooms used to store hazardous materials

AREAS WHERE HEAT DETECTORS ARE TO BE INSTALLED

- In every room in an area of a building classified as Group A Division 1, or Group B other than sleeping rooms
- Except in a hotel, in every suite, and every room not located within a suite, in portions

Group	Division	Description of Major Occupancies
A	1	Assembly occupancies intended for the production and viewing of the performing arts
A	2	Assembly occupancies not elsewhere classified in Group A
A	3	Assembly occupancies of the arena type
A	4	Assembly occupancies in which provision is made for the congregation or gathering of persons for the purpose of participating in or viewing open-air activities
B	1	Institutional occupancies in which persons are under restraint or incapable of self-preservation because of security measures not under their control
B	2	Institutional occupancies in which persons because of mental or physical limitations require special care or treatment
C	—	Residential occupancies
D	—	Business and personal services occupancies
E	—	Mercantile occupancies
F	1	High hazard industrial occupancies
F	2	Medium hazard industrial occupancies
F	3	Low hazard industrial occupancies

Fig. 6–4 Classification of major occupancies.

of buildings classified as Group C and more than three storeys in height

- In a hotel, in every room in a suite, and in every room not located within a suite, other than saunas and swimming pools or refrigerated areas

AREAS WHERE SMOKE DETECTORS ARE TO BE INSTALLED

The areas where smoke detectors are required are covered in the *National Building Code of Canada (N.B.C.)*. These are summarized as follows:

- Group A Division 1 — in every corridor
- Group B — every sleeping room including jail cells
  — every corridor or hallway serving as a means of egress from sleeping rooms
- Group C — every public corridor
  — every exit stair shaft
  — every room and corridor serving a contained use area
  — in the vicinity of draft stops, as per *Section 3.2.8.7 of the N.B.C.*

For the specific location on a ceiling or wall where the detector should be placed, refer to Fig. 6–5.

TYPES OF SYSTEMS

There are three building systems that protect occupants of buildings and the structure itself from fire.

1. Passive systems
2. Detection/signalling systems
3. Supression systems

Passive Systems

Passive systems include a variety of building components and choices made during the design

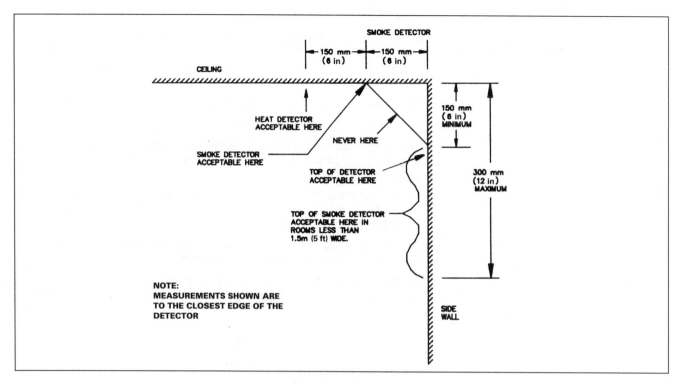

Fig. 6–5 Ceiling- or wall-mounted detectors.

and construction of a building. Many of these choices are mandated by the building and fire safety codes. These would include fire walls, separation between hazardous locations, fire resistance ratings, and flame spread ratings together with smoke control and fire stops.

There must be approved exit systems to give occupants a means of emergency egress from a building.

High-rise buildings pose special problems because of the stack effect. Created by wind and temperature differences, the stack effect can cause stairwells and corridors to fill with smoke during a fire.

Detection/Signalling Systems

A fire alarm system that is able to detect a fire and then warn the occupants and signal the fire department is another type of fire protection.

Fire alarm systems fall into two categories: local systems and proprietary systems. Most are local fire alarm systems. A local fire alarm system has three main components: (1) manual and automatic initiating and detection devices, (2) audible alarm signal devices, and (3) a central control unit to process the information. Refer to Fig. 6–6.

An annunciator panel at a remote location could be connected to the fire alarm system to indicate trouble or alarm in a zone. See Fig. 6–7.

Alarm detection and initiation can be manual or automatic. Manual pull stations provide for manual alarm initiation, Figs. 6–8A and 6–8B.

Heat and smoke detectors form a major category of automatic detection systems. Heat detectors can be either fixed temperature or a combination of fixed temperature and rate-of-rise. A dual action, fixed temperature and rate-of-rise heat detector is shown in Fig. 6–9. Smoke detectors can be of the ionization, photoelectric, or beam type. A typical ceiling-mount smoke detector is shown in Fig. 6–10.

Alarm signalling is usually in the form of bells, horns, and sirens. In areas where there is a high noise level or there are hearing-impaired occupants, visual signals such as strobe lights are used in addition to bells. Many modern systems use voice paging and tone signals on one-way speaker systems.

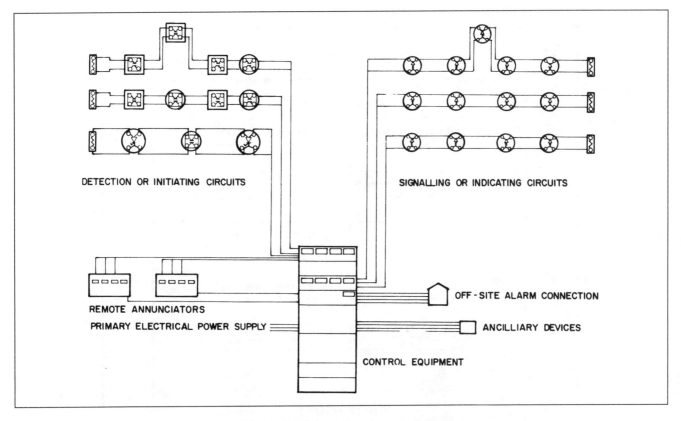

Fig. 6–6 Simple fire alarm system.

Supression Systems

The various fire extinguishing methods can be placed into four groups:

- diluting
- cooling
- smothering
- inhibiting chain-reaction

Among these methods there are a variety of extinguishing agents. Water is the most common because it is inexpensive and easily available. Wet pipe sprinklers constitute 97% of sprinkler system installations. They provide a very effective method of fire supression and are mandatory in new construction of almost all commercial buildings in Canada. The sprinkler system interconnects with the fire alarm system. Fig. 6–11 shows a typical sprinkler valve with initiating devices that are supervised by the fire alarm system. Various components of the sprinkler system could be monitored for the following indications of trouble:

- position of the supply valve
- loss of power to the fire pump
- low temperature in the supply tank
- high or low air pressure in pressure tanks
- failure of automatic fire pump controls

Other extinguishing systems include carbon dioxide, dry chemical, foam, and halon gas.

SMOKE DETECTORS

Two types of smoke detectors are commonly used: photoelectronic and ionization.

Photoelectronic Type

The photoelectronic type of smoke detector has a light sensor that measures the amount of light in a chamber. When smoke is present, a light is reflected off the smoke in the chamber, triggering the alarm. Refer to Fig. 6–12A. This type of sensor

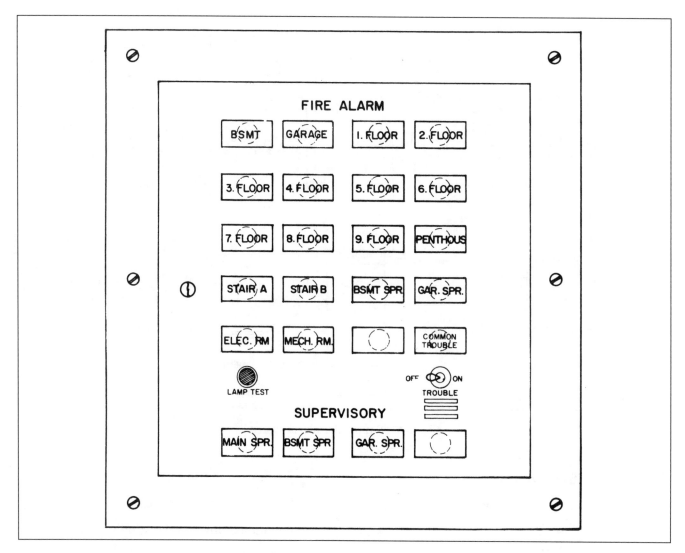

Fig. 6–7 Typical back-lit annunciator.

detects smoke from burning materials that produce great quantities of white or gray smoke, such as furniture, mattresses, and rags. This type of detector is less effective for gasoline and alcohol fires, which do not produce heavy smoke. Additionally, it is not as effective with black smoke, which absorbs light rather than reflecting it.

Ionization Type

The ionization type of detector contains a low-level radioactive source that supplies particles that ionize the air in the smoke chamber of the detector. Plates in this chamber are oppositely charged. Because the air is ionized, an extremely small amount of current (millionths of an ampere) flows between the plates. Smoke entering the chamber impedes the movement of the ions, reducing the current flow, which causes an alarm to go off. This type of detector is able to sense products of combustion (POC) in the incipient stage of a fire and is sometimes referred to as a POC detector. It can detect airborne particles (aerosols) as small as 5 microns in diameter.

The ionization type of detector is effective for detecting small amounts of smoke, as in gasoline and alcohol fires, which are fast-flaming. Also, it is more effective in detecting dark or black smoke than the photoelectric type of detector. A comparison of the effectiveness of the photoelectric smoke detector and the ionization smoke detector is shown in Fig. 6–12B.

Fig. 6–8A Manual fire alarm pull station.

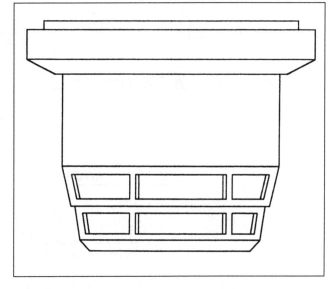

Fig. 6–10 Smoke detector.

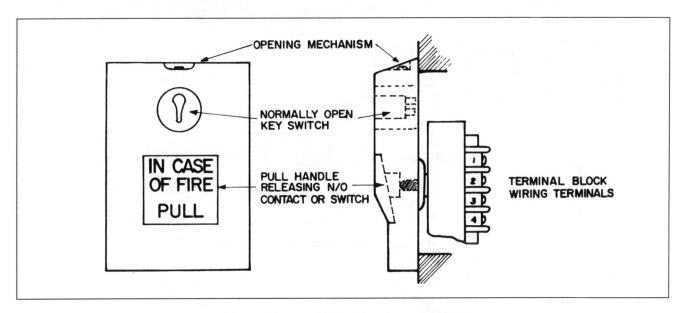

Fig. 6–8B Typical pull station assembly.

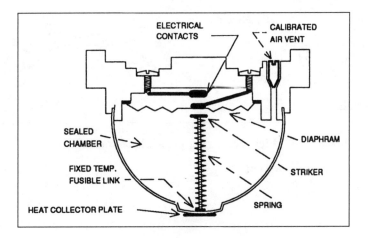

Fig. 6–9 Rate-of-rise heat detector.

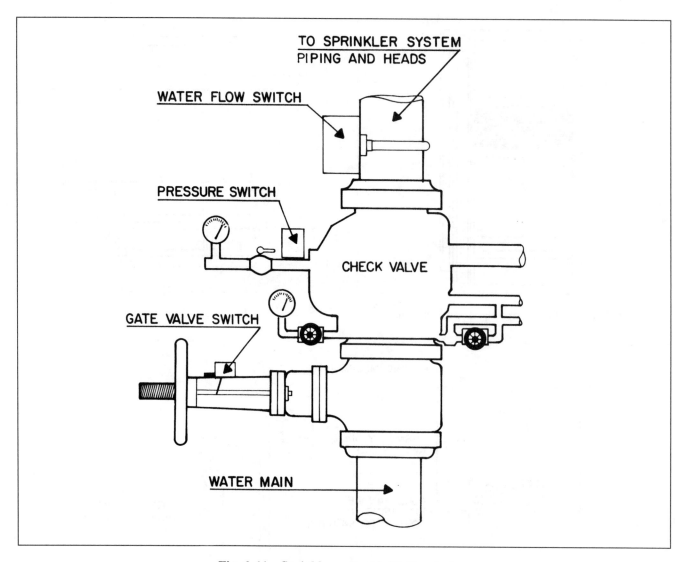

Fig. 6–11 Sprinkler system initiating device.

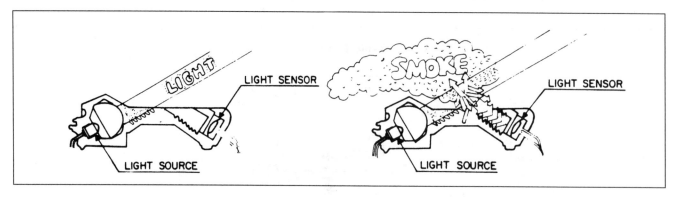

Fig. 6–12A Light scattering principle of smoke detector.

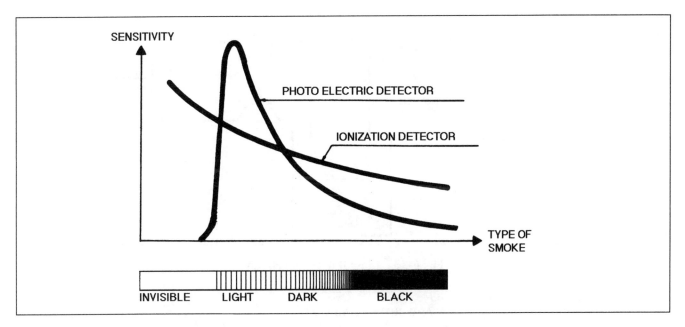

Fig. 6–12B Ionization vs. photoelectric smoke detector.

Features of Smoke Alarms

Smoke alarms may contain an indicator light to show that the unit is functioning properly. They may also have a test button. When the button is pushed, it is the circuitry and alarm that are tested, not the smoke-detecting ability.

Heat Detectors

Heat detectors are available that sense a specific *fixed* temperature, such as 57°C (135°F) or 90°C (194°F). Available also are *rate-of-rise* heat detectors that sense rapid changes in temperature (6°C (14°F) per minute), such as those caused by flash fires.

Fixed and rate-of-rise temperature detectors are available as a combination unit.

The spacing of heat detectors shall be as recommended by the manufacturer, because each detector is capable of sensing heat within a limited area in a given time limit.

Wiring Requirements

Class A and B Circuits

There are two classes of fire alarm system circuits—Class A and Class B. Class B is a two-wire system that has an end-of-line resistor and represents 80–90% of installed systems. A single break or ground fault in the wiring will initiate a trouble signal, but an alarm may not be sent in to the control panel.

A Class A fire alarm system is a four-wire system that originates and terminates in a terminal strip in the panel. It does not require an end-of-line resistor, since an end-of-line resistor is already mounted on a module located inside the panel. A single break or ground fault in the wiring will not prevent an alarm from being sounded, and a trouble signal will be initiated as soon as the ground fault or break in the wiring occurs.

Class B Circuits

Fig. 6–13 shows a typical Class B initiating circuit. Fig. 6–14 shows typical Class A and Class B signal circuits.

Because of the vital importance of smoke and fire detectors in preventing loss of life, it is suggested that the student obtain a copy of Underwriters Laboratories of Canada's standards.

Also, read very carefully the application and installation manuals that are prepared by the manufacturers of the smoke and heat detectors.

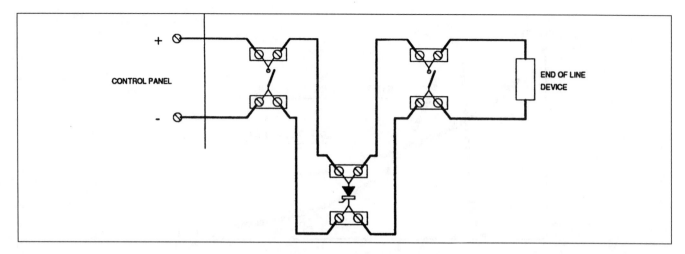

Fig. 6–13 Class B detection circuit.

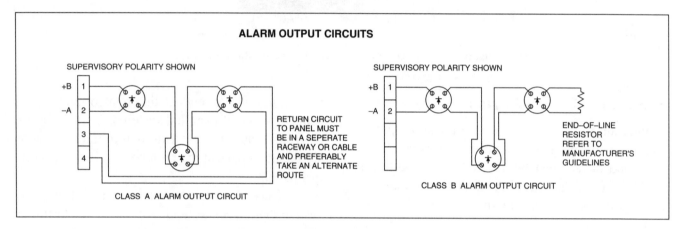

Fig. 6–14 Class A & B signal circuits.

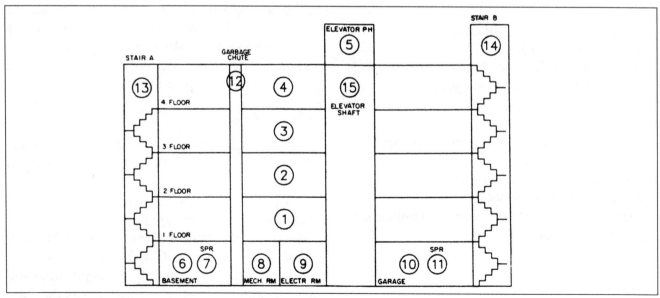

ZONING REQUIREMENTS

Alarm Receiving Zones: 1., 1st floor; 2. 2nd floor; 3. 3rd floor; 4. 4th floor; 5. Elev. penthouse; 6. Basement; 7. Bsmt. sprinkler; 8. Bsmt. mech. rm; 9. Bsmt. elec. rm.; 10. Garage; 11. Garage sprinkler; 12. Garage chute; 13. Stair A; 14. Stair B; 15. Elev. shaft. *Supervisory Receiving Zones:* 1. Sprinkler system; 2. Bsmt. sprinkler valve; 3. Garage sprinkler valve; 4. Bsmt. sprinkler low pressure; 5. Garage sprinkler low air pressure; 6. Standpipe main valve.

Fig. 6–15 Typical zoning of a five-storey building.

ZONING REQUIREMENTS OF FIRE ALARM SYSTEMS

If a building is not more than three storeys in height and has a total floor area of not more than 2000 m², then the fire alarm system does not require zoning. Even though this is the case, it is recommended that the fire alarm system installed in any building be zoned to provide quick location of a trouble or alarm signal in the building.

Fig. 6–15 shows a typical zoning of a five-storey building for the alarm initiating circuits. These would include smoke, heat, and rate-of-rise detectors in the required locations. In addition to this would be the signal circuit zoning to ensure adequate sound coverage to alert the occupants of an alarm.

FIRE CODES AND STANDARDS

Underwriters' Laboratories of Canada

The Underwriters' Laboratories of Canada (ULC) provides standards for the installation, verification, and inspection of fire alarm systems. These standards are:

- CAN/ULC S524–M91, "Standard for the Installation of Fire Alarm Systems"
- CAN/ULC S537–M86, "Standard for the Verification of Fire Alarm Systems"
- CAN/ULC S536–M86, "Standard for the Inspection of Fire Alarm Systems"

Minimum requirements are stated in the standards.

Fire warning devices commonly used in a commercial building are heat detectors, rate-of-rise detectors, and smoke detectors. These must be connected to a central fire alarm control panel that will process the information and send an output signal to alarm bells or buzzers.

In some provincial jurisdictions it is mandatory that qualified technicians install, verify, and maintain fire alarm systems. The certification to become a fire alarm technician is available through a variety of government approved courses, e.g., college courses, local electrical union (IBEW) courses, and Canadian Fire Alarm Association (CFAA) courses.

In recent years the standards of Underwriters' Laboratories of Canada together with the *National Building Code*, and associated provincial and municipal codes have made mandatory the installation of smoke alarms and heat and/or smoke detectors in commercial buildings that were built before these standards were in effect. A sufficient number of detectors must be installed so as to provide full coverage.

For more information please write to:

Underwriters' Laboratories of Canada
7 Crouse Road
Scarborough, ON M1R 3A9
Tel: (416) 757–3611

Canadian Fire Alarm Association

Organized in 1973, the Canadian Fire Alarm Association (CFFA) has the main objective of promoting a better understanding of the fire alarm industry. The CFAA's three specific goals are to foster and promote:

1. products and service standards for the fire alarm industry
2. the relationship among manufacturers, suppliers, and users of the products and services of the fire alarm industry
3. cooperation and understanding between CFAA members and those bureaus and agencies regulating the fire alarm industry

The CFAA offers courses to become certified as a Fire Alarm Technician and various other courses and seminars related to the fire alarm industry.

For more information please write to:

Canadian Fire Alarm Association
P.O. Box 262
Markham, ON L3P 2J7
Tel: (905) 513-5733
Fax: (905) 479-3639

C.E.C., PART I REQUIREMENTS (SECTION 32)

Section 32 covers the installation requirements of fire alarm systems and fire pumps. This section is an amendatory section, which means that its requirements override other sections of the *C.E.C., Part I* with regard to this type of installation.

Rule 32–102 covers the wiring methods required for the installation of conductors in a fire alarm system.

Rule 32–108 requires that the feed to the fire alarm system be as close to the load side of the service box as practicable, and that the disconnect be red and be lockable in the "On" position.

Rule 32–106 requires that wiring to dual terminals must have each wire must be independently terminated to each terminal.

It is recommended that the students familiarize themselves with the requirements of this section of the *C.E.C., Part I.*

FIRE PUMPS

Fire pumps and associated equipment may be connected to a separate service, according to *Rule 32–204*. This service may be placed at a different location from the main service, but must be labelled in a "conspicuous, legible and permanent manner" with the words "FIRE PUMP."

Consideration must be given to running the power feeds to fire pumps underground, *Rule 32–202, Appendix B*. Underground feeds are better protected from hazards such as flood, vandalism, or explosion, and so have fewer failures. All conductors must be copper, *Rule 32–200*. The wiring must be installed in a metal raceway or be an armored cable type listed in *Table 19*.

Overcurrent Protection for Fire Pumps

Overcurrent protection is based on the locked-rotor current (LRC) of the motor(s) in addition to the rated current of any control or power equipment on the circuit, *Rule 32–208*. No separate overload or overheating protection is required for fire pump motors; this protection is achieved by the overcurrent device, *Rule 32–210*. In a fire situation the fire pump is required to operate until the motor fails. Therefore reduced protection is permitted by *Rule 32–210*.

EXAMPLE: For a 10-horsepower, 575 volt, three-phase, Class H insulation, totally enclosed, nonventilated motor, the following is a calculation of disconnect, fuse, and wire size.

FLA = 11 amperes, *Table 44*

11 amperes $\times$ 6 = 66 amperes LRC, *Rule 32–208(2)*

Minimum overcurrent device is 70 amperes and minimum disconnect size is 100 amperes.

Minimum wire ampacity is 11 $\times$ 1.25 = 13.75 amperes, *Rule 32–200*.

Minimum wire size is No. 14 Type TWN75 copper, *Table 2*. Use $\frac{1}{2}$-in (16-mm) rigid or EMT conduit, *Table 6*.

From junction box to motor use No. 14 Type R90 silicone wire, *Rule 28–104(1), Table 37*, and *Table 19, Note 10*.

REVIEW

Note: Refer to the *C.E.C., Part I* or the plans where necessary.

1. List four stages of a fire.

   a. _____

   b. _____

   c. _____

   d. _____

2. Draw lines to connect the four classes of fire with the appropriate material.

   Class A   Electrical

   Class B   Paper

   Class C   Combustible materials

   Class D   Paint

3. Name two types of smoke detectors.

   _____

   _____

4. Which type is the most sensitive to black smoke?

   _____

5. What is the minimum clearance from the corner of a ceiling for a smoke detector?

   _____

6. List six types of circuits that may be connected to a fire alarm panel.

   a. _____

   b. _____

   c. _____

   d. _____

   e. _____

   f. _____

7. How does a rate-of-rise detector work?

   _____

   _____

   _____

8. With reference to Fig. 6–15, indicate
   a. the minimum number of detection or initiating circuits required.

   _____

   b. the minimum number of signalling circuits.

   _____

   c. where the remote annunciator panel will be located.

   _____

9. For a 15-horsepower, 575-volt, three-phase, Class H, totally enclosed, nonventillated fire pump, calculate:
   a. Wire size

   _____

   b. Minimum size of overcurrent device

   _____

   c. Minimum size of disconnect

   _____

   d. Minimum size of raceway

   _____

10. Draw a diagram of a Class B initiating circuit.

UNIT 7

Hazardous Materials in the Workplace

OBJECTIVES

After completing the study of this unit, the student will be able to
- describe the Workplace Hazardous Materials Information System (WHMIS)
- explain the responsibilities of suppliers and employers
- describe a typical Materials Safety Data Sheet (MSDS)
- explain the responsibilities of employees
- provide a list of relevant health hazards
- describe the major classes of hazardous materials
- provide a list of symbols for hazardous materials

WORKPLACE HAZARDOUS MATERIALS INFORMATION SYSTEM

The Workplace Hazardous Materials Information System (WHMIS) is a Canada-wide legislated system of providing information on workplace hazardous materials and how to safely use, store, and handle them.

Many of the materials in today's workplaces can cause injury, illness, or death due to fire, shock, explosion, or other serious accidents if they are not handled properly. WHMIS describes the dangers of these workplace materials and how workers can protect themselves from the hazards associated with them. WHMIS legislation also makes employers responsible for providing their workers with workplace-specific training and education.

The legislation was passed at the federal level (Bill C70) and states that workers "have the right to know" about hazardous materials to which they are exposed on the job. The legislation at the provincial level incorporates the federal legislation and introduces the concept of the Material Safety Data Sheet (MSDS) and the public's right to know concerning hazardous material in the public domain.

The major components of WHMIS legislation are as follows:

- warning labels must be placed on containers of hazardous materials.
- hazards of each material must be listed on an MSDS available in the workplace.
- employee education programs concerning WHMIS are mandatory in the workplace.

Compliance with WHMIS legislation falls into three categories and requires action by suppliers, employers, and employees.

1. *Responsibility of the supplier.* Suppliers have to determine which of their products intended for use in the workplace in Canada are hazardous materials under WHMIS. In addition, as a condition of sale or importation, suppliers have to provide information about the products that are hazardous materials. This information is provided in the form of a regulation label and an MSDS.

2. *Responsibility of the employer.* Employers must ensure that any container of potentially hazardous material received at the workplace has the required label. In addition, any hazardous material transferred from a supply container into another container must have a label fixed to the new container indicating its contents.

    An MSDS for each of these materials must be readily accessible at the worksite and be given to the joint health and safety committee. An MSDS must be provided for any hazardous material produced at the workplace as well.

    The employer also has the responsibility of training the employees in all matters related to WHMIS and MSDSs. The training program is to be reviewed on an annual basis in consultation with the health and safety committee.

3. *Responsibility of the employee.* The employee is required to participate in the training program and to use the information learned to maintain a safe and healthy workplace.

HEALTH

Hazardous materials are in use in all aspects of our lives. This includes many commonly-used cleaning supplies, paints, chemicals, solvents, and glues.

TYPICAL HAZARDS

- *Personal exposure (health).* Certain chemicals can cause harmful effects to the body. These can include burns, eye irritation, and allergic reactions. In addition, the breathing of chemical vapours into the body can harm various organs.

- *Reactivity.* Certain chemicals, when mixed with other chemicals or exposed to heat, water, or air, may burn, explode, or form toxic vapours. For example, mixing chlorine bleach with acid-based cleaners will produce poisonous chlorine gas. If this is breathed it will have a serious effect on the body.

- *Flammability.* Chemicals that may catch fire with improper use must be identified as being flammable.

WORKPLACE CONTROLS

The harmful effects of hazardous materials in the workplace can be minimized by the provision of proper storage areas with adequate ventilation, the implementation of safe work procedures, and the proper disposal of the hazardous materials.

PERSONAL PROTECTIVE EQUIPMENT

Appropriate personal protective equipment must be worn by employees when workplace controls are not possible. This may include adequate eye and face protection, gloves, proper footwear, safety mats, protective clothing, and breathing protection.

It must be mandatory that employees wear the appropriate personal protective equipment provided for them.

This is extremely important when the employee may be exposed to a hazardous material. For example, when taking an oil sample from a transformer that may contain PCBs or topping up liquid levels in batteries in battery rooms, the appropriate safety equipment must be worn by the employee. If an employee does not comply with this requirement, disciplinary action must be mandatory.

All staff must follow basic personal hygiene procedures, which include washing hands after chemical use and not eating or drinking where chemicals are used or stored.

MATERIAL SAFETY DATA SHEET

Employees have the right to know about the products they are using on the job. The information is found on the Material Safety Data Sheet (MSDS) for each product. See Fig. 7–1 on pages 112 to 114 for an example of an MSDS.

The MSDS must include the following information.

a. *Product identifier:* the common name and the chemical name, for example, household bleach/sodium hypochlorite

b. *Hazard symbols:* symbols that indicate materials that are flammable, corrosive, explosive, or poisonous (e.g., flammable symbol in Fig. 7–2)

c. *Physical data:* included in the MSDS

d. *Fire and explosion data:* includes the flash point of gases or vapours of the product

e. *Reactivity:* description of how the product reacts with other substances (For example, household bleach when mixed with acid forms chlorine gas—one of the types of corrosive poison gas that caused such devastation in World War I.)

f. *Toxicological properties:* describes the health effects of exposure by ingestion, inhalation, or direct skin or eye contact

g. *Preventive measures:* describes protective clothing required when using the product

h. *First aid measures:* describes emergency and first aid procedures (For example, certain products require vomiting to be induced while other products, such as acids, indicate that vomiting should not be induced.)

i. *Date and source of MSDS:* used to determine whether the information is the most up-to-date available.

SUPPLIER LABEL

Each package of a product must be provided with a label in both official languages. This label has a distinctive slashed border and depicts the applicable hazard symbols for the product. For an example of a supplier label, see Fig. 7–2.

Employers must ensure that all hazardous products bear a supplier label before they enter the workplace.

The supplier of the product should be identified along with an address and telephone number, and the information should be highlighted so that it stands out from the background colour of the container.

If a product arrives without a label, or is placed or poured into a new container, employers are responsible for ensuring that a workplace label is attached.

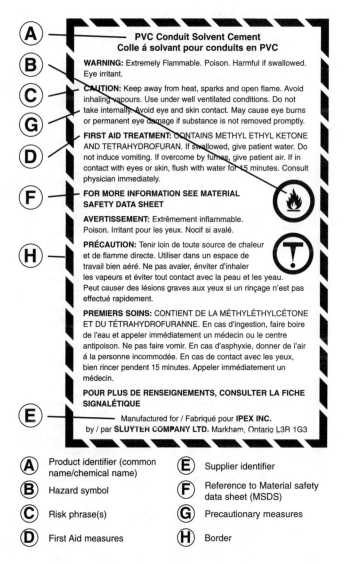

Fig. 7–2 Supplier label.

08/22/94	MATERIAL SAFETY DATA SHEET : 00000022	PAGE:1

IPEX Inc.
Port of Montreal Building
Wing #3, First Floor Cite Du Harve, Mtl, Quebec H3C 3R5

Product : PVC Conduit Cement (S100)

Section 01: Chemical Product and Company Identification

Manufacturer	IPEX Inc.
	Port of Montreal Building
	Wing #3, First Floor
	Cite Du Harve, Mtl, Quebec; H3C 3R5
Prepared By	Process Development
Preparation Date	Apr 07/93
Product Name	PVC Conduit Cement (S100)
Product Uses	Adhesives. For PVC pipe, conduit and fittings. Refer to label for information.
Chemical Family	Poly vinyl chloride based solvent cement.
Material Use	Adhesive.

Section 02: Composition/Information on Ingredients

Hazardous Ingredients	%	T.L.V.	C.A.S. #	LD/50, ROUTE, SPECIES	LC/50, ROUTE, SPECIES
Tetrahydrafuran	30–60	200 PPM	109-99-9	1650 MG/KG ORAL RAT	18000 PPM 4 HRS RAT
Methyl Ethyl Ketone (MEK)	30–60	200 PPM	78-93-3	3300 mg/kg ORAL RAT	not available
Cyclohexanone	3–7	25 ppm/200 mg/m^3	108-94-1	1535 MG/KG ORAL RAT	50 PPM 1 HR RAT

Section 03: Hazards Identification

Route of Entry	Skin contact. Inhalation. Ingestion. Eye contact.
Skin Contact	Can cause moderate irritation, defatting and dermatitis.
Skin Absorption	No data available.
Inhalation, Chronic	See "Effects of Chronic Exposure."
Inhalation	Breathing of high vapour concentrations may have results ranging from dizziness and headache to unconsciousness. May be anaesthetic and may cause other central nervous system effects.
Ingestion	Can cause gastro-intestinal irritation, nausea, vomiting and diarrhea. Small amounts of liquid aspirated into respiratory system can cause severe health effects (e.g., Bronchopneumonia or Pulmonary Edema).
Eye Contact	Contains materials that are severely irritating to the eyes.
Effects of Acute Exposure	As described above.
Effects of Chronic Exposure	May cause damage to the central nervous system. Prolonged or repeated skin contact may cause drying or cracking of skin.

Section 04: First Aid Measures

Eye Contact	Immediately flush with water for a minimum of 20 minutes. Get medical attention.
Skin Contact	Remove contaminated clothing. Wash affected area with water and soap. Seek medical attention if irritation occurs or persists.
Inhalation	Remove victim to fresh air. If not breathing, qualified personnel should administer artificial respiration. Get medical attention.
Ingestion	Do not induce vomiting. Get immediate medical attention.
Additional Information	Get in contact with your local poison control centre.

Section 05: Fire Fighting Measures

Flammable?	Yes
If Yes, under which conditions	Extremely flammable liquid. Dangerous fire hazard when exposed to heat, flame or temperatures above the flash point. Vapours are heavier than air and may travel to a source of ignition and flash back.
Special Procedures	A self-contained breathing apparatus is required for fire fighters. Use water spray to cool fire-exposed surfaces and to protect personnel.
Flash Point (C), Method	–14°C Tag closed cup.
Auto Ignition Temperature	321°C
Upper Flammable Limit (% VOL)	11.8
Lower Flammable Limit (% VOL)	2.0
Extinguishing Media	"Alcohol" foam, CO_2, dry chemical.
Hazardous Combustion Products	Carbon dioxide. Carbon monoxide, hydrogen chloride. Smoke.

Fig. 7–1 Material safety data sheet (MSDS).

08/22/94	MATERIAL SAFETY DATA SHEET : 00000022	PAGE:2

Product : PVC Conduit Cement (S100)

Section 05: Fire Fighting Measures

Sensitivity to Mechanical Impact ....Unknown.
Sensitivity to Static Discharge ....May be sensitive.

Section 06: Accidental Release Measures

Leak/Spill ....Ventilate. Remove all sources of ignition, open flames, sparks, etc. Wear protective gear. (See Section 8.) Large spills should be collected for disposal. Small spills may be wiped. Use a noncombustible absorbent inorganic material. Prevent runoff into drains, sewers, and other waterways.

Section 07: Handling and Storage

Handling Procedures ....Avoid skin and eye contact. Avoid breathing vapours. Use adequate ventilation. Keep away from heat, sparks and open flame.
Storage Needs...................................Store away from all sources of heat and ignition. Store in a well-ventilated area. Keep container closed when not in use.

Section 08: Exposure Controls/Personal Protection

Protective Equipment
    Eye/Type..................................Safety glasses.
    Respiratory/Type ....None required for normal use if adequate ventilation is maintained. Use NIOSH/MSHA-approved air supply respirator if TLV is exceeded.
    Gloves/Type............................Wear impervious gloves (neoprene or rubber).
    Clothing/Type..........................Not applicable.
    Footwear/Type........................Not applicable.
    Other/Type..............................Eye bath and safety shower.
Ventilation Requirements...................Natural or mechanical (explosion proof) ventilation to keep vapour concentration well below TLV.

Section 09: Physical and Chemical Properties

Physical State ....Liquid
Odour...Ketone odour
Specific Gravity.................................0.90 - 1.00
Odour Threshold (ppm) ....Not available
Vapour Pressure (mm Hg)................145 @ 20°C
Vapour Density (Air=1) ....> 1
Evaporation Rate ....6.0 (NBUAC = 1)
Boiling Point (deg C).........................65
pH ....Not applicable
Solubility in Water (% W/W) ....Negligible
Coefficient of Water/Oil.....................Not available
Distribution
Freezing Point...................................< 0°C
Melting Point (deg C)........................Not applicable
Molecular Weight ....Not applicable

Section 10: Stability and Reactivity

Incompatibility ....Strong acids and strong bases. Chlorinated solvents. Oxidizing agents
Reactivity Conditions ? ....Excessive heat, sparks and open flame
Hazardous Products of
    Decomposition.............................Oxides of Carbon (CO, CO_2). Toxic fumes. Smoke

Section 11: Toxicological Information

Exposure Limit of Material ....See Hazardous Ingredients Section (2)
Irritancy of Material ....Moderate
Sensitizing Capability of Material.......Not available
Material
Carcinogenicity ....Reversible liver damage at high dosages is possible.
Teratogenicity....................................No information is available and no adverse teratogenicity effects are anticipated.
Mutagenicity......................................No information is available and no adverse mutagenicity effects are anticipated.
Reproductive Effects.........................Not available
Synergistic Materials.........................Not available

Fig. 7–1 Material safety data sheet (MSDS) (continued).

08/22/94	MATERIAL SAFETY DATA SHEET : 00000022 PAGE:3

Product : PVC Conduit Cement (S100)

Section 12: Ecological Information

Environmental .... Not available
Biodegradability .... Not available

Section 13: Disposal Considerations

Waste Disposal .... Spilled material and water rinses are classified as chemical waste. Dispose of in accordance with current local, provincial and federal regulations.

Section 14: Transport Information

T.D.G. Classification .... Adhesives Class 3.2 U.N. 1133 P.G. II

Section 15: Regulatory Information

CPR Compliance .... This product has been classified in accordance with the hazard criteria of the CPR and the MSDS contains all the information required by the CPR.
WHMIS Classification .... Class B Div. 2 Flammable Liquid Class D DIV.2B Toxic Material

Section 16: Other Information

CANUTEC Emergency (613) 996-6666

Fig. 7–1 Material safety data sheet (MSDS) (concluded).

WORKPLACE LABEL

A label may be made up by the employer when using the product in small quantities. This label may be smaller than the supplier's label but must contain at least the following:

a. product identification.

b. preventive measures or precautions

c. a statement indicating that an MSDS is available

An example of a workplace label is shown in Fig. 7–3.

PVC GLUE
— *Wear safety gloves.*
— *Use in well-ventilated area.*
— *Refer to MSDS for further information.*

Fig. 7–3 Workplace label.

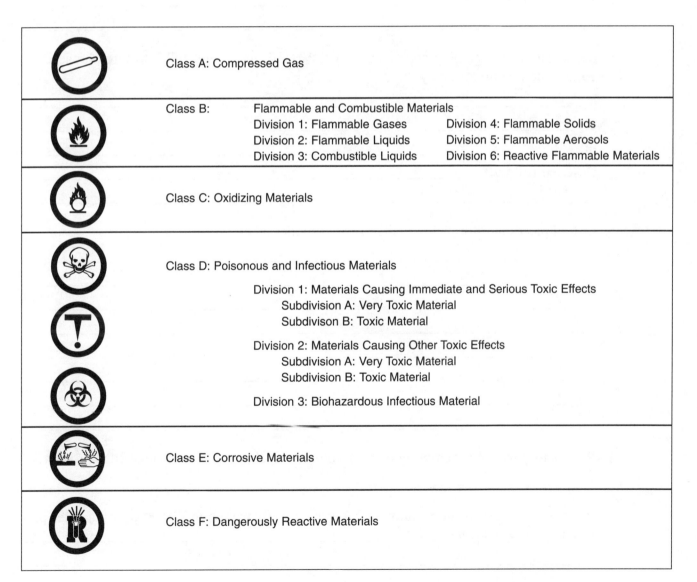

Fig. 7–4 Hazardous material symbols.

HAZARDOUS MATERIALS

There are six major classes of hazardous materials. These are shown with their symbols in Fig. 7–4. They are as follows:

Class A: Compressed gas
Class B: Flammable and combustible materials
Class C: Oxidizing materials
Class D: Poisonous and infectious materials
Class E: Corrosive materials
Class F: Dangerously reactive materials

REVIEW

1. What does WHMIS stand for? _____

2. What does MSDS stand for? _____

3. Responsibilities regarding compliance with WHMIS legislation fall into three broad categories, and require action by:
   i. _____
   ii. _____
   iii. _____

4. Describe Bill C70. _____

5. List all pieces of information required on a supplier label.
   _____

6. Who is responsible for ensuring that a workplace label appears on an unlabelled product? _____

7. What is the employee's obligation with regard to participation in WHMIS programmes provided by the employer? _____

8. What do the following hazard symbols stand for?

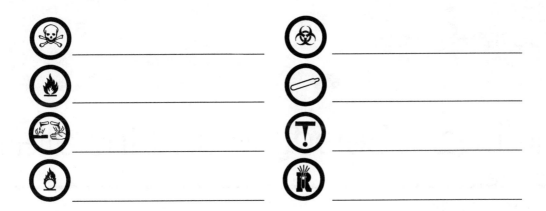

9. Using the MSDS in Fig. 7–1, answer the following questions.

   a. What are the chronic effects of exposure to this product?

   b. How should this product be handled?

   c. What are the storage requirements of this product?

   d. What materials should you avoid when using this product?

UNIT 8

Panelboard Selection and Installation

OBJECTIVES

After completing the study of this unit, the student will be able to

- select a panelboard, given the number of required branch circuits and their loads
- connect the branch circuits to a panelboard
- select the feeder size, given the loading of the panelboard

Changing needs for electrical energy in a commercial building will require the electrician to install new branch circuits. In addition, as the existing panelboards are filled, new panelboards must be installed. As a result, the electrician must be able to select and connect panelboards and the feeders to supply them.

PANELBOARDS

Separate feeders must be run from the main service equipment to each separate area of the commercial building. Each feeder will terminate in a panelboard that is located in the area to be served, Fig. 8–1.

A panelboard may be defined as a single panel or group of panel units designed for assembly in the form of a single panel; such a panel will include bus bars and automatic overcurrent devices, and may or may not include switches for the control of light, heat, or power circuits. A panelboard is designed to be installed in a cabinet or cutout box placed in or against a wall or partition. This cabinet (and panelboard) is to be accessible only from the front. *Rule 14–108* describes the enclosure of overcurrent devices. Panelboards are referred to as dead front because no live parts are accessible without removing the cover.

While a panelboard is accessible only from the front, a switchboard may be accessible from the rear as well.

Lighting and appliance branch-circuit panelboards generally are used on single-phase, three-wire systems and three-phase, four-wire electrical systems. This type of installation provides individual branch circuits for lighting and receptacle outlets and permits the connection of appliances such as the equipment in the bakery.

Panelboard Construction

In general, panelboards are constructed so that the main feed bus bars run the height of the panelboard. The branch-circuit protective devices are

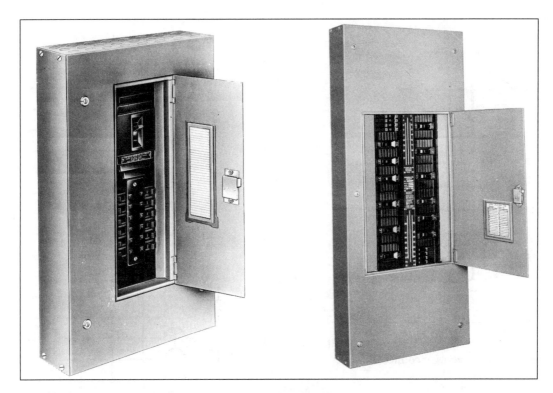

Fig. 8–1 Panelboards.

connected to alternate main buses as shown in Figs. 8–2A and 8–2B. In an arrangement of this type, the connections directly across from each other are on the same phase and the adjacent connections on each side are on different phases. As a result, multiple protective devices can be installed to serve the 208-volt equipment. The numbering sequence shown in Figs. 8–2A, 8–2B, and 8–3 is the system used for most panels. Figs. 8–4 and 8–5 show the phase arrangement requirements of *Rule 4–036(4,5)*.

Number of Circuits

The number of the overcurrent devices in a panelboard is determined by the needs of the area being served. Using the bakery as an example, there are 13 single-pole circuits and six 3-pole circuits. This is a total of 31 poles. When using a three-phase supply, the incremental number is six (a pole for each of the three phases on both sides of the panelboard). The minimum number of poles that could be specified for the bakery is 36.

Colour Code for Branch Circuits

The "hot" ungrounded conductors may be any colour *except* green or green with a yellow stripe (colours which are reserved for grounding purposes only), or white or natural gray (colours which are reserved for the grounded circuit conductors). See *Rule 4–036*. Table 8–1 provides suggestions for colour coding of conductors.

Table 8–1

Suggested Colour Coding of Conductors	
Two-wire circuits, 120 volts	One black, one white
Three-wire circuits, 120/240 volts	One black, one white, one red
Three-phase, four-wire wye circuits, 120/208 volts	One black, one white, one red, one blue
Three-phase, four-wire delta circuits, 240 volts	One black, one white, one blue, one red (for high leg A phase)
Three-phase, four-wire wye circuits, 347/600 volts	One black, one white, one red, one blue
Three-phase, three-wire delta circuits, 600 volts	One red, one black, one blue

120 Unit 8 Panelboard Selection and Installation

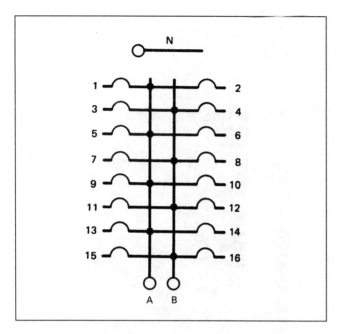

Fig. 8–2A Lighting and appliance branch-circuit panelboard; single-phase, three-wire connections.

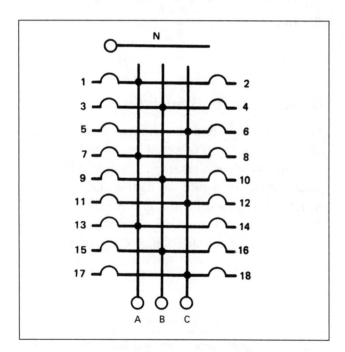

Fig. 8–2B Lighting and appliance branch-circuit panelboard; three-phase, four-wire connections.

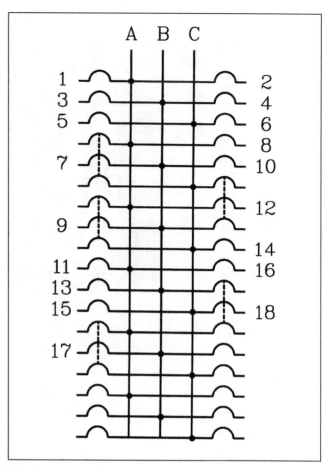

Fig. 8–3 Alternate panelboard circuit numbering scheme.

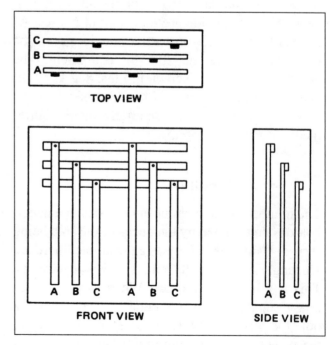

Fig. 8–4 Possible phase arrangement for switchboards and panelboards.

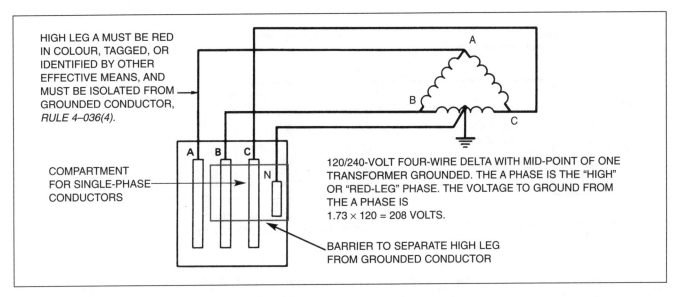

Fig. 8–5 Panelboards and switchboards supplied by four-wire, delta-connected systems shall have the A phase connected to the phase having the higher voltage to ground, *Rule 4–036(4)* and *(5)*.

Rule 4–036(3) requires that three-phase AC be identified with one red (phase A), one black (phase B), one blue (phase C), and, where neutral is required, one white. Where different voltages exist in a building, the ungrounded conductors for each system should be identified and posted at each branch-circuit panelboard.

This situation occurs where, for instance, the building is served with 347/600 volts and step-down transformers are used to provide 120/208 volts for lighting and receptacle outlets.

Table 8–2 Panelboard Summary

	Breakers 1p	2p	3p	Total Poles Used	Specified Poles
Drugstore	16	0	1	19	24
Bakery	13	0	6	31	36
Insurance office	9	0	1	22	30
Beauty salon	8	2	1	15	24
Doctor's office	7	2	0	11	18
Owner—EM Upper	5	0	0	5	12
Owner—EM Lower	12	0	0	12	18

Panelboard Sizing

Panelboards are available in various sizes, including 100, 125, 200, 225, 400, and 600 amperes. These sizes are determined by the current-carrying capacity of the main bus. After the feeder capacity is determined, the panelboard size is selected to carry the demand amperes at 80% or 100% of the panel ampere rating. The number of circuits required must also be considered. Refer to Table 8–2.

Panelboard Overcurrent Protection

In many installations, a single feeder may be sized to serve several panelboards, Fig. 8–6. In this situation, each panelboard must have a main overcurrent device with a trip rating not greater than the rating of the bus bars in the panelboard. In the case of the bakery installation, a main device is not required, since the panelboard is protected by the feeder protective device, Fig. 8–7. An example of a method of providing panelboard protection is shown in Fig. 8–8.

THE FEEDER

The size of a feeder is determined by the branch-circuit loads being served. In addition, a reasonable allowance for growth may be considered. Again using the bakery as an example, the connected load is 45 585 volt-amperes (see Table 10–1).

122 Unit 8 Panelboard Selection and Installation

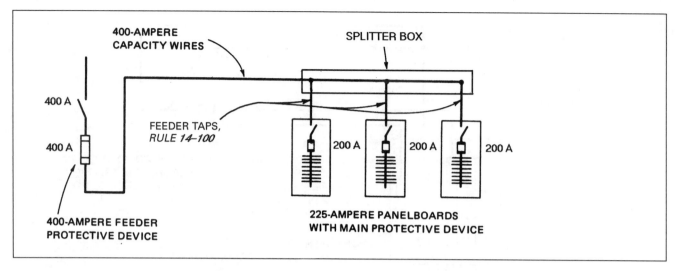

Fig. 8–6 Panelboards with main, *Rule 14–606*.

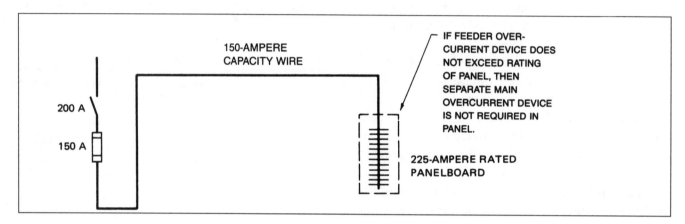

Fig. 8–7 Panelboards without main, *Rule 14–100(a)*.

Since we are using standard-rated equipment and conductors listed in *Table 2*, the ampacity of the feeder to the bakery will be based on 125% of the connected load. The connected load current is:

$$\frac{45\,585}{208 \times 1.73} = 126.6 \text{ A}$$

The ampere rating of the feeder to the bakery would be:

$$126.6 \text{ A} \times 1.25 = 158 \text{ A}$$

The conductor size using TWN75 copper conductors would be No. 2/0 AWG. The conduit size would be 53 mm (2 in).

Neutral Current

Under certain conditions, the neutral wire of a feeder may be smaller in size than the phase wires. This reduction in size is permissible because the neutral carries only the unbalanced portion of the load. However, on a three-phase, four-wire system, the load value is a vector sum (see Fig. 8–11). Table 8–3 shows several conditions that may exist. The neutral currents for conditions I, II, III, and VII are obvious, but conditions IV, V, and VI appear to be contradictory. To find the unbalanced neutral current in a three-phase, four-wire system, only the loads connected between the neutral and phase wires are considered. In addition,

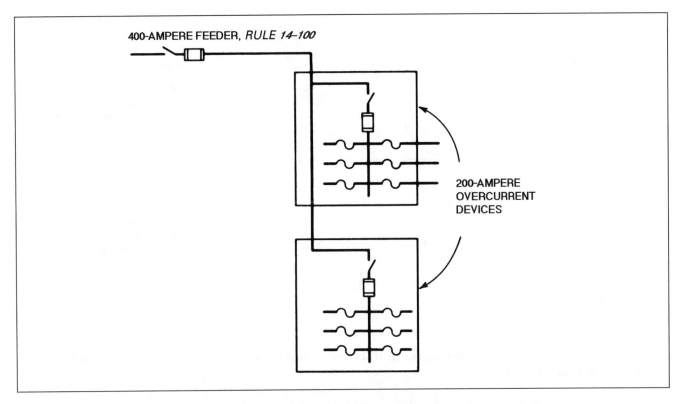

Fig. 8–8   Feedthrough panel with adequate gutter space, *Rule 14–100(b)*.

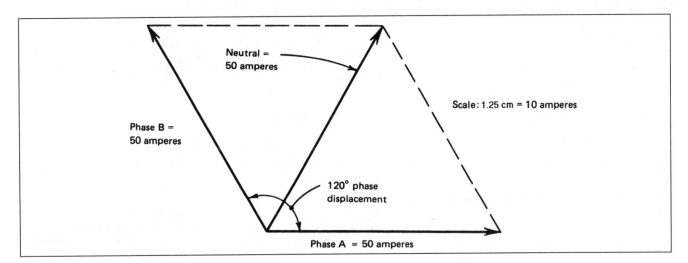

Fig. 8–9   Vector sum as shown in Table 8–3, Condition IV; loads are connected line-to-neutral.

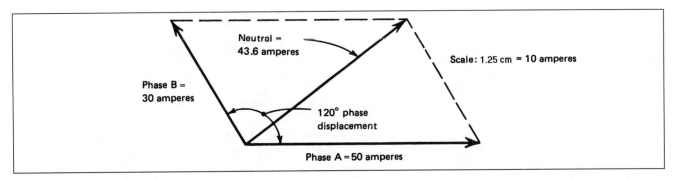

Fig. 8–10   Vector sum showing current in the neutral when Phase A is carrying 50 amperes and Phase B is carrying 30 amperes; neutral is carrying approximately 43.6 amperes; loads are connected line–to–neutral.

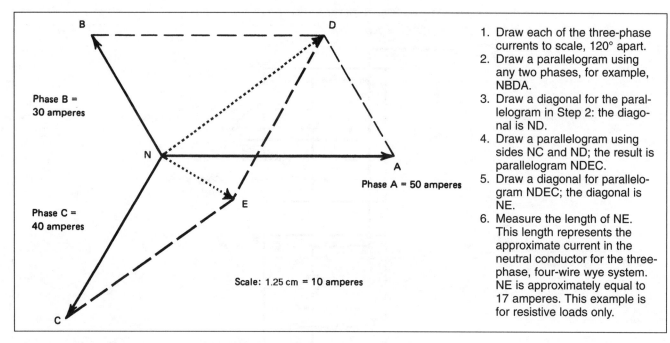

Fig. 8–11 Drawing vectors to determine approximate current in neutral of a three-phase, four-wire wye system; loads are connected line-to-neutral.

equal power factors are assumed. The equation used to find the unbalanced neutral current is:

$$N = \sqrt{A^2 + B^2 + C^2 - AB - BC - CA}$$

where:

N = current in the neutral

A = current in phase A

B = current in phase B

C = current in phase C

EXAMPLE: Compute the neutral current for Fig. 8–11.

$$N = \sqrt{30^2 + 40^2 + 50^2 - (30 \times 40) - (40 \times 50) - (50 \times 30)}$$
$$= \sqrt{900 + 1600 + 2500 - 1200 - 2000 - 1500}$$
$$= \sqrt{300}$$
$$= 17.32 \text{ amperes}$$

To calculate the neutral current when only two phases, from a three-phase four-wire system, are available.

$$N = \sqrt{A^2 + B^2 - AB}$$

EXAMPLE: Compute the neutral current for Fig. 8–9.

$$N = \sqrt{50^2 + 50^2 - (50 \times 50)}$$
$$= \sqrt{2500 + 2500 - 2500}$$
$$= \sqrt{2500}$$
$$= 50 \text{ amperes}$$

Another method for computing the neutral current is to tabulate the connected loads on the neutral. Do not include two- or three-pole loads.

Table 8–3

Condition	Phase A	Phase B	Phase C	Neutral
I	50	0	0	50
II	0	50	0	50
III	0	0	50	50
IV	50	50	0	50
V	50	0	50	50
VI	0	50	50	50
VII	50	50	50	0
VIII**	50	50	50	50 (approx.)

* Table is based on resistive loads.
**For inductive loads, such as fluorescent ballasts, the neutral conductor carries approximately the same value of current as the phase conductors. In this case, the neutral is considered to be a current-carrying conductor. See *Rules 4–004(4)* and *4–022(2)(a)*.

Table 8–3 Load in amperes, neutral unbalanced loading* three-phase, four-wire system.

Table 8–4 Doctor's Office Feeder Loading Schedule

Item	Count	VA/unit	Demand Load	Installed Load	Demand Factor Feeder	Demand Factor Service	Use Feeder	Use Service
Minimum basic load (C.E.C. Rule 8-210)	42 m^2	50 W/m^2	2100		1	0.9	2100	1890
Installed basic loads								
Style F luminaires	3	132		396			396	396
Style G luminaires	2	132		264			264	264
Style J luminaires	5	150		750			750	750
Style N luminaires	2	60		120			120	120
General receptacles	10	120		1200			1200	1200
							2130	2130
BASIC LOAD		Use the greater of the demand or installed loads, C.E.C. Rule 8–106(2).					**2130**	**2130**
SPECIAL LOADS								
Equipment outlet	1	2000		2000			2000	2000
Roof receptacle	1	1440		1440			1440	1440
Water heater	1	3800		3800			3800	3800
Motors	**Volts**	**FLA**	**Phase**					
Cooling unit								
• compressor	208	16.8	1	3494		1.25	4368	3494
• evap/cond. motor	208	3.7	1	770			770	770
TOTAL LOAD							**15 108**	**14 234**

$$I = \frac{P}{E} = \frac{15\,108}{208} = 72.63 \text{ A}$$

Assuming all loads are continuous, using standard-rated equipment and T90/TWN75 copper conductors from *Table 2, C.E.C.*, the ampere rating of the circuit would be:

$$\frac{72.63}{0.8} = 90.79 \text{ A}$$

The minimum size feeder would be 3 No. 3 AWG T90/TWN75 copper conductors in a 27-mm (1-in) conduit, supplied from a 100-A disconnect c/w 100-A fuses.

EXAMPLE: Assume the following volt-ampere loads.

	Phase A	Phase B	Phase C
	1305	609	1200
	900	264	1188
	348	1500	540
	1500	720	720
	900		
TOTALS	4953	3093	3648

In the worst-case scenario, only the loads on phase A would be operating. In this case the load on the neutral would be 4953 volt-amperes, or 41.3 amperes.

Figs. 8–9, 8–10, and 8–11 illustrate the use of vectors to determine the approximate current in the neutrals of a three-phase, four-wire system.

NEUTRAL SIZING *(RULE 4–022)*

Rule 4–022 sets forth a procedure for reducing the neutral size. The method is based on the loading as computed by *Section 8* (not the connected load) and makes the assumption that the unbalanced load is evenly distributed across the phases.

FEEDER LOADING SCHEDULE

The Doctor's office feeder loading schedule can be found in Table 8–4.

REVIEW

Note: Refer to the *C.E.C., Part I* or the plans as necessary.

1. Seven 120-volt branch circuits supplying incandescent lighting are run from a three-phase lighting panel to a pull box. The circuits are connected to breakers at the following positions in the panel: circuits 1, 2, 8, 17, 19, and 21. How many neutrals are required for the seven branch circuits?

2. A lighting panel has 225-ampere main buses. The load on the panel is 150 amperes. What is the maximum size overcurrent protection permitted for the panelboard?

3. A neutral of a three-phase, four-wire system will carry a current of _____ amperes when the currents in the phases are as follows: Phase A, 12 amperes; Phase B, 4 amperes; and Phase C, 2 amperes. All loads are connected neutral-to-phase.

4. The preferred colours to be used for the ungrounded conductors of a three-phase, four-wire circuit are _____, _____, and _____.

5. Using the information in the insurance office panel schedule shown in the drawings, determine the maximum unbalanced current the neutral supplying the panel will carry. Show all calculations.

UNIT 9

The Cooling System

OBJECTIVES

After completing the study of this unit, the student will be able to

- list the parts of a cooling system
- describe the function of each part of the cooling system
- make the necessary calculations to determine the sizes of the electrical components
- read a typical wiring diagram that shows the operation of a cooling unit

The electrician working on commercial construction is expected to install the wiring of cooling systems and troubleshoot electrical problems in these systems. Therefore, it is recommended that the electrician know the basic theory of refrigeration and the terms associated with it.

REFRIGERATION

Refrigeration is a method of removing energy in the form of heat from an object. When the heat is removed, the object is colder. An energy balance is maintained, which means that the heat must go somewhere. As long as the locations where the heat is discharged and where it is removed are separate from each other, it can be said that the space where the heat was absorbed is cooled. The inside of the household refrigerator or freezer is cold to the touch, but this cold cannot be used to cool the kitchen by having the refrigerator door open. Actually, leaving the door open causes the kitchen to become hotter overall. This situation demonstrates an important principle of mechanical refrigeration: to remove heat, it is necessary to add energy to it.

Mechanical refrigeration relies primarily on the process of evaporation. This process is responsible for the cool sensation that results when rubbing alcohol is applied to the skin or when gasoline is spilled on the skin. Body heat supplies the energy required to vaporize the alcohol or gasoline. It is the transfer of this energy from the body to the liquid that cools the skin. In refrigeration systems, such as those used in the commercial building, the evaporation process is controlled in a closed system. The purpose of this arrangement is to preserve the refrigerant so that it can be reused many times in what is known as the *refrigerant cycle*. As shown in Fig. 9–1, the four main components of the refrigerant cycle are:

- *Evaporator.* As the refrigerant evaporates, it absorbs energy from the air around the evaporator.
- *Compressor.* This device pressurizes the refrigerant and pumps it into the condenser. Although this adds more heat, the refrigerant can be condensed readily because of the increased pressure.

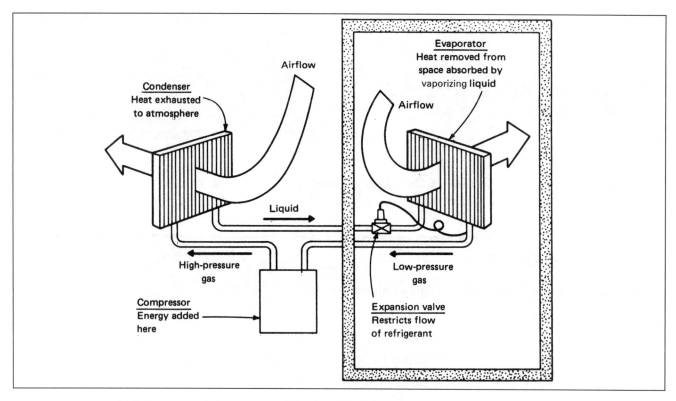

Fig. 9–1 The refrigerant cycle.

- *Condenser.* The compressed refrigerant condenses here as the heat is removed from it by the air blown past the condenser.
- *Expansion valve.* This control device maintains a lower pressure in the evaporator so the refrigerant can expand and vaporize.

EVAPORATOR

The evaporator in a commercial installation normally consists of a fin tube coil, Fig. 9–2, through which the building air is circulated by a motor-driven fan. A typical evaporator unit is shown in Fig. 9–3. The evaporator may be located inside or outside the building. In either case, the function of the evaporator is to remove heat from the interior of the building or from an enclosed space. The chilled air is usually circulated through pipes or duct-work to insure a more even distribution. The window-type air conditioner, however, discharges the air directly from the evaporator coil. In general, the cooling air from the evaporator is circulated through the space to be cooled and then passed again across the cooling coil. A certain

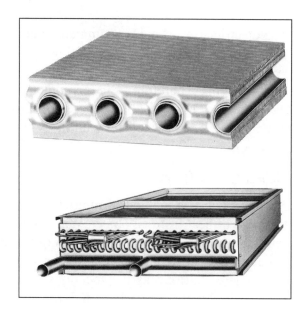

Fig. 9–2 Evaporator components.

Fig. 9–3 An evaporator with dual fans.

percentage of outside air is added to the circulating air to replace air lost through exhaust systems, fume hoods, or the gradual leakage through walls, doors, and windows.

COMPRESSOR

The compressor serves as a pump to draw the expanded refrigerant gas from the evaporator. In addition, the compressor boosts the pressure of the gas and sends it to the condenser, Fig. 9–4. This compression of the gas is necessary to condense it back into a liquid. When the temperature of the air or water surrounding the condenser is relatively warm, the gas pressure must be increased to ensure that the temperature around the condenser can liquefy the refrigerant.

Direct-drive compressors are usually used in large installations. For smaller installations, however, the trend is toward the use of hermetically sealed compressors. Due to several built-in electrical characteristics, these hermetic units cannot be used on all installations. (Restrictions on the use of hermetically sealed compressors are covered later in this text.)

CONDENSER

Condensers are generally available in three types: air-cooled units (Fig. 9–5), water-cooled units (Fig. 9–6), and evaporative cooling units in which water from a pump is sprayed on air-cooled coils to increase their cooling capacity (Fig. 9–7). The function of the condenser in the refrigerant cycle is to discharge both the heat extracted by the evaporator and the heat of compression. Thus, it can be seen that keeping the refrigerator door open causes the kitchen to become hotter because the condenser transfers the combined heat load to the condensing medium. In the case of the refrigerator in a residence, the condensing medium is the room air.

Air-cooled condensers use a motor-driven fan to drive air across the condensing coil. Water-cooled condensers require a pump to circulate the

Fig. 9–4 A motor-driven reciprocating compressor.

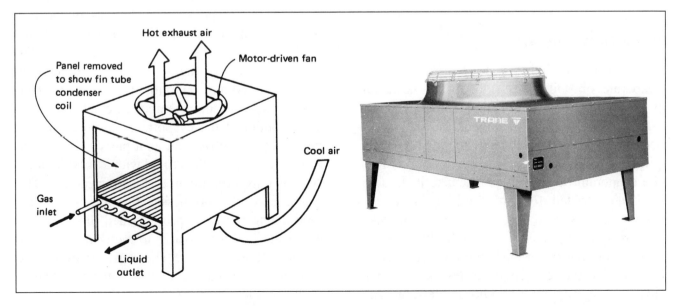

Fig. 9–5 An air-cooled condenser.

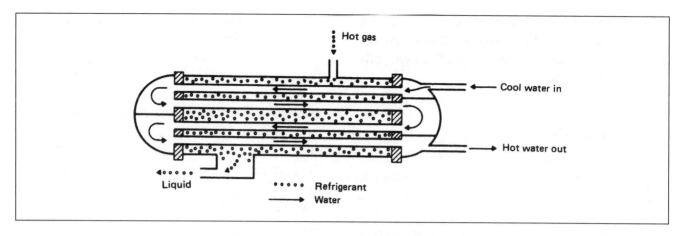

Fig. 9–6 A water-cooled condenser.

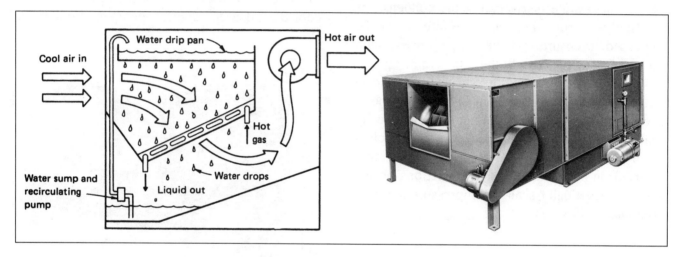

Fig. 9–7 An evaporative condenser.

water. Once the refrigerant gas is condensed to a liquid state, it is ready to be used again as a coolant.

EXPANSION VALVE

It was stated previously that the refrigerant must evaporate or boil if it is to absorb heat. The process of boiling at ordinary temperatures can occur in the evaporator only if the pressure is reduced. The task of reducing the pressure is simplified since the compressor draws gas away from the evaporator and tends to evacuate it. In addition, a restricted flow of liquid refrigerant is allowed to enter the high side of the evaporator. As a result, the pressure remains fairly low in the evaporator coil so that the liquid refrigerant entering through the restriction flashes into a vapour and fills the evaporator as a boiling refrigerant.

The restriction to the evaporator may be a simple orifice. However, commercial systems generally use a form of automatic expansion valve that is responsive to changes in the heat load. The expansion valve is located in the liquid refrigerant line at the inlet to the evaporator. The valve controls the rate at which the liquid flows into the evaporator. The liquid flow rate is determined by installing a temperature-sensitive bulb at the outlet of the evaporator to sense the heat gained by the refrigerant as it passes through the evaporator. The bulb is filled with a volatile liquid (which may be similar to the refrigerant). This liquid expands and passes through a capillary tube connected to a spring-loaded diaphragm to cause the expansion valve to meter more or less refrigerant to the coil. The delivery of various amounts of refrigerant compensates for changes in the heat load of the evaporator coil.

HERMETIC COMPRESSORS

The tremendous popularity of mechanical refrigeration for household use in the 1930s and 1940s stimulated the development of a new series of refrigerants that were nonexplosive and less toxic. These refrigerants, known as chloro-fluorocarbons (CFCs) were relatively expensive at the time of their initial production. Because of their expense, it was no longer feasible to permit the normal leakage that occurred around the shaft seals of reciprocating, belt-driven units. As a result, the hermetic compressor was developed, Fig. 9–8. This unit consists of a motor-compressor completely sealed in a gas-tight, steel casing. The refrigerant gas is circulated through the compressor and over the motor windings, rotor, bearings, and shaft. The circulation of the expanded gas through the motor helps to cool the motor.

Recent research has developed a CFC-free refrigerant that, it is hoped, will reduce the harmful effects of CFCs on the ozone layer in the upper atmosphere.

The initial demand for hermetic compressors was for use in residential-type refrigerators. Therefore, most of the hermetic compressors were constructed for single-phase service. In other words, it was necessary to provide auxiliary winding and starting devices for the compressor installation. Because the refrigerant gas surrounded and filled the motor cavity, it was necessary to remove the centrifugal switch commonly provided to disconnect the starting winding at approximately 85% of the full speed. This switch was removed because any arcing in the presence of the refrigerant gas caused the formation of an acid from the hydrocarbons in the gas. This acid attacked and etched the finished surfaces of the shafts, bearings, and valves. The acid also carbonized the organic materials used to insulate the motor winding and caused the eventual breakdown of the insulation. To replace the switch, the relatively heavy magnetic winding of a relay was connected in series with the main motor winding. The initial heavy inrush of current caused the relay to lift its magnetic core and energize the starting winding. As the motor speed increased, the main winding current decreased causing the relay to disconnect the starting or auxiliary winding.

A later refinement of this arrangement was the use of a voltage-sensitive relay that was wound to permit pickup at values of voltage greater than the line voltage. The coil of this voltage-sensitive relay was connected across the starting winding. This connection scheme was based on the principle that the rotor of a single-phase induction motor induces in its own starting winding a voltage that is approximately in quadrature (phase) with the main winding voltage and has a value greater than the main voltage. As long as the motor was running the voltage-sensitive relay broke the circuit to the starting winding.

Since the most maintenance problems in hermetic compressors were due to starting relays and

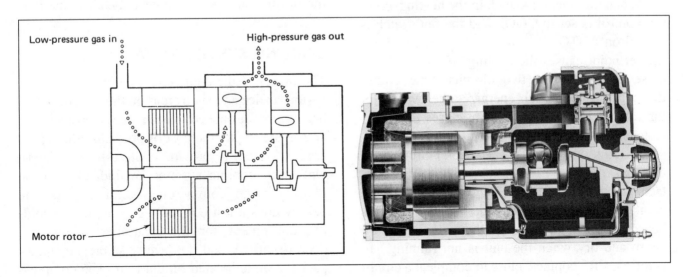

Fig. 9–8 The hermetic compressor.

capacitors, it was desirable to eliminate as many as possible of these devices. It was soon realized that small- and medium-size systems could make use of a different form of refrigerant metering device, thus eliminating the automatic expansion valve. Recall that this valve has a positive shutoff characteristic, which means that the refrigerant is restricted when the operation cycle is finished as indicated by the evaporator reaching the design temperature. As a result, when a new refrigeration cycle begins, the compressor must start against a head of pressure. If a small-bore, open capillary tube is substituted for the expansion valve, the refrigerant is still metered to the evaporator coil, but the gas continues to flow after the compressor stops until the system pressure is equalized. Therefore, the motor is required only to start the compressor with the same pressure on each side of the pistons.

This ability to decrease the load led to the development of a new series of motors that contained only a running capacitor (no relay was installed). This type of motor furnished sufficient torque to start the unloaded compressor and, at the same time, greatly improved the overall power factor.

COOLING SYSTEM CONTROL

Fig. 9–9 shows a wiring diagram that is representative of standard cooling units.

When the selector switch in the heating–cooling control is set to COOL and the fan switch is placed on AUTO, any rise in temperature above the set point causes the cooling contacts (TC) to close and complete two circuits. One circuit through the fan switch energizes the evaporator fan relay (EFR), which, in turn, closes EFR contacts to complete the 208-volt circuit to the evaporator fan motor (EFM). The second circuit is to the control relay (CR) and causes the CR1 contacts to open. When the CR1 contacts open, the crankcase heater (CH) is de-energized. (This heater is installed to keep the compressor oil warm and dry when the unit is not running.) In addition, CR2 contacts close to complete a circuit to the condenser fan motor (CFM) and another

circuit through low-pressure switch 2 (LP2), the high-pressure switch (HP), the thermal contacts (T), and the overload contacts (OL) to the compressor motor starter coil (CS). When the CR2 contacts close, the three CS contacts in the power circuit to the compressor motor (CM) also close.

Contacts LP1 and LP2 open when the refrigerant pressure drops below a set point. When contacts LP2 open, CS is de-energized. When contacts LP1 open, CR is de-energized, the circuit to CFM opens, and the circuit to CH is completed. The high-pressure switch (HP) contacts open and de-energize the compressor motor starter (CS) when the refrigerant pressure goes above the set point of the switch

The low-pressure control (LP1) is the normal operating control and the high-pressure control (HP) and LP2 act as safety devices. The T contacts shown in Fig. 9–9 are located in the compressor motor and open when the winding temperature of the motor is too high.

The OL contacts are controlled by the OL elements installed in the power leads to the motor. The OL elements are sized to correspond to the current draw of the motor. That is, a high current causes the overload elements to overheat and open the OL contacts. As a result, the compressor starter is de-energized and the compressor stops. The evaporator fan motor (EFM) used to circulate the cooled air in the store area can be run continuously if the fan switch is turned to FAN. Actually, in many situations, it is recommended that the fan motor run continuously to keep the air in motion.

COOLING SYSTEM INSTALLATION

The owner of the commercial building leases the various office and shop areas on the condition that heat will be furnished to each area. However, tenants agree to pay the cost of operating the refrigeration system to provide cooling in their areas. Only four cooling systems are indicated in the plans for the commercial building since the bakery does not use a cooling system. The cooling equipment for the insurance office, the doctor's office, and the beauty salon are single-package units located on the roof. The compressor, condenser, and evaporator for each of these

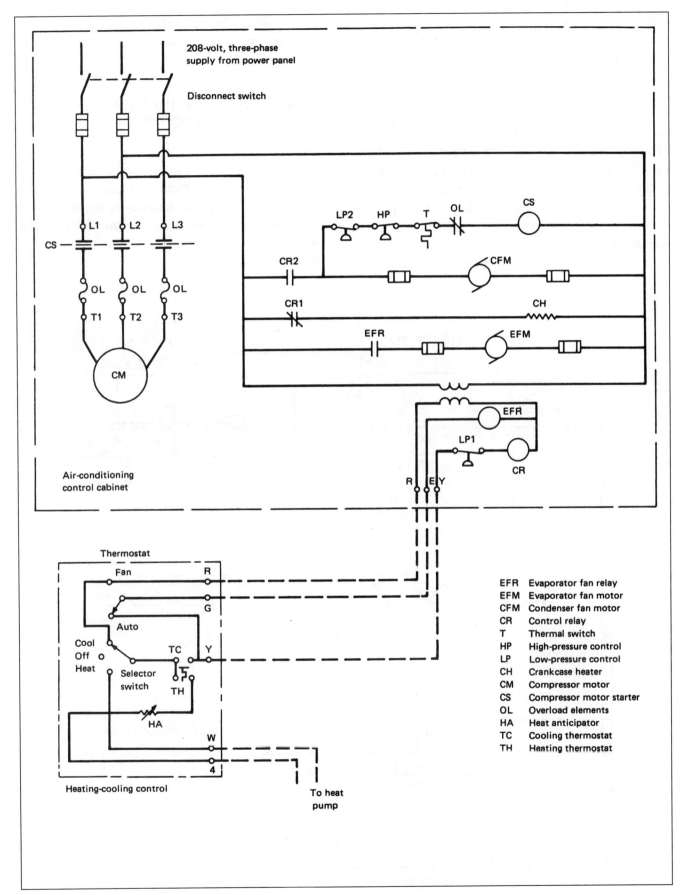

Fig. 9-9 A cooling system control diagram.

units are constructed within a single enclosure. The system for the drugstore is a split system with the compressor and condenser located on the roof and the evaporator located in the basement, Fig. 9–10. Regardless of the type of cooling system installed, a duct system must be provided to connect the evaporator with the area to be cooled. The duct system shown diagrammatically in Fig. 9–10 does not represent the actual duct system, which will be installed by another contractor.

The electrician is expected to provide a power circuit to each air-conditioning unit as shown on the plans. In addition, it is necessary to provide wiring to the thermostat in each area, and, in the case of the drugstore, wiring must be provided to the evaporator located in the basement.

ELECTRICAL REQUIREMENTS FOR AIR-CONDITIONING AND REFRIGERATION EQUIPMENT

The *C.E.C., Part I* rules covering air-conditioning and refrigeration equipment are found in *Section 28*. For air-conditioning and refrigeration equipment that does not incorporate hermetically sealed motor-compressor(s), *Rules 28–106, 28–200,* and *28–304* apply.

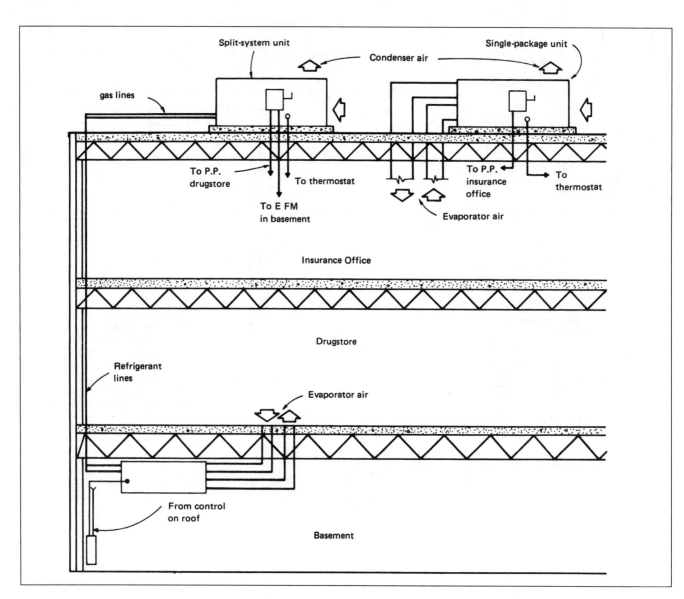

Fig. 9–10 Single-package and split-system cooling units.

For the insurance office, an air-conditioning unit is located on the roof of the commercial building. The data furnished with this air-conditioning unit supply the following information:

Voltage: 208 volts, three-phase, three-wire, 60 Hz

Hermetic refrigerant compressor–motor: Rated-load amperes 20.2 at 208 volts three-phase

Evaporator motor: Full-load amperes 3.2 at 208 volts single-phase

Condenser motor: Full-load amperes 3.2 at 208 volts single-phase

Minimum circuit ampacity: 31.65 amperes[1]

Maximum overcurrent protection: 70 amperes, time-delay fuse. 50% LRC, *Rule 28–708*.

Locked-rotor current (LRC): 140 amperes

When the nameplate on equipment specifies fuses, then only fuses are permitted to be used as the branch-circuit overcurrent protection. Circuit breakers are not permitted.

Air-conditioning and refrigeration equipment is almost always marked with the minimum supply circuit ampacity, maximum ampere rating, and type of overcurrent protective device permitted.

SPECIAL TERMINOLOGY

Understanding the following definitions is important when installing air-conditioners, heat pumps, and other equipment utilizing hermetic refrigerant motor–compressors. These terms are found in *Rule 28–010* and in various manufacturers' data.

Rated-Load Current (RLA). The RLA is determined by the manufacturer of the hermetic refrigerant motor–compressor through testing at rated refrigerant pressure, temperature conditions, and voltage.

In most instances, the RLA is at least equal to 64.1% of the hermetic refrigerant motor–compressor's maximum continuous current (MCC).

EXAMPLE: The nameplate on an air-conditioning unit is marked:
Compressor RLA...17.8 amperes

Locked-Rotor Current or Locked-Rotor Amperes. (LRA). This is marked on the equipment nameplate. If it is not marked it can be calculated by multiplying the full-load current or full-load amperes (FLA) by six. If the FLA is not marked it can be obtained from *Table 44* or *45*.

Branch-Circuit Selection Current (BCSC). Some hermetic refrigerant motor–compressors are designed to operate continuously at currents greater than 156% of the RLA. In such cases, the unit's nameplate is marked with "Branch-Circuit Selection Current." The BCSC will be no less than 64.1% of the maximum continuous current rating (MCC) of the hermetic refrigerant motor–compressor.
Note:
156% and 64.1% have an inverse relationship:
$1 / 1.56 = 0.641$ and $1 / 0.641 = 1.56$

EXAMPLE: The maximum continuous current of a hermetic refrigerant motor–compressor is 31 amperes. The BCSC will be no less than:

31 amperes $\times$ 0.641 = 19.9 amperes

Since the BCSC value, when marked, is always equal to or greater than the unit nameplate marked RLA value, the manufacturer of the air-conditioning unit must use the BCSC value, instead of the RLA value, to determine the minimum circuit ampacity (MCA) and the maximum overcurrent protection (MOP). For installation of individual hermetic refrigerant motor–compressors, where the electrician must select the conductors, the controller, the disconnecting means, and the short-circuit and ground-fault protection, the electrician must use the BCSC, if given, instead of the RLA.

Maximum Continuous Current (MCC). The MCC is determined by the manufacturer of the hermetic refrigerant motor–compressor under specific test conditions. The MCC is needed to

---

[1] $[(20.2 \times 1.25) + 3.2 + 3.2] = 31.65$ amperes, *Rule 28–106* (see Unit 12).

properly design the end-use product. The installing electrician is not directly involved with the MCC.

The MCC is not on the nameplate of the packaged air-conditioning unit. The MCC is recognized as being no greater than 156% of the RLA or the BCSC, for the hermetic refrigerant motor–compressor.

Except in special conditions, the overload protection must cut out if the current exceeds the 156% limit.

EXAMPLE: A hermetic refrigerant motor–compressor is marked:
Maximum Continuous Current. . .31 amperes

Minimum Circuit Ampacity (MCA). The manufacturer of an air-conditioning unit is required to mark the nameplate with this value. This is what the electrician needs to know. The manufacturer determines the MCA by multiplying the RLA, or the BCSC, of the hermetic refrigerant motor–compressor by 125%. The current ratings of all other concurrent loads, such as fan motors, transformers, and relay coils, are then added to this value, *Rule 28–108*.

EXAMPLE: An air-conditioning unit's nameplate is marked:
Minimum Circuit Ampacity. . .26.4 amperes

This was derived as follows:

BCSC of 19.9 amperes × 1.25 = 24.9 amperes
plus $\frac{1}{4}$ horsepower fan motor @ 1.5 amperes
MCA = 26.4 amperes

The electrician must install the conductors, the disconnect switch, and the overcurrent protection based upon the MCA value. No further calculations are needed; the manufacturer of the equipment has done it all.

Maximum Overcurrent Protective Device (MOP). The electrician should always check the nameplate of an air-conditioning unit for this information.

The manufacturer is required to mark this value on the nameplate. This value is determined by multiplying the RLA, or the BCSC, of the hermetic refrigerant motor–compressor by 225%, then adding all concurrent loads such as electric heaters and motors, *Rule 28–204(1)*. This is not to exceed 50% LRC, *Rule 28–708(1)*.

EXAMPLE: The nameplate of an air-conditioning unit is marked:
Maximum Time Delay Fuse. . . 45 amperes

This was derived as follows:

BCSC of 19.9 amperes × 2.25 = 44.775 amperes
plus 1/4 horsepower fan motor @ 1.5 amperes
Therefore, the MOP = 46.275 amperes

Since 46.275 is the maximum, the next lower standard size time-delay fuse, 45 amperes, must be used.

The electrician installs branch-circuit overcurrent devices with a rating **not to exceed** the MOP as marked on the nameplate of the air-conditioning unit. The electrician makes no further calculations; all calculations have been done by the manufacturer of the air-conditioning unit.

Overcurrent Protection Device Selection. How does an electrician know if fuses or circuit breakers are to be used for the branch-circuit overcurrent device? The electrician reads the nameplate and the instructions carefully. *Rule 2–024* requires that "Electrical equipment...shall be approved and shall be of a kind or type and rating approved for the specific purpose for which it is to be employed."

EXAMPLE: If the nameplate of an air-conditioning unit reads "Maximum Size Time-Delay Fuse: 45 amperes," fuses are required for the branch-circuit protection. To install a circuit breaker would be a violation of *Rule 2–100(1)(m)*.

If the nameplate reads "Maximum Fuse or HACR Type Breaker. . . ," then either a fuse or a HACR-type circuit breaker is permitted.

Maximum Overload Protection (MOLP). Referring to *Rule 28–710*, the maximum rating is not to exceed 140% of the RLC of the motor–compressor.

Disconnecting Means Rating. The *C.E.C., Part I* rules for determining the size of disconnecting means required for HVAC equipment are covered in *Rule 28–714* (see below).

Because disconnect switches are horsepower rated, it is sometimes necessary to convert the horsepower rating to a locked-rotor current rating. *Rule 28–704(1)* indicates this to be six times the FLA from the nameplate of the compressor. If this information is not available, the values may be obtained from *Tables 44* or *45* in the *C.E.C., Part I*.

Air-Conditioning and Refrigeration Equipment Disconnecting Means

- The disconnecting means for an individual hermetic motor–compressor must not be less than 115% of the rated-load current. See *Rule 28–714(1)(a)*.

- The disconnecting means for an individual hermetic motor–compressor shall have an interrupting capacity or LRC rating equal to the LRC of the motor–compressor.

- For equipment that has at least one hermetic motor–compressor, and other loads such as fans, heaters, solenoids, and coils, the disconnecting means must not be less than 115% of the largest RLA plus the sum of the currents of all of the components. See *Rule 28–714(2)(a)*.

In our example, the full-load current equals 31.65 amperes. Checking *Table 44*, we find that a full-load ampere rating of 31.65 amperes is slightly greater than the 30.8 amperes that a 10-horsepower, 208-volt motor draws. Thus a 15-horsepower disconnect would be required.

- The total locked-rotor current must also be checked to be sure that the horsepower rating of the disconnecting means is capable of safely disconnecting the full locked-rotor current, *Rule 28–714(2)(b)*.

In our example, the locked-rotor current equals 140 amperes. The equivalent LRC rating of the 15-horsepower disconnect, as determined by *Rule 28–704*, would be 277.2 amperes (46.2 amperes × 6 = 277.2 amperes). This would fall well within the requirements of *Rule 28–714(2)(b)*.

Table 45

(See Rules 28–010 and 28–704)

Single-Phase AC Motors

Single-Phase AC Motors Full-Load Current in Amperes (See Notes (1) to (4))

hp Rating	115 V	230 V
1/6	4.4	2.2
1/4	5.8	2.9
1/3	7.2	3.6
1/2	9.8	4.9
3/4	13.8	6.9
1	16	8
1 1/2	20	10
2	24	12
3	34	17
5	56	28
7 1/2	80	40
10	100	50

Notes:
(1) For full-load currents of 208- and 200-volt motors, increase the corresponding 230-volt motor full-load current by 10 and 15%, respectively.
(2) These values of motor full-load current are to be used as guides only. Where exact values are required (e.g., for motor protection), always use those appearing on the motor nameplate.
(3) These values of full-load current are for motors running at usual speeds and motors with normal torque characteristics. Motors built for especially low speeds or high torques may have higher full-load currents, and multi-speed motors will have full-load current varying with speed, in which case the nameplate current ratings shall be used.
(4) The voltages listed are rated motor voltages. Corresponding Nominal System Voltages are 120 and 240 volt. Refer to CSA Standard CAN3–C235.

Table 9–1 *Table 45 of the C.E.C., Part I.*

If one of the selection methods (FLA or LRA) indicates a larger-size disconnect switch than the other method does, then the larger switch shall be installed.

- The disconnecting means must be within sight of, and not more than 3 metres from, the air-conditioning and refrigeration equipment. See *Rule 28–604(5)*.

- The disconnecting means must be accessible. See *Rule 28–604(6)*.

Air-Conditioning and Refrigeration Equipment Branch-Circuit Conductors

- For individual motor–compressor equipment, the branch-circuit conductors must have an ampacity of not less than 125% of the rated-load current or branch-circuit selection current, whichever is greater. See *Rules 28–706* and *28–106*.

- For motor–compressor(s) plus other loads, such as a typical air-conditioning unit that contains a motor–compressor, fan, heater, coils, etc., add all of the individual current values, plus 25% of that of the largest motor or motor–compressor. See *Rule 28–108*.

One can readily see the difficulty encountered when attempting to compute the actual currents flowing in phases A, B, and C for a typical three-phase unit, Fig. 9–11.

Rather than attempting to add the currents vectorially, for simplicity, one generally uses the arithmetic sum of the current. However, the student must realize than an ammeter reading taken while the air-conditioning unit is operating will not be the same as the calculated current values.

For the air-conditioning unit in the insurance office, the calculation is:

(20.2 + 3.2 + 3.2) amperes	=	26.60 amperes
plus 25% of 20.2 amperes	=	5.05 amperes
Total	=	31.65 amperes

The minimum ampacity for the conductors serving the above air-conditioning unit is 31.65 amperes, with the calculation done as above.

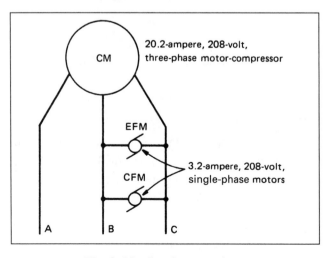

Fig. 9–11 Load connections.

The nameplate of the air-conditioning unit also calls for a minimum circuit ampacity of 31.65 amperes. Refer to *Table 2*, which shows that No. 8 is the smallest size with adequate ampacity.

Air-Conditioning and Refrigeration Equipment Branch-Circuit Overcurrent Protection

- For individual units, the branch-circuit overcurrent protection shall not exceed 50% of the locked-rotor current (LRC). But if the 50% sizing will not allow the equipment to start, then the *C.E.C., Part I* permits sizing not to exceed 65% LRC. See *Rule 28–708(1)*.

- Where the circuit has more than one hermetic refrigerant motor–compressor, or one hermetic refrigerant motor–compressor and other motors or other loads, refer to *Rule 28–206*.

The branch-circuit overcurrent protection must also meet the requirements of *Rule 28–200*. Simply stated, this means that overcurrent protection must be provided that does not exceed the protection required for the motor–compressor that draws the largest amount of current, plus the sum of the rated-load current or branch-circuit selection currents of other motor–compressors, plus the ampere ratings of any other loads.

In our example:

$[(20.2 \times 1.75) + 3.2 + 3.2]$ amps = 41.75 amps

The nameplate for the air-conditioning unit supplying the cooling for the drugstore calls for maximum 40-ampere, dual-element, time-delay fuses. This value must not be exceeded.

If the feeder or branch circuit originates from a circuit-breaker panel, then the manufacturer's requirements can be met by installing a fusible disconnect switch near, or on, the air-conditioning equipment.

Air-Conditioning and Refrigeration Equipment Motor–Compressor Overload Protection

This topic is covered in *Rule 28–710*. The manufacturer provides the overload protection for the

Table 44

(See Rules 28–010 and 28–704)

Three-Phase AC Motors

3-Phase	AC Motor Full-Load Current in Amperes (See Notes (1), (2), (3), and (5))								
	Induction Type, Squirrel Cage and Wound Rotor Amperes					Synchronous Type, Unity Power Factor (See Note (4)) Amperes			
Motor Rating hp	115 V	230 V	460 V	575 V	2300 V	230 V	460 V	575 V	2300 V
$\frac{1}{2}$	4	2	1	0.8	—	—	—	—	—
$\frac{3}{4}$	5.6	2.8	1.4	1.1	—	—	—	—	—
1	7.2	3.6	1.8	1.4	—	—	—	—	—
$1\frac{1}{2}$	10.4	5.2	2.6	2.1	—	—	—	—	—
2	13.6	6.8	3.4	2.7	—	—	—	—	—
3	—	9.6	4.8	3.9	—	—	—	—	—
5	—	15.2	7.6	6.1	—	—	—	—	—
$7\frac{1}{2}$	—	22	11	9	—	—	—	—	—
10	—	28	14	11	—	—	—	—	—
15	—	42	21	17	—	—	—	—	—
20	—	54	27	22	—	—	—	—	—
25	—	68	34	27	—	54	27	22	—
30	—	80	40	32	—	65	33	26	—
40	—	104	52	41	—	86	43	35	—
50	—	130	65	52	—	108	54	44	—
60	—	154	77	62	16	128	64	51	12
75	—	192	96	77	20	161	81	65	15
100	—	248	124	99	26	211	106	85	20
125	—	312	156	125	31	264	132	106	25
150	—	360	180	144	37	—	158	127	30
200	—	480	240	192	49	—	210	168	40

Notes:
(1) For full-load currents of 208- and 200-volt motors, increase the corresponding 230-volt motor full-load current by 10 and 15%, respectively.
(2) These values of motor full-load current are to be used as guides only. Where exact values are required (e.g., for motor protection), always use those appearing on the motor nameplate.
(3) These values of full-load current are for motors running at speeds usual for belted motors and motors with normal torque characteristics. Motors built for especially low speeds or high torques may require more running current, and multi-speed motors will have full-load current varying with speed, in which case the nameplate current ratings shall be used.
(4) For 90 and 80% power factor multiply the above figures by 1.1 and 1.25, respectively.
(5) The voltages listed are rated motor voltages. Corresponding Nominal System Voltages are 120, 240, 480, and 600 volts. Refer to CSA Standard CAN3–C235.

Table 9–2 *Table 44 of the C.E.C., Part I.*

various internal components that make up the complete air-conditioning unit. The manufacturer submits the equipment to CSA for testing of the product. Overload protection of the internal components is covered in the CSA Standards. This protection can be in the form of fuses, thermal protectors, certain types of inverse-time circuit breakers, overload relays, etc.

Rule 28–710 discusses overload protection for motor–compressor appliances. This rule requires that the motor–compressor be protected from overloads and failure to start by:

- a separate overload relay that is responsive to current flow and trips at not more than 140% of the rated-load current, or

- a time-delay fuse or inverse-time circuit breaker rated at not more than 125% of the rated-load current, or

- an integral thermal protector arranged so that its action interrupts the current flow to the motor–compressor.

It is rare for the electrician to have to provide additional overload protection for these components. The electrician's job is to make sure the equipment is fed by the proper-size conductors with the proper size and type of branch-circuit overcurrent protection and the proper horsepower-rated disconnecting means.

REVIEW

Note: Refer to the *C.E.C., Part I* or the plans as necessary.

1. Refrigeration is a method of _____ from an object.

2. Match the following items by inserting the appropriate letter from column 2 in the blank in column 1.

   a. This device raises the energy
      level of the refrigerant.   _____   A. Evaporator

   b. The refrigerant condenses here
      as heat is removed.   _____   B. Compressor

   c. This is a metering device.   _____   C. Condenser

   d. The refrigerant absorbs energy
      here in the form of heat.   _____   D. Expansion valve

3. Name three types of condensers.
   _____   _____   _____

4. A 230-volt, single-phase hermetic compressor with an FLA of 21 and an LRA of 250 requires a disconnect switch with a rating of _____ horsepower.

5. A motor–compressor unit with a rating of 32 amperes is protected from overloads by a separate overload relay selected to trip at not more than _____ ampere(s).

6. Should the electrician be required to furnish overload protection for a motor–compressor refrigeration unit, he or she can meet the requirements of *Rule 28–306* by selecting fuses sized at not more than ____% of the motor–compressor rated-load current.

7. Does the *C.E.C., Part I* permit the disconnect switch for a large rooftop air-conditioner to be mounted inside the unit? Give the appropriate reference. _____
   _____

8. Where the nameplate of an air-conditioning unit specifies the minimum ampacity required for the supply conductors, must the electrician increase this ampacity by 25% to determine the correct ampacity for the conductors that will supply the air-conditioning unit? Explain.
   _____
   _____
   _____

9. Where appliances are required to be grounded, what section of the *C.E.C., Part I* would you refer to for specific grounding requirements?
   _____

10. Which current rating of an air-conditioning unit is greater:

    a. rated-load current

    b. branch-circuit selection current

    Circle the letter of the correct answer.

UNIT 10

Reading Electrical Drawings—Bakery, and Switches and Receptacles

Part I: Reading Electrical Drawings—Bakery

OBJECTIVES

After completing the study of Part I of this unit, the student will able to
- determine loading requirements
- select branch-circuit conductors

PRINTS

The electrician should check the electrical drawings long before the actual installation begins. Many decisions are required even when there is a complete set of drawings. Few plans provide full information on (1) the exact paths that the conduits or cables shall take, and (2) the actual sizes of the conductors and conduits. The circuit routings are very project-dependent and the sizings depend upon the routings.

Frequently, it is possible to install short connecting conduits between wall boxes while the slabs are being poured. Later, longer conduits may be needed for the same connections. If longer runs are necessary, the electrician may need to increase the conductor sizes and also the conduit sizes.

This section of the unit will give the student experience in determining the proper conductor size.

THE BAKERY PRINTS

Several different applications are illustrated in the bakery. See Fig. 10–1. The luminaires are surface mounted as individual units. Raceway connections are made to all the luminaires. Connections are

Table 10–1 Bakery Feeder Loading Schedule

Item	Count	VA/unit	Demand Load	Installed Load	Demand Factor Feeder	Demand Factor Service	Use Feeder	Use Service
MINIMUM BASIC LOAD								
(C.E.C. Rule 8-210)	218 m²	30 W/m²	6540		1	1	6 540	6 540
INSTALLED BASIC LOADS								
Style C luminaires	15	87		1 305			1 305	1 305
Style B luminaires	4	87		348			348	348
Style E luminaires	2	144		288			288	288
Style L luminaires	12	87		1 044			1 044	1 044
Style N luminaires	2	60		120			120	120
General receptacles	20	120		2 400			2 400	2 400
Installed basic loads							5 505	5 505
BASIC LOAD		Use the greater of the demand or installed loads, C.E.C. Rule 8–106(2).					6 705	6 705
SPECIAL LOADS								
Show win. lts. style C	3	87		261			261	261
Show win. rec.	4	720		2 880			2 880	2 880
Sign	1	1 200		1 200			1 200	1 200
Bake oven	1	16 000		16 000			16 000	16 000
Doughnut mach. htr	1	2 000		2 000			2 000	2 000
Motors	**Volts**	**FLA**	**Phase**					
Dishwasher	208	23.9	3	8 600	1.25		10 750	8 600
Disposer	208	7.44	3	2 677			2 677	2 677
Dough divider	208	2.2	3	792			792	792
Doughnut machine	208	2.2	3	792			792	792
Exhaust fan	120	2.9	1	348			348	348
Multimixer	208	7.48	1	2 692			1 556	1 556
Multimixer	208	3.96	1	1 426			824	824
Total load							45 585	43 435

$$I = \frac{P}{E \times 1.73} = \frac{45\ 585}{208 \times 1.73} = 126.6 \text{ A}$$

Assuming all loads are continuous, using standard-rated equipment, and RW90–XLPE copper conductors from *Table 2, C.E.C.*, the ampere rating of the circuit would be:

$$\frac{126.6}{0.8} = 158.3 \text{ A}$$

The minimum size feeder would be 4 No. 2/0 RW90–XPLE copper conductors in a 53-mm (2-in) raceway, supplied by a 200-A switch c/w 175-A fuses.

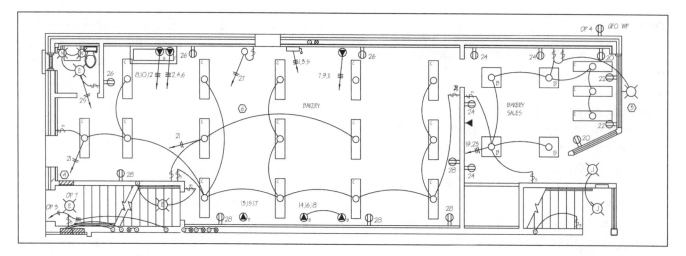

Fig. 10–1 **Electrical drawing for a bakery.** *Note:* **For complete blueprint, refer to blueprint E3 at back of text.**

made from an electrical box located flush in the ceiling or by a raceway installed between the ends.

Special attention should be paid to the branch circuits as these plans are being reviewed. The branch-circuit panelboard schedule shown in Table 8–2 (page 121) provides important information about these circuits. However, project conditions may require the electrician to make changes.

The electrician is responsible for checking voltage drop and conductor ampacity. Once these have been determined, the electrician must check to see if any of the items on the schedule need updating. A review of what can, what should not, and what cannot be changed will be useful.

- Changing the circuit load should be avoided. Changing the load could require the development of entirely new loading and panelboard schedules.
- Overcurrent protection should be carefully checked if there are changes in load or load type. Otherwise, it should not be changed.
- Reassigning phase connections can be useful if there is an advantage in grouping circuits to share neutrals.
- Continuous and unbalanced load designations are usually self-evident and should not be revised except in conjunction with other changes.
- The wire size may need to be increased if voltage drop is excessive or the conductor ampacity is derated.
- When the wire size is changed it may be possible to increase the circuit capacity.

Part II: Switches and Receptacles

OBJECTIVES

After completing the study of Part II of this unit, the student will be able to
- select switches and receptacles with the proper rating for a particular application
- install various types of receptacles correctly
- connect single-pole, three-way, four-way, and double-pole switches into control circuits

During the course of the work, an electrician selects and installs numerous receptacles and switches. Therefore, it is essential that the electrician know the important characteristics of these devices, and how they are to be connected into the electrical system.

RECEPTACLES

The National Electrical Manufacturers Association (NEMA) and the Canadian Standards Association (CSA) have developed standards for the physical appearance of locking and nonlocking plugs and receptacles. The differences in the plugs and receptacles are based on the ampacity and voltage rating of the device. For example, the most commonly used receptacle is the CSA 5–15R, Fig. 10–2. The CSA 5–15R receptacle has a 15-ampere, 125-volt rating and has two parallel slots and a ground-pin hole. This receptacle will accept the CSA 5–15P plug only, Fig. 10–3. The CSA 5–20R (Fig. 10–4) receptacle has two slots at right angles. This receptacle is rated at 20 amperes, 125 volts. The CSA 5–20R will only accept a CSA 5–20P plug, *Rule 26–700, Appendix B*, Fig. 10–5. A CSA 6–20R receptacle is shown in Fig. 10–6. This receptacle has a rating of 20 amperes at 250 volts. CSA 6–15P and 6–20P are shown in Fig. 10–7. Another receptacle that is specified for the commercial building is the CSA 6–30R, Fig. 10–8. This receptacle is rated at 30 amperes, 250 volts.

The CSA standards for general-purpose, non-locking and locking receptacles are shown in Figs. 10–9 and 10–10. Following is the key to the catalogue numbers for these standards.

- A/ Single receptacle
- B/ Duplex receptacle
- C/ Plug
- D/ Connector
- E/ Flanged inlet
- F/ Flanged outlet
- + Hospital Grade
- * Available in hospital grade; add suffix HG
- ++ Available in isolated ground; add prefix IG

A special note should be made of the differences between the 125/250-volt (CSA 14) devices and the three-phase, 250-volt (CSA 15) devices. The connection of the 125/250-volt receptacle requires a neutral, a grounding wire, and two-phase connections. For the three-phase, 250-volt receptacle, a grounding wire, and three-phase connections are required.

Fig. 10–3 CSA 5–15P. A 15-ampere, 125-volt plug.

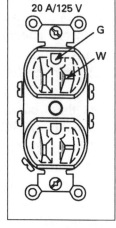

Fig. 10–4 CSA 5–20R. A 20-ampere, 125-volt receptacle

Fig. 10–5 CSA 5–20P. A 20-ampere, 125-volt plug.

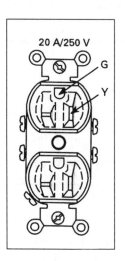

Fig. 10–6 CSA 6–20R. A 20-ampere, 250-volt receptacle.

Fig. 10–8 CSA 6–30R. A 30-ampere, 250-volt receptacle.

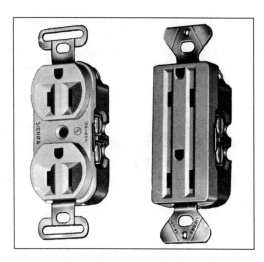

Fig. 10–2 CSA 5–5R. A 15-ampere, 125-volt receptacle.

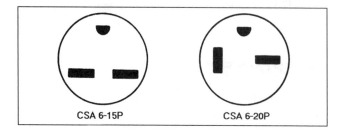

Fig. 10–7 CSA 6–15P, a 15-ampere, 250-volt plug; and CSA 6–20P, a 20-ampere, 250-volt plug.

Diagram 1

(See *Rules 26–700, 26–702, 26–746, 78–052, 78–102,* and *82–014* and *Appendix B.*)

CSA Configurations for Nonlocking Receptacles

	DESCRIPTION		15 AMPERE RECEPTACLE	20 AMPERE RECEPTACLE	30 AMPERE RECEPTACLE	50 AMPERE RECEPTACLE	60 AMPERE RECEPTACLE
2-POLE 3-WIRE GROUNDING	125V	5	5-15R	5-20R	5-30R	5-50R	
	125V	5A		5-20RA ALTERNATE			
	*250V	6	6-15R	6-20R	6-30R	6-50R	
	*250V	6A		6-20RA ALTERNATE			
	277V AC	7	7-15R	7-20R	7-30R	7-50R	
	347V AC	24	24-15R	24-20R	24-30R	24-50R	
3-POLE 4-WIRE GROUNDING	125/250V	14	14-15R	14-20R	14-30R	14-50R	14-60R
	3∅ 250V	15	15-15R	15-20R	15-30R	15-50R	15-60R

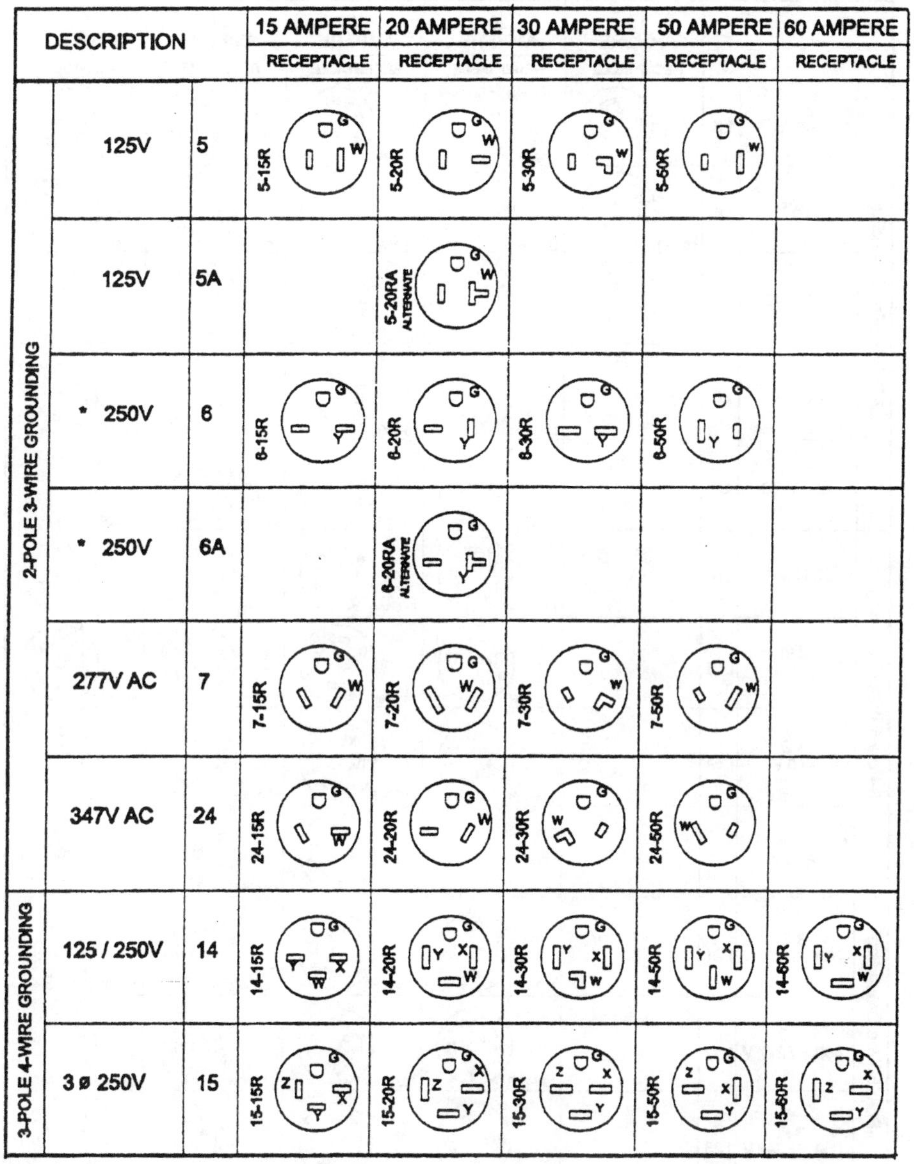

*For configurations 6–15R, 6–20R, 6–30R, and 6–50R, Y denotes identified terminal when used on circuits derived from three-phase, four-wire 416-volt circuits.

Fig. 10–9 CSA Diagram 1.

Two new CSA configurations—5-20RA and 6-20RA—are designed to accept both 15- and 20-ampere plugs. The introduction of these configurations will make the CSA configurations for nonlocking receptacles identical to those used in the United States.

HOSPITAL-GRADE RECEPTACLES

In locations where severe abuse or heavy use is expected, hospital-grade receptacles are recommended. These are high-quality products. They meet CSA requirements as hospital-grade receptacles. These receptacles are marked with a small dot, Fig. 10–11.

Diagram 2

(See *Rules 12–020, 26–700, 78–052, 78–102,* and *82–014* and *App. B.*)

CSA Configurations for Locking Receptacles

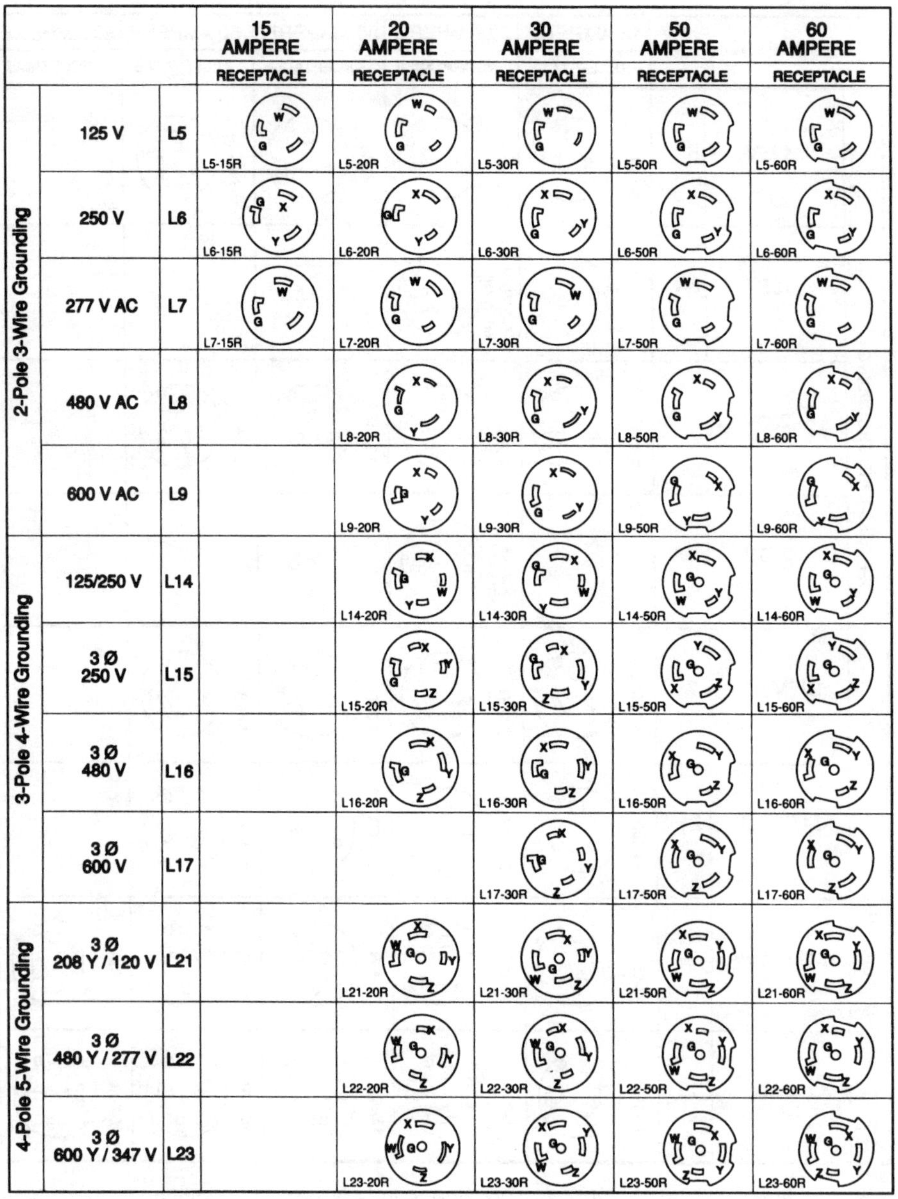

			15 AMPERE RECEPTACLE	20 AMPERE RECEPTACLE	30 AMPERE RECEPTACLE	50 AMPERE RECEPTACLE	60 AMPERE RECEPTACLE
2-Pole 3-Wire Grounding	125 V	L5	L5-15R	L5-20R	L5-30R	L5-50R	L5-60R
	250 V	L6	L6-15R	L6-20R	L6-30R	L6-50R	L6-60R
	277 V AC	L7	L7-15R	L7-20R	L7-30R	L7-50R	L7-60R
	480 V AC	L8		L8-20R	L8-30R	L8-50R	L8-60R
	600 V AC	L9		L9-20R	L9-30R	L9-50R	L9-60R
3-Pole 4-Wire Grounding	125/250 V	L14		L14-20R	L14-30R	L14-50R	L14-60R
	3Ø 250 V	L15		L15-20R	L15-30R	L15-50R	L15-60R
	3Ø 480 V	L16		L16-20R	L16-30R	L16-50R	L16-60R
	3Ø 600 V	L17			L17-30R	L17-50R	L17-60R
4-Pole 5-Wire Grounding	3Ø 208 Y / 120 V	L21		L21-20R	L21-30R	L21-50R	L21-60R
	3Ø 480 Y / 277 V	L22		L22-20R	L22-30R	L22-50R	L22-60R
	3Ø 600 Y / 347 V	L23		L23-20R	L23-30R	L23-50R	L23-60R

Fig. 10–10 CSA Diagram 2.

ELECTRONIC EQUIPMENT RECEPTACLES

Circuits with receptacle outlets that serve microcomputers, solid-state cash registers, or other sensitive electronic equipment should be served by receptacles that are specially designed. These receptacles may be constructed for an isolated ground, Fig. 10–12, they may have transient voltage surge protection, they may be hospital grade, or any combination of these three. Where the isolated grounding is desired, an insulated grounding conductor is permitted to be run from the ground terminal at the distribution panel to the isolated terminal on the receptacle, *Rule 10–906(9)*. These

Fig. 10–11 Hospital-grade receptacle.

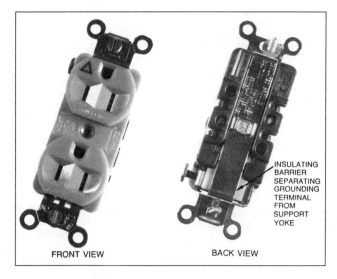

Fig. 10–12 Isolated grounding receptacle.

receptacles are required to have an orange triangle on their faces, except where the receptacle is orange. In this case, the triangle is black.

Surge protection will absorb high voltage surges on the line and further protect the equipment, Fig. 10–13. Surge suppressors are highly recommended in areas of the country where lightning strikes are common. Compared to the cost of replacing damaged equipment and lost data, and the cost of the receptacle, probably under $50, seems small.

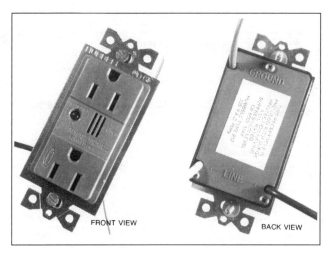

Fig. 10–13 Surge suppressor receptacle.

GROUND-FAULT CIRCUIT INTERRUPTER RECEPTACLES (*RULE 26–700(13)*)

Rule 26–700(11) requires ground-fault circuit interrupters (GFCI) be installed in circuits where receptacles are installed within 3 m of washbasins, bathtubs, and shower stalls. The *C.E.C., Part I* also requires the installation of GFCI protection on temporary wiring for construction sites and in a limited number of other applications. GFCI protection is required for the bathrooms of commercial buildings. For further information concerning GFCI application, the student is referred to *Electrical Wiring: Residential*, Third Canadian Edition (Nelson, 2002).

Deaths and personal injuries have been caused by electrical shock from appliances such as radios, shavers, and electric heaters. This shock hazard exists whenever a person touches both a defective appliance and a grounded conducting surface such as a water pipe or metal sink. To protect against this possibility of shock, the 15-ampere branch circuits can be protected with GFCI receptacles or GFCI circuit breakers. The GFCI receptacles are the most commonly used form of protection.

The CSA requires that Class A GFCIs trip on ground-fault currents of 4 to 6 milliamperes (0.004 to 0.006 ampere). Figs. 10–14 and 10–15 illustrate the principles of how a GFCI operates.

148 Unit 10 Reading Electrical Drawings—Bakery, and Switches and Receptacles

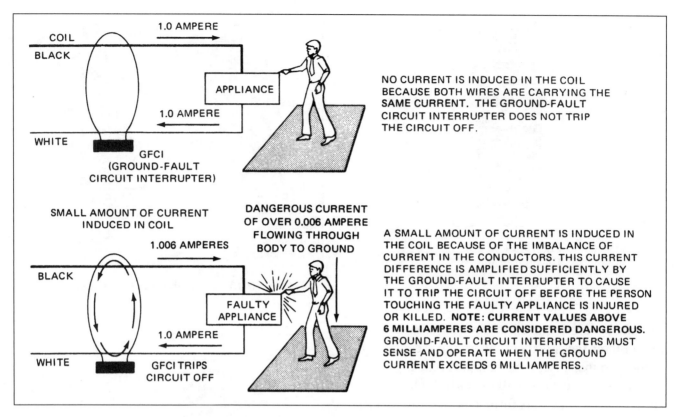

Fig. 10–14 Basic principle of how a ground-fault circuit interrupter operates.

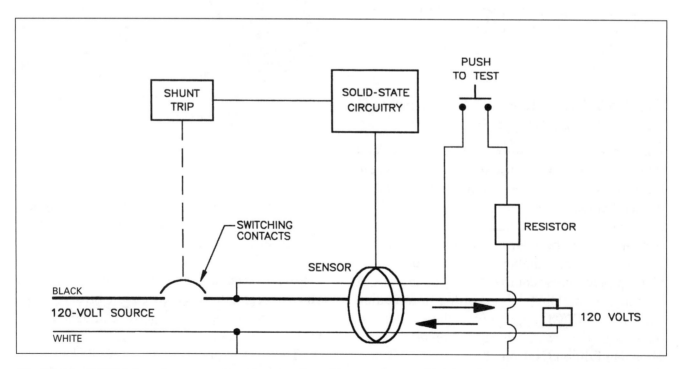

Fig. 10–15 GFCI internal components and connections. Receptacle-type GFCIs switch both the black and white conductors. Note that when the test button is pushed, the test current passes through the sensor to the test button but bypasses the sensor on the way back to the opposite circuit conductor. This is how an unbalance is created and then monitored by the solid-state circuitry to signal the GFCI's contacts to open. Note that since both load currents pass through the sensor, no unbalance is present under normal receptacle use.

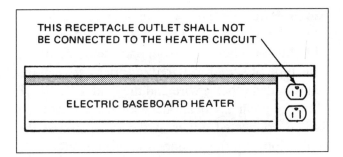

Fig. 10–16 Factory-mounted receptacle outlets on permanently installed electric baseboard heaters may be counted as the required outlets for the space occupied by the baseboard unit.

RECEPTACLES IN ELECTRIC BASEBOARD HEATERS

Electric baseboard heaters are available with or without receptacle outlets. Fig. 10–16 shows how a receptacle outlet can be installed in an electric baseboard heater. *Section 62* details the requirements for fixed electric space heating equipment.

SWITCHES (*RULE 14–500*)

Most electricians refer to switches as toggle switches or wall switches. Refer to Fig. 10–17. These switches are divided into two categories.

Category 1 contains those ac/dc general-use switches that are used to control

- ac or dc circuits
- resistive loads not to exceed the ampere rating of the switch at rated voltage
- inductive loads not to exceed one-half the ampere rating of the switch at rated voltage
- tungsten filament lamp loads not to exceed the ampere rating of the switch at rated voltage if the switch is marked with the letter "T" (A tungsten filament lamp draws a very high current at the instant the circuit is closed. As a result, the switch is subjected to a severe current surge.)

The ac/dc general-use switch normally is not marked "ac/dc." However, it is always marked with the current and voltage rating, such as "10A–125V, 5A–250V–T."

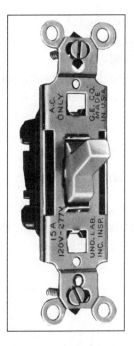

Fig. 10–17 General-use switch.

Category 2 contains those ac general-use switches that are used to control

- ac circuits only
- resistive, inductive, and tungsten-filament lamp loads not to exceed the ampere rating of the switch at 120 volts
- motor loads not to exceed 80% of the ampere rating of the switch at rated voltage

Ac general-use switches may be marked "ac only," or they may also be marked with their current and voltage ratings. A typical switch marking is "15A, 120–277V AC" or "120–277–347V AC." The 347-V rating is required on 347/600-volt systems. Refer to *Section 30* for the special requirements of 347-V switches for lighting circuits. See *Rule 14–512* for additional information pertaining to maximum voltage limitations.

Terminals of switches rated at 20 amperes or less, when marked "CU/AL" or "CO/ALR" are suitable for use with aluminum, copper, and copper-clad aluminum conductors. Switches not marked "CU/AL" or "CO/ALR" are suitable for use with copper and copper-clad conductors only.

Many switches have screwless pressure terminals where a connection can be made by simply inserting the bared end of the conductor. These terminals may be used with copper and copper-clad aluminum conductors only. Push-in terminals are not suitable for use with ordinary aluminum conductors.

Further information on switch ratings is given in *Rules 14–508* and *14–510*.

Switch Types and Connections

Switches for branch circuit lighting are readily available in four basic types: single-pole, three-way, four-way, and double-pole.

Single-Pole Switch. A single-pole switch is used where it is desired to control a light or group of lights, or other load, from one switching point. This type of switch is wired in series between the ungrounded ("hot") wire and the load. Fig. 10–18 shows typical applications with the feed to the switch connected either at the switch or at the light. Note that the grounded wire goes directly to the load and the ungrounded wire is broken at the single-pole switch. Three diagrams are shown for each switching connection. The first diagram is a schematic drawing and is valuable when visualizing the current path. The second diagram repre-

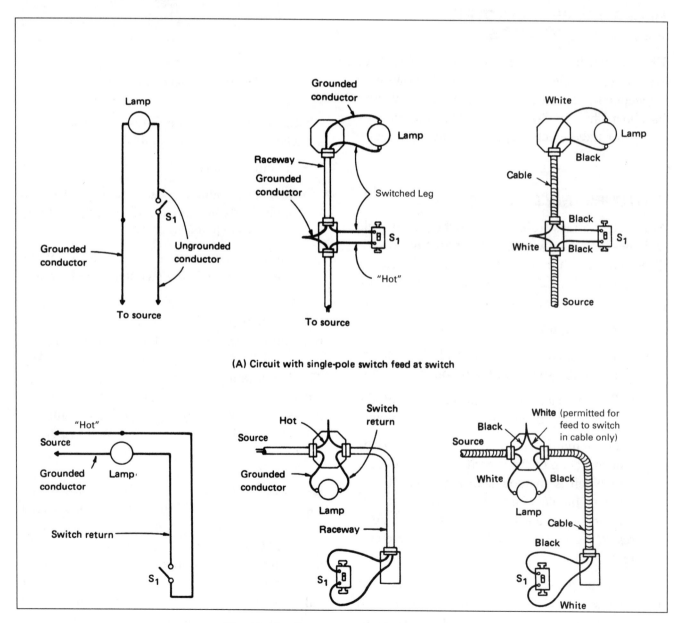

Fig. 10–18 Single-pole switch connection.

sents the situation where a raceway will be available for installing the conductors. The third diagram illustrates the connections necessary when using armoured cable (AC) or nonmetallic-sheathed cable (NMSC), *Rule 4–034(2)*.

Three-Way Switch. A three-way switch has a *common terminal* to which the switch blade is always connected. The other two terminals are called the *traveller terminals*, Fig. 10–19. In one position, the switch blade is connected between the common terminal and one of the traveller terminals. In the other position, the switch blade is connected between the common terminal and the second traveller terminal. The three-way switch can be identified readily because it has no *On* or *Off* position. Note that *On* and *Off* positions are not marked on the switch handle in Fig. 10–19. The three-way switch is also identified by its three terminals. The common terminal is darker in colour than the two traveller terminals, which have a natural brass colour. Fig. 10–20 shows the application of three-way switches to provide control of the light from two points.

Four-Way Switch. A four-way switch is similar to the three-way switch in that it does not have *On* and *Off* positions. However, the four-way switch has four terminals. Two of these terminals are connected to traveller wires from one three-way switch and the other two terminals are connected to traveller wires from another three-way switch, Fig. 10–21. Notice that terminals A1 and A2 are connected to one three-way switch and terminals B1 and B2 are connected to the other three-way switch. In position 1, the switch connects A1 to B2 and A2 to B1. In position 2, the switch connects A1 to B1 and A2 to B2.

The four-way switch is used when a light or a group of lights, or other load, must be controlled from more than two switching points. The switches that are connected to the source and the load are three-way switches. At all other control points, however, four-way switches are used. Fig. 10–22 illustrates a typical circuit in which a lamp is controlled from any one of three switching points. Care must be used to ensure that the traveller wires are connected to the proper terminals of the four-way switch. That is, the two traveller wires from one three-way switch must be connected to the two terminals on one end of the four-way switch. Similarly, the two traveller wires from the other three-way switch must be connected to the two terminals on the other end of the four-way switch.

Double-Pole Switch. A double-pole switch is rarely used on lighting circuits. As shown in Fig. 10–23, a double-pole switch can be used for those installations where two separate circuits are to be controlled with one switch. Both conductors of a circuit may be disconnected at the same time using a two-pole switch, *Rule 14–016(a)*.

SWITCH AND RECEPTACLE COVERS

A cover that is placed on a recessed box containing a receptacle or a switch is called a faceplate, and a cover placed on a surface mounted box is called a raised cover.

Faceplates come in a variety of colours, shapes, and materials, but for our purposes they can be placed in two categories, insulating and metal. Metal faceplates can become a hazard because they can conduct electricity; they are considered to be effectively grounded through the 6-32 screws that fasten the faceplate to the grounded yoke of a receptacle or switch. If the box is nonmetallic, the yoke must be grounded. Both switches and receptacles are available with a grounding screw for grounding the metal yoke.

It is a requirement that two screws or another approved method be used to fasten a receptacle to a raised cover. Receptacles that are installed outdoors in damp or wet locations must be covered with weatherproof covers that maintain their integrity when the receptacle is in use. These covers would be deep enough to shelter the attachment plug, *Rule 26–702*.

152 Unit 10 Reading Electrical Drawings—Bakery, and Switches and Receptacles

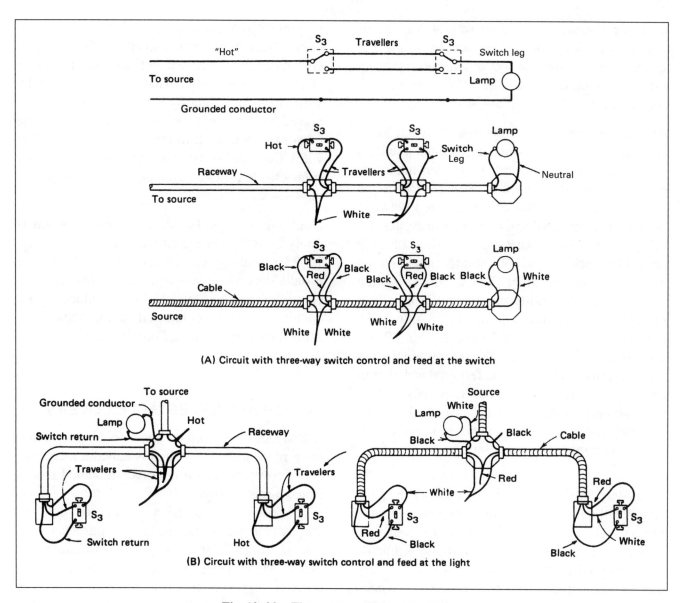

Fig. 10–20 Three-way switch connections.

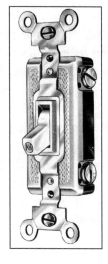

Fig. 10–19 Three-way switch.

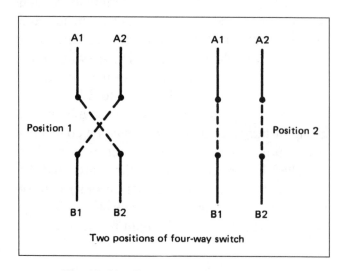

Fig. 10–21 Four-way switch operation.

Unit 10 Reading Electrical Drawings—Bakery, and Switches and Receptacles 153

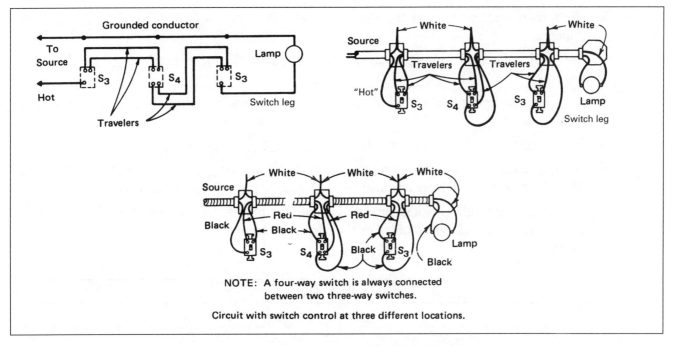

Fig. 10–22 Four-way switch connections.

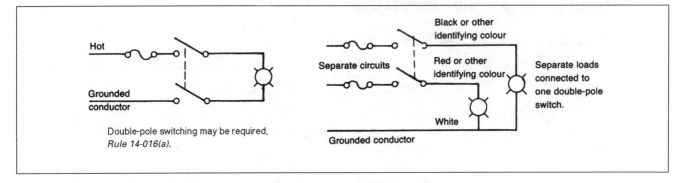

Fig. 10–23 Double-pole switch connections.

REVIEW

PART I: READING ELECTRICAL DRAWINGS—BAKERY

Note: Use the following questions to review the loading schedule for the bakery. Use the drugstore loading schedule, the bakery plans, and the *C.E.C, Part I* as necessary.

1. What is the basic demand load for the drugstore?

   _____
   _____

2. The total connected load of the "special loads" (i.e., not basic load) is _____ volt-amperes.

3. The total connected lighting load on the main floor in the bakery is _____ volt-amperes.

4. To supply the lighting tabulated in the previous question, a total of _____ branch circuits are used.

5. The connected load on circuit number 1 is _____ amperes.

6. The maximum allowable load on branch circuit 10 will be _____ volt-amperes if No. 10 AWG copper conductors are installed to increase the voltage at the load.

7. The electrician might consider installing a single circuit for branch circuits 12 and 9. Is this a good idea? Explain in detail. Do the changes required by this strategy overshadow its advantages?

_____

_____

_____

PART II: SWITCHES AND RECEPTACLES

Note: Refer to the *C.E.C., Part I* or the plans as necessary.

1. Indicate which of the following switches may be used to control the loads listed below.

   A. Ac/dc   10 A–125 V   5 A–250 V
   B. Ac only   10 A–120 V
   C. Ac only   15 A–120/277–347 V
   D. Ac/dc   20 A–125 VT   10 A–250 V

   a. A 120-volt incandescent lamp load (tungsten filament) consisting of ten 150-watt lamps. _____

   b. A 120-volt fluorescent lamp load (inductive) of 1500 volt-amperes. _____

   c. A 347-volt fluorescent lamp load of 625 volt-amperes. _____

   d. A 120-volt motor drawing 10 amperes. _____

   e. A 120-volt resistive load of 1250 watts. _____

   f. A 120-volt incandescent lamp load of 2000 watts. _____

   g. A 230-volt motor drawing 3 amperes. _____

2. Using the information in Figs. 10–9 and 10–10, select by number the correct receptacle for each of the following outlets.

   a. 120/208-volt, single-phase, 20-ampere outlet. _____

   b. 208-volt, three-phase, 50-ampere outlet. _____

   c. 208-volt, single-phase, 30-ampere outlet. _____

3. Show all wiring for the following circuit. The circuit consists of two three-way switches controlling two luminaires with the feed at one luminaire. Indicate conductor colours.

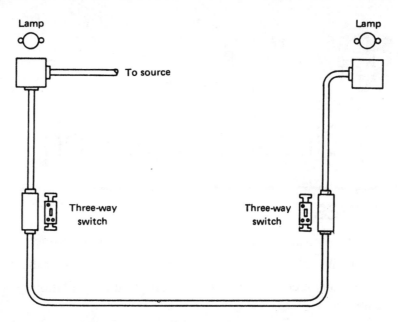

Two three-way switches controlling two luminaires with feed at one luminaire. Show wiring scheme.

4. Show all wiring for the following circuit. The circuit consists of two three-way switches and one four-way switch controlling a light. One of the three switches is to be selected as the four-way switch; then complete the wiring accordingly. Indicate conductor colours.

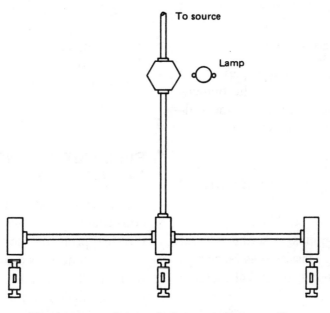

Two three-way switches and a four-way switch controlling a light. Choose which switch should be the four-way and show wiring scheme.

UNIT 11

Branch-Circuit Installation

OBJECTIVES

After completing the study of this unit, the student will be able to

- complete the installation of various raceway products, using the correct materials
- select the correct raceway size for a branch-circuit installation, using appropriate calculations
- select the correct size and type of box for the installation, and the conductors to be installed to the box
- specify the proper raceway support
- demonstrate the proper method of connecting copper and aluminum conductors

BRANCH-CIRCUIT INSTALLATION

In a commercial building, the major part of the electrical work is the installation of the branch-circuit wiring. The electrician must have the ability to select and install the correct materials to ensure a successful job.

The term *raceway*, which is used in this unit as well as others, is defined by the *C.E.C., Part I* as any channel that is designed and used expressly for holding wires, cables, or bus bars.

The following paragraphs describe several types of materials that are classified as raceways, including rigid metal conduit, electrical metallic tubing, liquidtight flexible conduit, flexible conduit, rigid nonmetallic conduit, and electrical nonmetallic tubing.

RIGID METAL CONDUIT (*RULES 12–1000* TO *12–1014*)

Rigid metal conduit, Fig. 11–1, is of heavy-wall construction to provide a maximum degree of physical protection to the conductors run through it. Rigid conduit is available in either steel or aluminum. The conduit can be threaded on the job, or nonthreaded fittings may be used where permitted by local codes, Fig. 11–2.

Rigid conduit bends can be purchased or they can be made using special bending tools. Bends in 16-mm ($\frac{1}{2}$-in), 21-mm ($\frac{3}{4}$-in), and 27-mm (1-in) conduit can be made using hand benders or hickeys, Fig. 11–3. Hydraulic benders must be used to make bends in larger sizes of conduit.

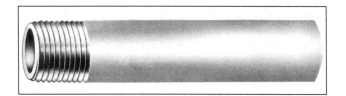

Fig. 11–1 Rigid steel conduit.

Fig. 11–2   Rigid metal conduit fittings.

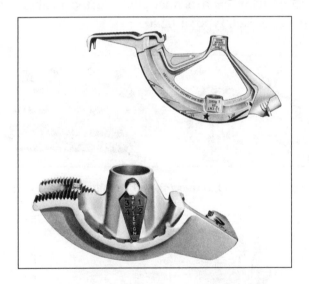

Fig. 11–3   Conduit benders (without handles).

Installation of Metallic Raceway

Rigid metal conduit and electrical metallic tubing are installed according to the requirements of *Rules 12–1000* to *12–1014* and *12–1400* to *12–1412*. The following points summarize these rules. All conduit runs should be level, straight, plumb, and neat with a good standard of workmanship, *Rule 2–108*.

The rigid types of metal conduit

- may be installed in concealed and exposed work

- may be installed in or under concrete when of the type approved for this purpose

- must be threaded as specified in *Table 40, Rule 12–1006* (This rule is to ensure a minimum thread length as well as to prevent a running thread that would prevent the joints from being securely tightened.)

- may contain up to four quarter bends (for a total of 360 degrees) in any run

- must be supported at least every 1.5 metres (4.9 ft) for 16-mm ($\frac{1}{2}$-in) and 21-mm ($\frac{3}{4}$-in) trade size, every 2.0 metres (6.6 ft) for 27-mm (1-in) and 35-mm ($1\frac{1}{4}$-in), and every 3.0 metres (9.8 ft) for 41-mm ($1\frac{1}{2}$-in) to 155-mm (6-in).

- may be installed in wet or dry locations if the conduit is of the type approved for this use

- must have the ends reamed to remove rough edges and a bushing installed on the end of the conduit

- must be run through drilled, bored, or punched holes in framing members, such as studs or joists, in order for conduit to be considered adequately supported; conduit may also be supported with straps approved for the purpose

ELECTRICAL METALLIC TUBING (*RULES 12–1400* TO *12–1412*)

Electrical metallic tubing (EMT) is a thin-wall metal raceway that is not to be threaded, Fig. 11–4. Several types of nonthreaded fittings and connectors are available for use with EMT. The specifications for the commercial building permit the use of EMT under 53-mm (2-in) trade size for several types of installations. EMT must not be installed where subject to mechanical damage.

Fig. 11–4 Electrical metallic tubing.

EMT Fittings

Since EMT is not to be threaded, tubing sections are joined together and connected to boxes and cabinets by fittings called *couplings* and *connectors*. Several styles of EMT fittings are available, including setscrew, compression, and indenter styles. Indenter-style fittings are not commonly used.

Setscrew. When used with this type of fitting, the EMT is pushed into the coupling or connector and is secured in place by tightening the setscrews, Fig. 11–5. This type of fitting is classified as concrete-tight.

Compression. EMT is secured in these fittings by tightening the compressing nuts with a wrench or pliers, Fig. 11–6. These fittings are classified as raintight and concrete-tight types.

Indenter. A special tool is used to secure EMT in this style of fitting. The tool places an indentation in both the fitting and the tubing. It is a standard wiring practice to make two sets of indentations at each connection. This type of fitting is classified as concrete-tight. Fig. 11–7 shows a straight indenter connector.

The efficient installation of EMT requires the use of a bender similar to the one used for rigid conduit, Fig. 11–8. This tool is commonly available in hand-operated models for EMT in sizes from 16-mm ($\frac{1}{2}$-in) to 35-mm ($1\frac{1}{4}$-in) and in power-operated models for EMT in sizes greater than 27-mm (1-in).

Three kinds of bends can be made with the use of the bending tool. The stub bend, the back-to-back bend, and the angle bend are shown in Fig. 11–9. The manufacturer's instructions that accompany each bender indicate the method of making each type of bend.

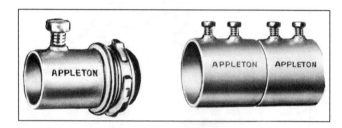

Fig. 11–5 EMT connector and coupling, setscrew type.

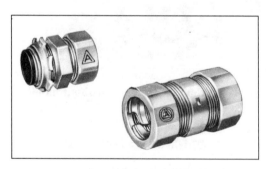

Fig. 11–6 EMT connector and coupling, compression type.

Fig. 11–7 EMT connector, indenter type.

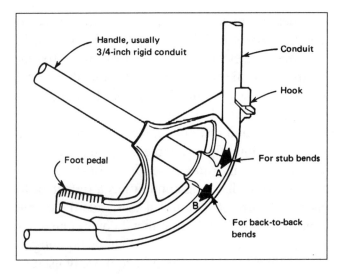

Fig. 11–8 EMT bender.

ELECTRICAL NONMETALLIC TUBING (*RULES 12–1500* TO *12–1516*)

This type of raceway is very popular where some degree of flexibility is required. It is referred to as ENT and is not to be confused with EMT. ENT is discussed later in this Unit.

FLEXIBLE CONNECTIONS (*RULES 12–1010* AND *12–1300* TO *12–1306*)

The installation of certain equipment requires flexible connections, both to simplify the installation and to stop the transfer of vibrations.

The two basic types of material used for these connections are flexible metal conduit, Fig. 11–10, and liquidtight flexible metal conduit, Fig. 11–11.

Flexible metal conduit shall not be less than 16-mm ($\frac{1}{2}$-in) trade size, *Rule 12–1004*, except

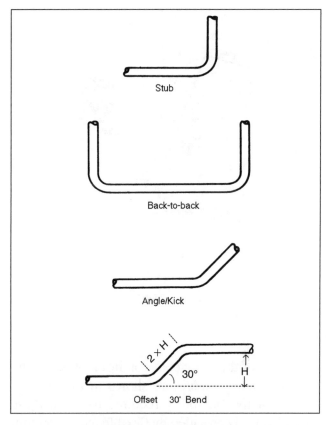

Fig. 11–9 Conduit bends.

when connecting equipment, in which case 12-mm ($\frac{3}{8}$-in) to 14-mm ($\frac{7}{16}$-in) trade size may be used used in lengths of 1.5 m or less.

Rule 12–1000 to *12–1014* covers the use and installation of flexible metal conduit. This type of conduit is similar to armored cable, except that the conductors are installed by the electrician. For armored cable, the cable armor is wrapped around the conductors at the factory to form a complete cable assembly.

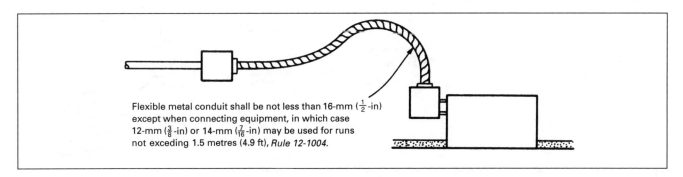

Fig. 11–10 Flexible metal conduit.

160 Unit 11 Branch-Circuit Installation

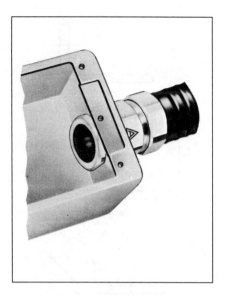

Fig. 11–11 Liquidtight flexible conduit and fitting.

Some of the more common installations using flexible metal conduit are shown in Fig. 11–12. Note that the flexibility required to make these installations is provided by flexible metal conduit. The figure calls attention to the requirement to install a separate bonding conductor in the raceway, *Rule 12–1306*.

Flexible metal conduit is commonly used to connect recessed lighting fixtures. When used in this manner, electricians refer to it as a "fixture whip" or a "fixture drop." *Note*: Flexible metal conduit always requires a separate bonding conductor in the conduit.

The use and installation of liquidtight flexible metal and liquidtight flexible nonmetallic conduit are described in *Rules 12–1300* to *12–1306*. Liquidtight flexible metal conduit has a tighter fit to its spiral turns as compared to standard flexible metal conduit. Liquidtight metal conduit also has a thermoplastic outer jacket that is liquidtight. Liquidtight flexible conduit is commonly used as the flexible connection to a central air-conditioning unit located outdoors, Fig. 11–13.

Liquidtight flexible nonmetallic conduit may be used:

- in exposed or concealed locations
- where flexibility is required
- where not subject to damage
- where the combination of ambient and conductor temperature does not exceed that for which the flexible conduit is approved

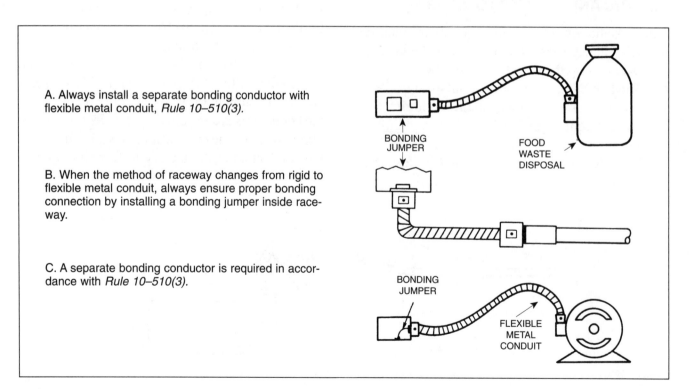

Fig. 11–12 Installations using flexible metal conduit.

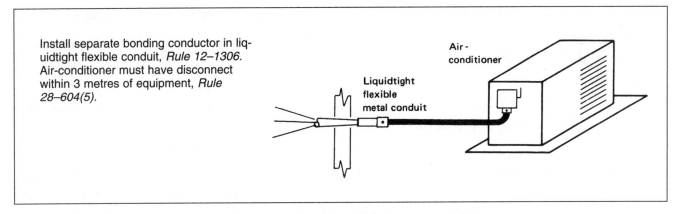

Fig. 11–13   A use of liquidtight flexible conduit.

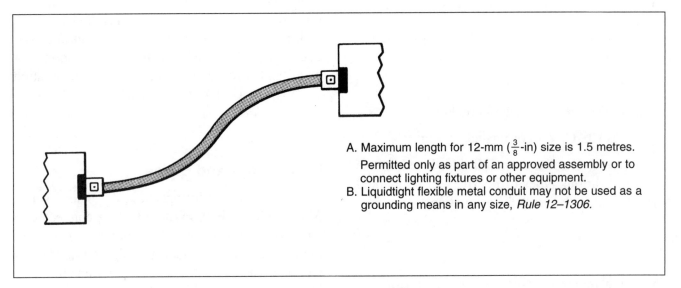

Fig. 11–14   Flexible liquidtight flexible metal conduit, *Rule 12–1300*.

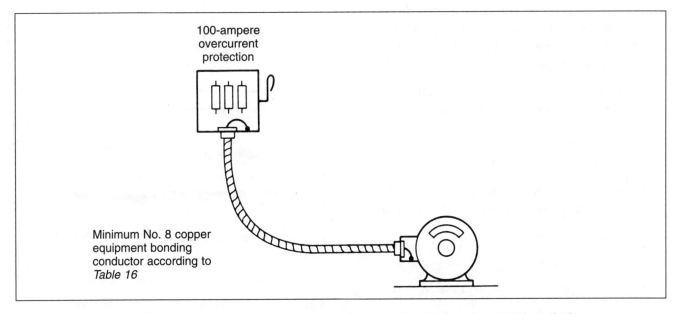

Fig. 11–15   Separate equipment bonding conductor in flexible metal conduit installation.

- in lengths not over 1.5 metres (4.9 ft) for 12-mm ($\frac{3}{8}$-in) conduit
- only with fittings identified for use with the flexible conduit
- in sizes 16-mm ($\frac{1}{2}$-in) to 53-mm (2-in); 12-m ($\frac{3}{8}$-in) is premissible for enclosing motor leads
- for flexible connection to equipment, such as a motor

A separate equipment bonding conductor must be installed. This conductor is sized according to *Table 16*. Fig. 11–15 illustrates the application of this table.

There are three types of liquidtight flexible nonmetallic conduit, which are marked:

- "FNMC-A" for layered conduit
- "FNMC-B" for integral conduit
- "FNMC-C" for corrugated conduit

Both metal and nonmetal fittings listed for use with the various types of FNMC will be marked with the "A," "B," or "C" designations.

RIGID PVC CONDUIT (*RULE 12–1100*)

Rigid PVC conduit (PVC) is made of polyvinyl chloride, which is a plastic material. Solvent-welded fittings may be used for PVC connections and terminations, Fig. 11–16.

PVC may be used:

- concealed in walls, floors, and ceilings, provided no thermal insulation is present
- in cinder fill
- in wet and damp locations
- for exposed work where not subject to physical damage unless identified for such use
- for underground installations

PVC may **not** be used:

- in hazardous locations, *Section 18*
- for support of lighting fixtures
- where exposed to physical damage
- where subject to ambient temperatures exceeding 75°C (167°F)
- when conductor ampacities are based on values which exceed those permitted for a 90°C conductor
- where enclosed in thermal insulation

On outdoor exposed runs of PVC, it is important that expansion joints (Fig. 11–17) be installed to comply with the requirements set forth in *Rule 12–1118* and *Appendix B,* where the expansion characteristics of PVC are given. An equipment grounding conductor must be installed in PVC, and, if metal boxes are used, the grounding conductor must be connected to the metal box. For additional information regarding the installation of PVC, see Unit 4 of *Electrical Wiring: Residential* and *Rule 12–1100*.

PVC Fittings and Boxes

A complete line of fittings, boxes, and accessories is available for PVC. Where PVC is connected to a box or enclosure, an offset connector is recommended over a male adapter because the offset is stronger and provides a gentle radius for wire pulling, as shown in Fig. 11–18, item (c).

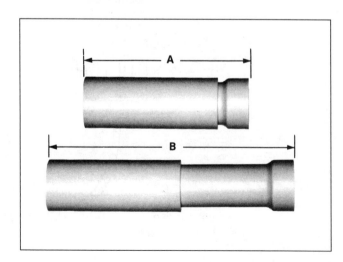

Fig. 11–17 An example of a PVC "O" ring expansion joint. *Rule 12–1118* and *Appendix B.*

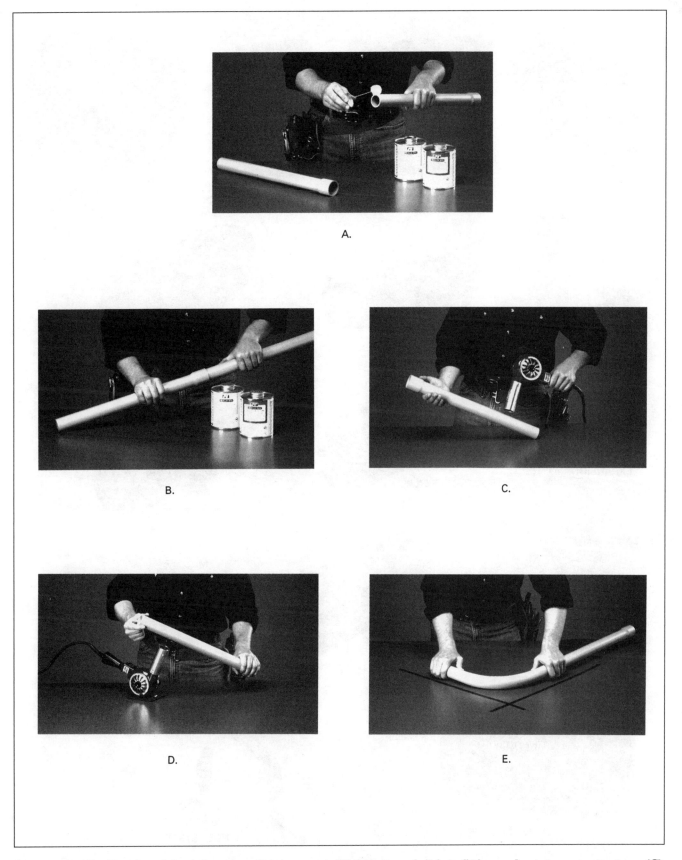

Fig. 11–16 (A) Clean conduit and apply solvent cement. (B) Insert conduit into fitting and rotate a quarter turn. (C) Heat conduit till soft using approved heater or heat gun (D) for bending conduit. (E) Use a template for 90° bends. *Rule 12–1108 and Appendix B.*

164 Unit 11 Branch-Circuit Installation

Fig. 11–18 Various PVC fittings.

Electrical Nonmetallic Tubing (*Rules 12–1500* to *12–1516*)

Electrical nonmetallic tubing (ENT) is a pliable, corrugated raceway that is made of polyvinyl chloride, a plastic material that is also used to make PVC. ENT is hand bendable and is available in coils and reels to facilitate a single length installation between pull points. ENT is designed for use within building walls, encased in concrete, or buried directly in the earth. It is not intended for outdoor use. The specification for the commercial building allows the use of ENT for feeders, branch circuits, and telephone wiring where a raceway size of 53-mm (2-in) trade size or less is required. ENT is available in a variety of colours, which allows the colour coding of the various systems; e.g., electrical circuits could be blue, telephone, red, etc.

ENT may be used:

- concealed in masonry walls, see Table 11–1
- in corrosive environments (check manufacturer's recommendations)
- in damp and wet locations, with fittings identified for the use
- for direct burial, if approved for the purpose
- embedded in poured concrete, including slab-on-grade or below grade, where fittings identified for the purpose are used

ENT may **not** be used:

- in hazardous locations
- for lighting fixture or equipment support, *Rule 12–1512*
- subject to ambient temperature in excess of 75°C (167°F)
- when conductor ampacities are based on values in excess of those permitted for a 90°C conductor
- where exposed, unless provided with protection from mechanical damage

ENT Fittings

Two styles of listed mechanical fittings are available for ENT; one piece snap-on and a clam-shell variety. Solvent cement PVC fittings are approved for use with ENT. The package label will indicate which of these fittings are concrete tight without tape.

	Rigid Metal Conduit	Flexible Metallic Conduit	Rigid PVC and HTF Conduit	Electrical Metallic Tubing	Electrical Nonmetallic Tubing
C.E.C., Part I	Subsection 12–1000	Subsection 12–1000	Subsection 12–1100	Subsection 12–1400	Subsection 12–1500
Min.– Max. Size	16–155 mm (1/2–6 in)	12–53 mm (3/8–2 in)	16–155 mm (1/2–6 in)	16–103 mm (1/2–4 in)	16–27 mm (1/2–1 in)
Min. Radii of Bends, *Rule 12–922*	6 × internal diameter.		6 × internal diameter.	6 × internal diameter.	6 × internal diameter.
Maximum Spacing of Supports	1.5 to 3 m depending on size of conduit, *Rule 12–1010*	300 mm from box/outlet. 1.5 m apart. *Rule 12–1010(3)*	.75 to 2.5 m depending on size of conduit. *Rule 12–1114*	1 m from box/outlet. 1.5 m to 3m apart depending on size of conduit. *Rules 12–1404* and *12–1010*	1 m from box/outlet. 1 m apart.
Uses Allowed	All conditions. Under cinder fill if protected from corrosion, *Rule 12–936*	Dry, accessible locations.	All conditions except in hazardous. *Rule 12–1102*	As permitted. *Rule 12–1402*	Concealed in masonry walls, encased in concrete, or direct burial, if approved.
Uses Not Allowed	See *Rule 12–936*	Hazardous. Underground in concrete.	Hazardous. Thermal insulation.	Where subject to damage. Burial. Hazardous. Wet. See *Rule 12–1402*	Hazardous. Support of equipment. Exposed. Thermal Insulation.
Miscellaneous	Grounding. Threaded with taper die. See *Rule 12–1002* for damp and wet locations	Liquidtight with nonmetallic jacket. See *Rule 12–1304* for fill of 12-mm (3/8-in)	Expansion joints required. Nongrounding— bonding conductor required.	Not threaded.	Pliable. Corrugated. Nongrounding.

Table 11–1 **Summary of conduit and tubing types.**

ENT Boxes and Accessories

Wall boxes and ceiling boxes are available with knockouts for 16-mm ($\frac{1}{2}$-in) to 27-mm (1-in) ENT. ENT can also be connected to any PVC box or access fitting. ENT mud boxes listed for fixture support are available, with knockouts for 16-mm ($\frac{1}{2}$-in) to 27-mm (1-in) trade sizes.

RACEWAY SIZING

The conduit size required for an installation depends upon three factors: (1) the number of conductors to be installed, (2) the cross-sectional area of the conductor, and (3) the permissible raceway fill. These factors are described in *Rule 12–1014*. After the student examines the plans and determines the number of conductors to be installed to a certain point, either of the following procedures can be used to find the conduit size.

If all of the conductors have the same insulation and are of the same wire size, the conduit size can be determined directly by referring to *Table 6* (Table 11–2). For example, three No. 8 T90 Nylon conductors may be installed to the air-conditioning unit in the drugstore in the commercial building. Refer to *Table 6* to find the insulation type (T90 Nylon) and the conductor size (8 AWG). It can be seen that 16-mm ($\frac{1}{2}$-in) conduit or tubing is allowed.

Now assume that in addition to the three No. 8 T90 Nylon conductors, a No. 10 T90 Nylon equipment grounding conductor is to be installed. This type of installation will meet the require-

TABLE 6
(See Rule 12–014(5))

MAXIMUM NUMBER OF CONDUCTORS OF ONE SIZE IN TRADE SIZES OF CONDUIT OR TUBING

Size of Conduit or Tubing		16 ($^1/_2$)	21 ($^3/_4$)	27 (1)	35 (1$^1/_4$)	41 (1$^1/_2$)	53 (2)	63 (2$^1/_2$)	78 (3)	91 (3$^1/_2$)	103 (4)	116 (4$^1/_2$)	129 (5)	155 (6)
Conductor														
Type	Size AWG, kcmil													
TWN75	14	12	22	36	62	85	140	200	200	200	200	200	200	200
	12	9	16	26	45	62	102	145	200	200	200	200	200	200
T90 Nylon	10	4	8	13	23	31	52	74	114	153	197	200	200	200
	8	3	5	9	16	22	36	52	80	100	138	173	200	200
	6	1	4	6	11	16	26	37	58	80	100	125	157	200
	4	1	2	4	7	9	16	23	35	47	61	77	96	140
	3	1	1	3	6	8	13	19	30	40	52	65	82	118
	2	1	1	2	5	7	11	16	25	34	43	55	69	99
	1	1	1	2	3	5	8	12	18	25	32	40	51	73
	1/0	0	1	1	3	4	7	10	15	21	27	34	42	62
	2/0	0	1	1	2	3	6	8	13	17	22	28	35	51
	3/0	0	1	1	1	3	5	7	11	14	19	23	29	43
	4/0	0	0	1	1	1	4	5	9	12	15	19	24	35
	250	0	0	1	1	1	3	4	7	10	12	16	20	29
	300	0	0	1	1	1	2	4	6	8	11	13	17	25
	350	0	0	0	1	1	2	3	5	7	9	12	15	22
	400	0	0	0	1	1	1	3	5	6	8	10	13	19
	450	0	0	0	1	1	1	2	4	6	7	9	12	17
	500	0	0	0	1	1	1	2	4	5	7	9	11	16

NOTES: (1) *The calculated values in the foregoing table are based on conventional concentric Class B stranded conductors.*
(2) *The calculated values in the foregoing table are based on metallic conduit. Other types of raceway of the same nominal size may have different dimensions.*
(3) *Some raceways are required to contain a separate bonding or grounding conductor. No allowance is made for extra conductors in the foregoing table.*

Table 11–2 A section of *Table 6,* showing TWN75 and T90 Nylon conductors' conduit fill.

ments of *Rule 10–510(3)* if a flexible connection is used to connect to the air-conditioning unit. The No. 10 equipment bonding conductor must be included in the determination of the required conduit size, as specified by *Rule 12–1014*.

Thus, for the case of three No. 8 T90 Nylon conductors and one No. 10 T90 Nylon equipment bonding conductor, refer to *Table 10* (Table 11–4) to determine the conductor area. The total cross-sectional area of all of the conductors can now be calculated.

1 No. 10 T90 Nylon	13.63 mm²
3 No. 8 T90 Nylon @ 23.66 mm²	70.98 mm²
Total conductor cross-sectional area	84.61 mm²

Next refer to *Table 8* (Table 11–3) in the category "Conductors (not lead-sheathed)" and the subheading "4" conductors. This shows a maximum fill of 40%. Now refer to *Table 9 C.E.C.* (Table 11–5) under the 40% column to find a value equal to or greater than 84.61. This takes us to the second number in the column, which is 137.6. Read across to the left to find the conduit size. For this situation, 21-mm ($\frac{3}{4}$-in) trade size conduit is required.

SPECIAL CONSIDERATIONS

Following the selection of the circuit conductors and the branch-circuit protection, there are a number of special considerations that generally are factors for each installation. The electrician is usually given the responsibility of planning the routing of the conduit to ensure that the outlets are connected properly. As a result, the electrician must determine the length and the number of conductors in each raceway. In addition, the electrician must derate the conductors' ampacity as required, make allowances for voltage drops, recognize the various receptacle types, and be able to install these receptacles correctly on the system.

BOX STYLES AND SIZING

The style of box required on a building project is usually established in the specifications. However, the sizing of the boxes is usually one of the decisions made by the electrician. Box sizing should be determined by *Tables 22* and *23* (Tables 11–7 and 11–8) and *Rule 12–3036*.

Switch Boxes (Device Boxes)

Switch boxes are 76 × 51 mm (2 by 3 inches) in size and are available with depths ranging from 38 to 89 mm ($1\frac{1}{2}$ inches to $3\frac{1}{2}$ inches), Fig. 11–19. These boxes can be purchased with knockouts for 16-mm ($\frac{1}{2}$-in) or 21-mm ($\frac{3}{4}$-in) conduit, or with cable clamps. Each side of the switch box has holes through which a suitable nail can be inserted for nailing the box to wood studs. The boxes can be ganged by removing the common sides of two or more boxes and connecting the boxes together. Plaster ears may be provided for use on plasterboard or for work on old installations. Device boxes are also available with the sides welded in place to provide a stronger construction.

Masonry Boxes

Masonry boxes are designed for use in masonry block or brick. These boxes do not require an extension cover; they will accommodate devices directly, Fig. 11–20. Masonry boxes are available with depths of 64 mm ($2\frac{1}{2}$ in) and 90 mm ($3\frac{1}{2}$ in) and in through-the-wall depths of 90 mm ($3\frac{1}{2}$ in), 140 mm ($5\frac{1}{2}$ in), and 190 mm ($7\frac{1}{2}$ in). Boxes of this type are available with knockouts up to 27-mm (1 inch) trade size.

Utility boxes, Fig. 11–21, are generally used in exposed surface-mount installations and are available with 16-, 21-, and 27-mm ($\frac{1}{2}$-, $\frac{3}{4}$-, or

Table 8
Maximum Allowable Percent Conduit and Tubing Fill

	Maximum Conduit and Tubing Fill Percent				
	Number of Conductors or Multi-Conductor Cables				
	1	2	3	4	Over 4
Conductors or multi-conductor cables (not lead-sheathed)	53	31	40	40	40
Lead-sheathed conductors or multi-conductor cables	55	30	40	38	35

Table 11–3 *Table 8 C.E.C.* showing maximum allowable conduit and tubing fill. See *Rule 12–1014*.

TABLE 10 (Part 1)

(See Rule 12–1014)

DIMENSIONS OF CABLE FOR CALCULATING CONDUIT AND TUBING FILL (MM)

Conductor Size AWG/kcmil	R90XLPE*, RW75XLPE*, RW90XLPE* 600 V		R90XLPE*, RW75XLPE*, RW90XLPE* 1000 V		R90XLPE†, RW75XLPE†, R90EP†, RW75EP†, RW90XLPE†, RW90EP‡ 600 V		TWN75 T90 NYLON		TW, TW75		TWU, TWU 75 RWU90XLPE*	
	Dia. (mm)	Area (mm²)	Dia. (mm)	Area (mm²)	Dia. (mm)	Area (mm²)	Dia. (mm)	Area (mm²)	Dia. (mm)	Area (mm²)	Dia. (mm)	Area (mm²)
14	3.36	8.89	4.12	13.36	4.12	13.36	2.80	6.18	3.36	8.89	4.88	18.70
12	3.84	11.61	4.60	16.55	4.60	16.75	3.28	8.47	3.84	11.61	5.36	22.56
10	4.47	15.67	5.23	21.45	5.23	21.45	4.17	13.63	4.47	15.67	5.97	27.99
8	5.99	28.17	5.99	28.17	6.75	35.77	5.49	23.66	5.99	28.17	7.76	47.29
6	6.95	37.98	7.71	46.73	8.47	56.39	6.45	32.71	7.71	46.73	8.72	59.72
4	8.17	52.46	8.93	62.67	9.69	73.79	8.23	53.23	8.93	62.67	9.95	77.76
3	8.88	61.99	9.64	73.05	10.40	85.01	8.94	62.83	9.64	73.05	10.67	89.42
2	9.70	73.85	10.46	85.88	11.22	98.82	9.76	74.77	10.46	85.88	11.48	103.5
1	11.23	99.10	12.49	122.6	13.51	143.4	11.33	100.9	12.49	122.6	13.25	137.9
1/0	12.27	118.3	13.53	143.9	14.55	166.4	12.37	120.3	13.53	143.9	14.28	160.2
2/0	13.44	141.9	14.70	169.8	15.72	194.2	13.54	144.0	14.70	169.8	15.45	187.5
3/0	14.74	170.6	16.00	201.0	17.02	227.5	14.84	172.9	16.00	201.0	16.76	220.6
4/0	16.21	206.4	17.47	239.7	18.49	268.5	16.31	209.0	17.47	239.7	18.28	262.4
250	17.90	251.8	19.17	288.5	21.21	353.2	18.04	255.7	19.43	296.4	20.20	320.5
300	19.30	292.6	20.56	332.1	22.60	401.2	19.44	296.9	20.82	340.5	21.54	364.4
350	20.53	331.0	21.79	372.9	23.83	446.0	20.67	335.6	22.05	381.9	22.81	408.6
400	21.79	373.0	23.05	417.3	25.09	494.5	21.93	377.8	23.31	426.8	24.07	455.0
450	22.91	412.2	24.17	458.8	26.21	539.5	23.05	417.3	24.43	468.7	25.19	498.4
500	23.95	450.5	25.21	499.2	27.25	583.2	24.09	455.8	25.47	509.5	26.24	540.8
600	26.74	561.7	27.24	582.9	30.04	708.8	—	—	28.26	627.3	29.02	661.4
700	28.55	640.0	29.05	662.6	31.85	796.5	—	—	30.07	710.0	30.82	746.0
750	29.41	679.3	29.91	702.6	32.71	840.3	—	—	30.93	751.3	31.69	788.7
800	30.25	718.7	30.75	742.6	33.55	884.0	—	—	31.77	792.7	32.53	831.1
900	31.85	796.6	32.35	821.8	35.15	970.2	—	—	33.37	874.5	34.13	914.9
1000	33.32	872.0	33.82	898.4	36.62	1053	—	—	34.84	953.4	35.60	995.4
1250	37.56	1108	38.32	1153	42.38	1411	—	—	39.08	1200	39.08	1199
1500	40.68	1300	41.44	1349	45.50	1626	—	—	42.20	1399	42.96	1449
1750	43.58	1492	44.34	1544	48.40	1840	—	—	45.10	1598	45.86	1652
2000	46.27	1681	47.03	1737	51.09	2050	—	—	47.79	1794	48.55	1851

* Unjacketed
† Jacketed
‡ Includes EPCV

NOTE: Dimensions for aluminium conductors are subject to the range of sizes for which they are certified.

Table 11–4 *Table 10, C.E.C., Part I* showing dimensions of conductors.

1-inch) knockouts. Since these boxes range in depth from 32 to 64 mm ($1\frac{1}{4}$ to $2\frac{1}{2}$ inches), they can accommodate a device without the use of an extension cover.

102-mm (4-Inch) Square Boxes

A 102-mm (4-inch) square box, Fig. 11–22, is used commonly for surface or concealed installations. Extension covers of various depths are available to accommodate devices in those situations where the box is surface mounted. This type of box is available with knockouts up to 27-mm (1-in) trade size.

Octagonal Boxes

Boxes of this type are used primarily to install ceiling outlets. Octagonal boxes are available for mounting in concrete, surface mounting, or concealed installations, Fig. 11–23. Extension covers are available, but are not always required. Octagonal boxes are commonly used in depths of 38 mm ($1\frac{1}{2}$ in) and 54 mm ($2\frac{1}{8}$ in). These boxes are available with knockouts up to 27-mm (1-in) trade diameter.

TABLE 9
(See Rule 12–1014)
CROSS-SECTIONAL AREAS OF CONDUIT AND TUBING (mm²)

Nominal Conduct Size (Metric Designator)	Internal Diameter mm	Cross-Sectional Area of Conduit and Tubing (mm²)							
		100%	55%	53%	40%	38%	35%	31%	30%
16 (½)	15.8	196	107.8	103.9	78.41	74.49	68.61	60.77	58.81
21 (¾)	20.9	344	189.2	182.3	137.6	130.7	120.4	106.7	103.2
27 (1)	26.6	557.6	306.7	295.5	223	211.9	195.2	172.9	167.3
35 (1¼)	35.1	965	530.7	511.4	386	366.7	337.7	299.1	289.5
41 (1½)	40.9	1313	722.4	696.1	525.4	499.1	459.7	407.2	394
53 (2)	52.5	2165	1191	1147	866	822.7	757.7	671.1	649.5
63 (2½)	62.7	3089	1699	1637	1236	1174	1081	957.5	926.7
78 (3)	77.9	4769	2623	2528	1908	1812	1669	1479	1431
91 (3½)	90.1	6379	3508	3381	2551	2424	2233	1977	1914
103 (4)	102.3	8213	4517	4353	3285	3121	2875	2546	2464
116 (4½)	114.5	10288	5659	5453	4115	3910	3601	3189	3086
129 (5)	128.2	12907	7099	6841	5163	4905	4517	4001	3872
155 (6)	154.1	18639	10251	9879	7456	7083	6524	5778	5592

NOTE: The dimensions shown are typical of metallic conduit and tubing. Other figures more accurately representing the actual dimensions of a particular product may be substituted, when known. Dimensions of other circular raceways may be obtained from the approved standard to which they are manufactured.

Table 11–5 *Table 9, C.E.C., Part I. See Rule 12–1014.*

TABLE 22
(See Rule 12–3036)
Space for Conductors in Boxes

Size of Conductor	Usable Space Required for Each Insulated Conductor
AWG	Millilitres
14	24.6
12	28.7
10	36.9
8	45.1
6	73.7

Table 11–6 *Table 22, C.E.C., Part I. See Rule 12–3036.*

4 11/16-Inch (119-mm) Boxes

These more spacious boxes are used where the larger size is required. Available with 16-, 21-, 27-, and 35-mm (½-, ¾-, 1-, and 1¼-inch) trade size size knockouts, these boxes require an extension cover or a raised cover to permit the attachment of devices, Fig. 11–24. These are primarily used for stove and dryer outlets and junction boxes.

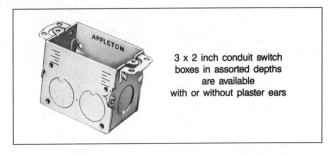

Fig. 11–19 Device box for use with conduit.

Box Sizing

Rule 12–3036 indicates that outlet boxes, switch boxes, and device boxes must be large enough to provide ample room for the wires in them so that the wires do not have to be jammed in or crowded.

When conductors are the same size, the proper box size can be determined by referring to *Table 23*. When conductors are of different sizes, refer to *Table 22*.

Tables 22 and *23* do not consider fittings or devices such as fixture studs, hickeys, switches, pilot lights, or receptacles which may be in the box. Refer to Figs. 11–25 to 11–29 for information on determining the proper box size.

TABLE 23
NUMBER OF CONDUCTORS IN BOXES

	Box Dimensions Inches Trade Size	Capacity Millilitres (Cubic Inch)	Maximum Number of Conductors Size AWG				
			14	12	10	8	6
Octagonal	4 x 1½	245 (15)	10	8	6	5	3
	4 x 2⅛	344 (21)	14	12	9	7	4
Square	4 x 1½	344 (21)	14	12	9	7	4
	4 x 2⅛	491 (30)	20	17	13	10	6
	4¹¹⁄₁₆ x 1½	491 (30)	20	17	13	10	6
	4¹¹⁄₁₆ x 2⅛	688 (42)	28	24	18	15	9
Round	4 x ½	81 (5)	3	2	2	1	1
Device	3 x 2 x 1½	131 (8)	5	4	3	2	1
	3 x 2 x 2	163 (10)	6	5	4	3	2
	3 x 2 x 2¼	163 (10)	6	5	4	3	2
	3 x 2 x 2½	204 (12.5)	8	7	5	4	2
	3 x 2 x 3	245 (15)	10	8	6	5	3
	4 x 2 x 1½	147 (9)	6	5	4	3	2
	4 x 2⅛ x 1⅞	229 (14)	9	8	6	5	3
	4 x 2⅜ x 1⅞	262 (16)	10	9	7	5	3
Masonry	3¾ x 2 x 2½	229 (14)/gang	9	8	6	5	3
	3¾ x 2 x 3½	344 (21)/gang	14	12	9	7	4
	4 x 2¼ x 2⅜	331 (20.25)/gang	13	11	9	7	4
	4 x 2¼ x 3⅜	364 (22.25)/gang	14	12	9	8	4
Through Box	3¾ x 2	3.8/mm (6/inch)	4	3	2	2	1
Concrete Ring	4	7.7/mm (12/inch)	8	6	5	4	2
FS	1 Gang	229 (14)	9	8	6	5	3
	1 Gang Tandem	557 (34)	22	19	15	12	7
	2 Gang	426 (26)	17	14	11	9	5
	3 Gang	671 (41)	27	23	18	14	9
	4 Gang	917 (56)	37	32	24	20	12
FD	1 Gang	368 (22.5)	15	12	10	8	5
	2 Gang	671 (41)	27	23	18	14	9
	3 Gang	983 (60)	40	34	26	21	13
	4 Gang	1392 (85)	56	48	37	30	18

Table 11–7   *Table 23, C.E.C., Part I. See Rule 12–3036.*

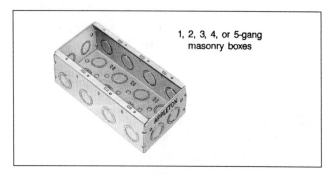

Fig. 11–20   Four-gang masonry device box.

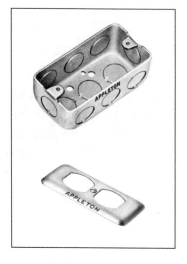

Fig. 11–21   Utility box and receptacle cover.

Unit 11 Branch-Circuit Installation 171

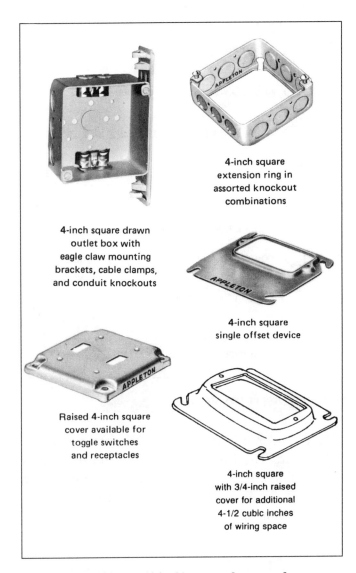

Fig. 11–22 102-mm (4-inch) square boxes and covers.

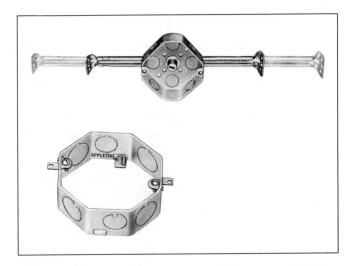

Fig. 11–23 Octagonal box on telescopic hanger and box extension.

EXAMPLE: A box contains one fixture stud and two wire connectors. The number of conductors permitted in the box shall be two less than shown in the table. (Deduct one conductor for the fixture stud; deduct one conductor for the two wire connectors.) See Fig. 11–25.

When the box contains different size wires, *Rule 12–3036,* do the following:

- Size the box based upon the total cubic-inch volume required for the conductors, according to *Table 22.*

- Then make the adjustments necessary for devices, wire connectors, and fixture studs.

- When conductors of different sizes are connected to device(s) on one yoke or strap, use the size of the largest conductor connected to the device(s) in computing the box fill.

- Count insulated grounding conductors the same as any other insulated conductor when computing the box fill.

Note: 1 cubic centimetre (cm^3) = 1 millilitre (mL)

EXAMPLE: A box is to contain two devices: a duplex receptacle and a toggle switch. Two No. 12 conductors are connected to the receptacle, and two No. 14 conductors are connected to the toggle switch. There are two grounding conductors in the box, a No. 12 insulated and a No. 14 bare. There are two wire connectors in the box. The minimum box size would be computed as follows:

Fig. 11–24 4-in (102-mm) box.

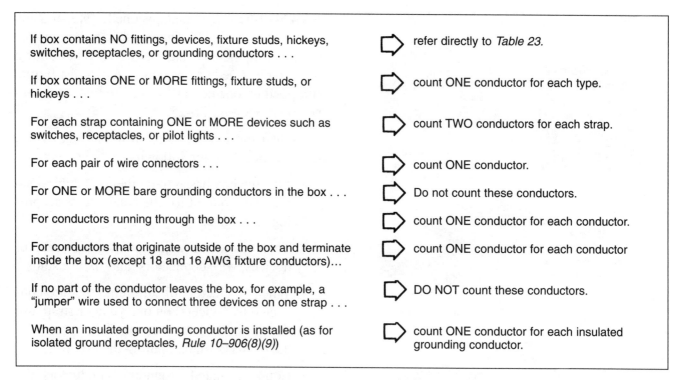

Fig. 11-25 Checklist for determining proper box sizes.

2 No. 12 conductors @ 28.7 cm³	= 57.4 cm³
2 No. 14 conductors @ 24.6 cm³	= 49.2 cm³
1 wire connector @ 28.7 cm³	= 28.7 cm³
(based on the size of the largest conductor in the box)	
1 equipment grounding conductor @ 28.7 cm³	= 28.7 cm³
1 duplex receptacle @ 2 × 28.7 cm³	= 57.4 cm³
(based on the size of the connecting conductors)	
1 toggle switch @ 2 × 24.6 cm³	= 49.2 cm³
(based on the size of the connecting conductors)	
Minimum box volume	= 270.6 cm³ or mL

Therefore, select a box having a *minimum* volume of 270.6 mL. The cubic-inch volume may be marked on the box; otherwise, refer to the second column of *Table 23*, entitled "Capacity."

Two 76 × 51 × 51-mm (3 × 2 × 2-in) boxes with a volume of 163 mL each would be adequate. When sectional boxes are ganged together, the volume to be filled is the total volume of the assembled boxes. Fittings may be used with the sectional boxes, such as plaster rings, raised covers, and extension rings.

When these fittings are marked with their volume, or have dimensions comparable to those boxes shown in *Table 23*, then their volume may be considered in determining the total volume to be filled. The following example illustrates how the volume of a plaster ring or of a raised cover, Fig. 11-27, is added to the volume of the box to increase the total volume.

EXAMPLE: How many No. 12 conductors are permitted in this box and raised plaster ring?

ANSWER: See *Rule 12-3036* and *Tables 22* and *23* for the volume required per conductor. This box and cover will take:

418 mL space/28.7 mL per No. 12 conductor = 14 No. 12 conductors maximum, less the deductions for devices and connectors, per *Rule 12-3036*.

Many devices, such as GFCI receptacles, dimmers, and timers, are much larger than a conventional receptacle or switch. To determine actual volume of a box where outlet devices are to be installed, subtract the volume of each outlet device from the box total. This will give a more accurate volume for wire fill in the box. It is always good practice to install a box that has ample room for the conductors, devices, and

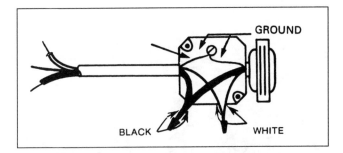

Fig. 11–26 Transformer leads No. 18 AWG or larger must be counted when selecting the proper size box, *Rule 12–3036(1)*. In the example shown, the box is considered to contain four conductors.

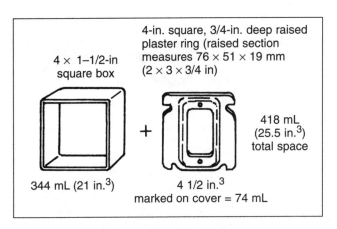

Fig. 11–27 Area of box and plaster ring.

fittings rather than forcibly crowding the conductors into the box.

Insulated bonding conductors are the same size as the circuit conductors in cables having No. 14, No. 12, and No. 10 circuit conductors. Thus, box sizes can often be calculated using *Table 23* when the bonding conductor is insulated.

Another example of the procedure for selecting the box size is the installation of a four-way switch in the bakery. Four No. 12 conductors and one device are to be installed in the box. Therefore, the box selected must be able to accommodate six conductors (two for the device). A $76 \times 51 \times 64$-mm ($3 \times 2 \times 2\frac{1}{2}$-in) switch box is adequate.

When the box contains fittings such as fixture studs or hickeys, the number of conductors permitted in the box is reduced by one. For example, if a box contains one fixture stud, deduct one conductor from the allowable amount as shown in *Table 23*.

An example of this situation occurs in the bakery at the box where the conduit leading to the four-way switch is connected.

- There are two sets of travellers (one from each of the three-way switches) passing through the box on the way to the four-way switch (four conductors).
- There is a neutral, from the panelboard, that is spliced to a neutral that serves the luminaire attached to the box and to neutrals entering the two conduits leaving the box that go to other luminaires (four conductors).
- There is a switch return that is spliced in the box to a conductor that serves the luminaire that is attached to the box and to switch returns that enter the conduits leaving the box to serve other luminaires (4 conductors).
- A fixture stud–stem assembly is used to connect the luminaire directly to the box (one conductor).
- This is a total of 13 conductors.

From *Table 23*, a 102×54-mm ($4 \times 2\frac{1}{8}$-in) square or a 119×38-mm ($4\frac{11}{16} \times 1\frac{1}{2}$-in) square box would qualify. Since these boxes require plaster rings to accept finishing devices, a smaller size box could be used if the plaster ring volume added to the box volume were sufficient for the 13 conductors.

When necessary, conduit fittings, Fig. 11–30, or pull boxes, Fig. 11–31, are used. If pull boxes are used, they must be sized according to *Rule 12–3038*, as shown in Fig. 11–32. The following *C.E.C., Part I* requirements apply to the installation of pull boxes and junction boxes.

- For straight pulls, the box must be at least eight times the diameter of the largest raceway when the raceways contain No. 4 or larger conductors, *Rule 12–3038(2)(a)*.
- For angle or U pulls, Fig. 11–33, the distance *between* the raceways and the opposite wall of the box must be at least six times the diameter of the largest raceway. To this value, it is necessary to add the diameters of all other raceways entering in any one row in

174 Unit 11 Branch-Circuit Installation

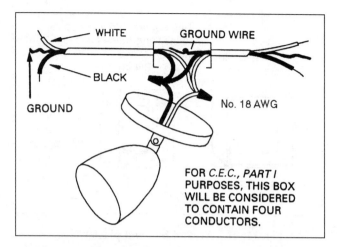

Fig. 11–28 Fixture wires are not counted when determining correct box size. *Rule 12–3036 (1)(d)* specifies that fixture wires No. 18 and No. 16 AWG shall not be counted.

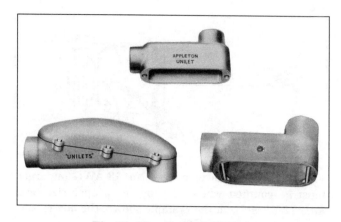

Fig. 11–30 Conduit fittings.

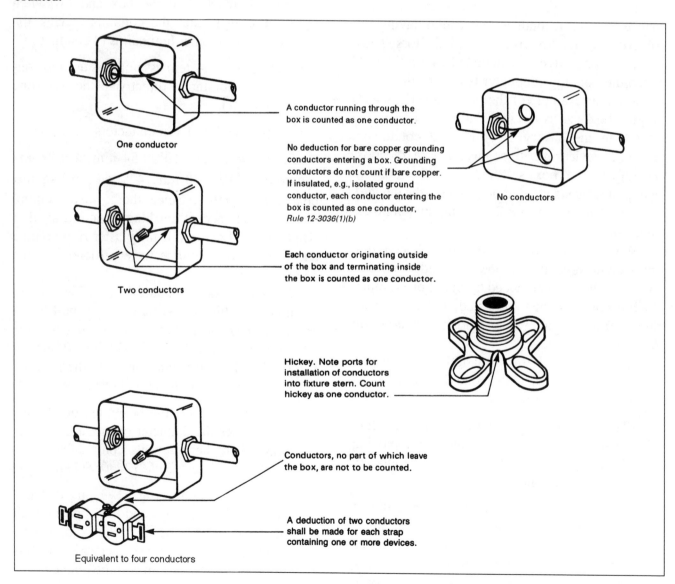

Fig. 11–29 Illustration of *Rule 12–3038* requirements.

Fig. 11–31 Threaded pull box.

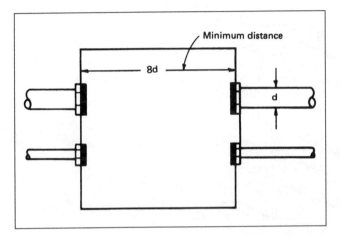

Fig. 11–32 Straight pull box, *Rule 12–3038*.

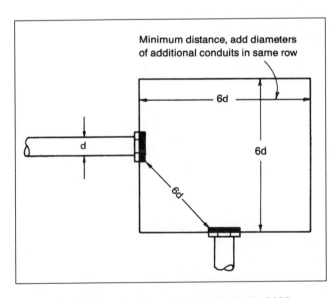

Fig. 11–33 Angle or U pulls, *Rule 12–3038*.

the same wall of the box, Fig. 11–34. When there is more than one row, the row with the maximum diameters must be used in sizing the box. This requirement applies when the raceways contain No. 4 or larger conductors. See *Rule 12–3038*.

- Boxes may be smaller than the preceding requirements when approved and marked with the size and number of conductors permitted.
- Boxes must have an approved cover.
- Boxes must be accessible.

RACEWAY SUPPORT

There are many devices available that are used to support conduit. The more popular devices are shown in Figs. 11–35 and 11–36.

The *C.E.C., Part I* section that deals with a particular type of raceway also gives the requirements for supporting that raceway. Fig. 11–37 shows the support requirements for rigid metal conduit, electrical metallic tubing, flexible metal conduit, and liquidtight flexible metal conduit.

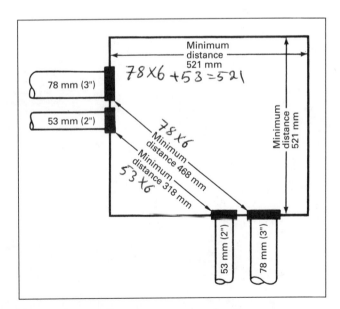

Fig. 11–34 An example of angle pull when more than one raceway enters same wall of box.

176 Unit 11 Branch-Circuit Installation

Fig. 11–35 Raceway support devices.

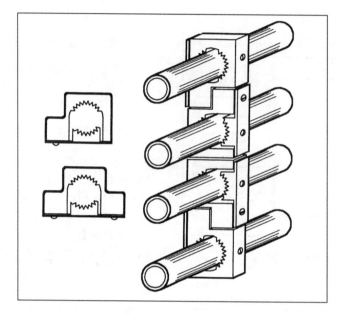

Fig. 11–36 Nonmetallic conduit and straps.

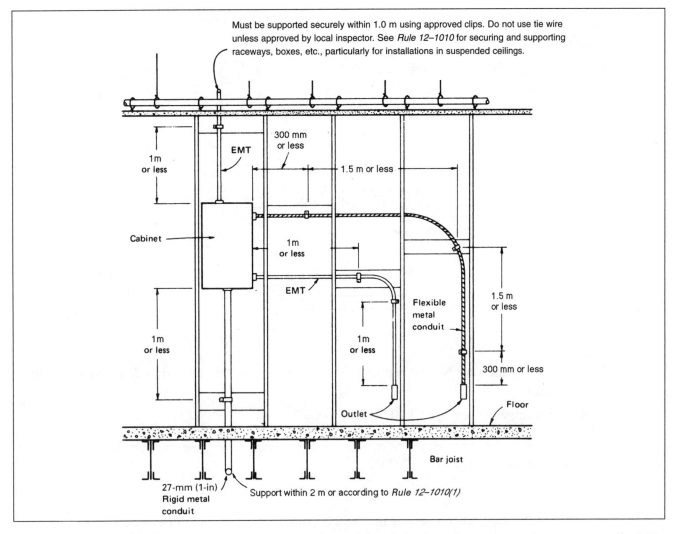

Fig. 11–37 Supporting raceway. See *Rule 12–1010(1)* for securing and supporting raceways, boxes, etc., particularly for installation in suspended ceilings.

REVIEW

Note: Refer to the *C.E.C., Part I* or the plans as necessary.

1. From the diagram that follows, determine the following dimensions.
    a. The minimum acceptable length dimension is __8×63__ mm.
    b. The minimum acceptable width dimension is _____ mm.

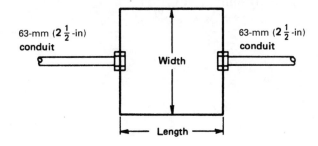

2. From the diagram below, determine the following dimensions.
   a. Dimension a must be at least __624__ mm. — 78 × 8 = 624
   b. Dimension b must be at least __318__ mm. — 6 × 53 = 318
   c. Dimension c must be at least __371__ mm. 6 × 53 + 53 = 371

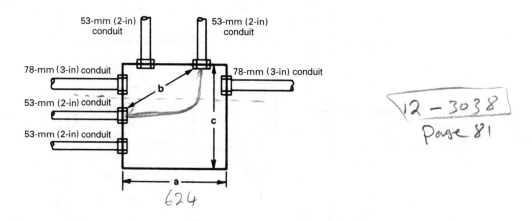

12 – 3038
Page 81

3. Four No. 12 and four No. 10 conductors are to be installed in a conduit. What size conduit is required if
   a. the conductors are Type T90 Nylon?   Table 10
   4 × 8.47 + 4 × 13.63 = 21 mm   (3/4")
   b. the conductors are Type RW90?

4. What is the smallest square box that will accommodate four No. 12 conductors and four No. 10 conductors if
   a. these conductors are pulled through without a splice?   Table 22
   4 × 28.7 + 4 × 36.9 = 262.4 mL   →   4 × 1½
   b. these conductors are all spliced in the box?   28.7 × 10 + 36.9 × 10 = 656
   4-11/16 × 2-1/8   From T23

5. For each of the following raceways, how many supports are required to be installed? Both ends of each raceway terminate in outlet boxes.
   a. 15.2 m (50 ft) of 16-mm (½-in) EMT
   10
   1m   15.2 − 2 = 13.2
   13.2 / 1.5 = 9 spaces = 8 supports
   2 + 8 = 10 supports
   b. 6.09 m (20 ft) of 16-mm (½-in) EMT
   4

6.09 − 2 = 4.09 / 1.5 = 2.7 ≈ 3 spaces
3 + 1 = 4 supports

6. What is the minimum volume that is permissible for the outlet box shown below? Show your calculations.

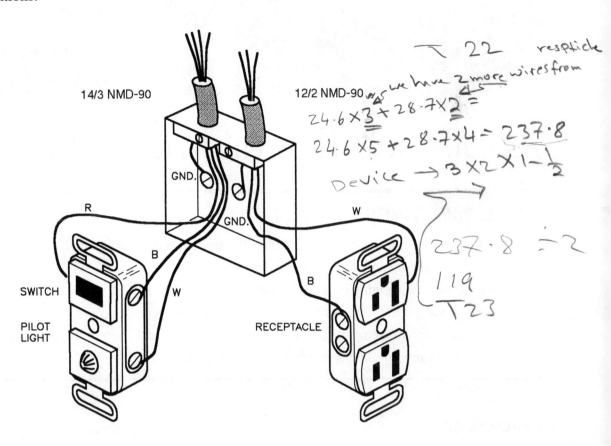

7. Rigid metal conduit must be secured within __1.5__ m (__5__ ft) of a junction box.

8. Must additional support be provided for a conduit that is run through holes that have been bored in a series of studs? Explain your answer.

_____
_____
_____
_____

UNIT 12

Appliance Circuits (Bakery)

OBJECTIVES

After completing the study of this unit, the student will be able to
- define the meaning of the word "appliance"
- define the meaning of the term "utilization equipment"
- understand the different wiring methods for connecting appliances
- determine branch-circuit ratings, conductor sizes, and overcurrent protection for appliances
- understand grounding requirements for appliances

APPLIANCES

Most of the rules that relate to the installation and connection of electrical appliances are found in *Section 26* and *Rule 8–302*. For motor-operated appliances, reference will also be made to *Section 28—Motors and Generators*. Where the appliance is equipped with hermetic refrigerant motor–compressor(s), such as refrigeration and air-conditioning equipment, *Subsection 28–700* applies.

An appliance could be defined as utilization equipment, generally other than industrial, of standardized sizes or types, that is installed or connected as a unit to perform such functions as air conditioning, washing, food mixing, and cooking.

The *C.E.C., Part I* defines utilization equipment as equipment that utilizes electric energy for mechanical, chemical, heating, lighting, or similar purposes.

The most common equipment types discussed in the *C.E.C., Part I* are

- permanently connected
- cord connected
- fixed

In commercial buildings, one of the most common tasks performed by the electrician is to make the electrical connections to appliances. The branch-circuit sizing may be specified by the consulting engineer in the specifications or on the electrical plans. In other cases, it is up to the electrician to do all of the calculations, and/or make decisions based upon the nameplate data of the appliance.

The bakery in the commercial building provides examples of several different methods that can be used to connect electrical appliances.

THE EXHAUST FAN (CONNECTED TO A LIGHTING BRANCH CIRCUIT)

The exhaust fan is a permanently connected motor-operated appliance. *Permanently connected* means "hard wired" with flexible conduit rather than being "cord connected." The exhaust fan is supplied by a 15-ampere branch circuit.

The rating of any one cord-and-plug connected piece of utilization equipment shall not exceed 80% of the branch-circuit ampere rating.

This particular exhaust fan has a small, $\frac{1}{10}$-horsepower, 120-volt motor having a full-load ampere rating of 2.9 amperes. Because of its low current draw the exhaust fan motor may be controlled by a standard single-pole switch, *Rule 28–500(3)(b)*, and supplied by the 15-ampere lighting branch circuit.

THE BASICS OF MOTOR CIRCUITS

Disconnecting Means

The *C.E.C., Part I* requires that all motors be provided with a means to disconnect them from their electrical supply. These requirements are found in *Rule 28–600*.

For most applications, the disconnect switch for a motor must be horsepower rated, *Rule 28–602*. There are some exceptions, the most common being

- A manual across-the-line starter can serve as both the starter and disconnect if marked "suitable for motor disconnect."

- For three-phase stationary motors over 100 horsepower, a general-use disconnect switch or isolating switch is permitted when the switch is marked "Do not open under load," *Rule 28–602(3)(a)*.

- An attachment plug can serve as the disconnecting means for a portable motor if the attachment plug cap is rated not less than the ampacity of the conductors of the motor branch circuit.

Both the nameplate on all disconnect switches and the manufacturer's technical data, furnish the horsepower rating, the voltage rating, and the ampere rating of the disconnect switch.

For those motors that are turned "off" and "on" by a controller (often referred to as a motor starter), a disconnecting means must be in sight of and within 9 metres of the controller and shall disconnect the controller, *Rule 28–604(3)(b)*, Fig. 12–1. That is why combination starters are so popular. Combination starters have the controller and the disconnecting means in one enclosure, as permitted in *Rule 28–602*. The electrician does not have to mount a separate disconnect switch and a separate controller. This results in labour savings.

According to *Rule 28–604(3)(a)*, Fig. 12–1, the disconnecting means must also be in sight of and within 9 metres of the motor location and the machinery driven by it. However, if the disconnect is for air-conditioning equipment it must be within sight of and within 3 metres of the equipment, *Rule 28–604(5)*. In the case of large units the disconnect may be installed on or within the unit.

For a motor and the motor-disconnecting means to meet the *C.E.C., Part I* definition of being "in sight" of and "within 9 metres" of each other, both of the following conditions must be met:

- the disconnecting means must be visible from the motor, and

- the disconnecting means must not be more than 9 metres from the motor.

To turn the exhaust fan on and off, an ac, general-use single-pole (toggle) switch is installed on the wall below the fan. The electrician will install an outlet box approximately 300 mm (1 ft) from the ceiling (see the electrical plan) from which he or she will run a short length of flexible metal conduit to the junction box on the fan. The fan exhausts the air from the bakery. The fan is a through-the-wall installation, approximately 450 mm (18 in) below the ceiling, Fig. 12–2.

Rules 28–500(3)(b) and *28–602(3)(e)* permit this switch to serve as both the controller and the required disconnecting means.

Motor Circuit Conductors

The conductors that supply a single motor shall rated at not less than 125% of the full-load current rating of the motor, *Rule 28–106*. The branch-circuit conductors supplying the exhaust fan are No. 12 T90 Nylon conductors which have an

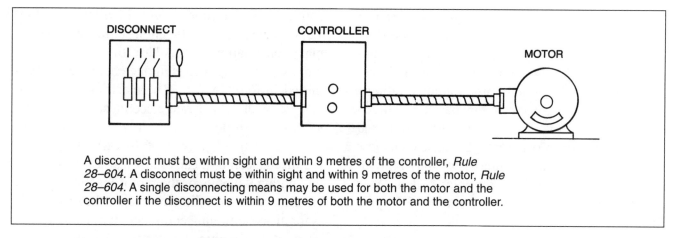

Fig. 12–1 Location of motor disconnect means.

A disconnect must be within sight and within 9 metres of the controller, *Rule 28–604*. A disconnect must be within sight and within 9 metres of the motor, *Rule 28–604*. A single disconnecting means may be used for both the motor and the controller if the disconnect is within 9 metres of both the motor and the controller.

ampacity well above the required 125% times the FLA of the motor. Refer to *C.E.C., Table 2* for conductor ampacities.

Motor Overload Protection

Subsection 28–300 covers motor overload protection. The exhaust fan in the bakery draws 2.9 amperes, is less than 1 horsepower, and is not automatically started. According to *Rule 28–308(a)*, "any motor rated at 1 horsepower or less that is continuously attended while in operation and that is on a branch circuit having overcurrent protection rated or set at not more than 15 A does not require overload protection."

Unit 17 covers in detail the subject of overcurrent protection using fuses and circuit breakers. Time-delay fuses are an excellent choice for overload protection of motors. They can be sized close to the ampere rating of the motor. When the motor has integral overload protection or the motor controller has thermal overloads, time-delay fuses are selected to provide backup overload protection to the thermal overloads. See Table 12–1.

Table 12–1

Motor Nameplate Rating	Overload Protection as a Percentage of Motor Nameplate Full-Load Current Rating
Service factor not less than 1.15	125%
All other motors	115%

The momentary inrush current when a motor starts can be quite high, Fig. 12–3. For example, a 4-ampere, full-load rated motor could have a starting current as high as 24 amperes. The fuse should not open needlessly because of this inrush current.

In this situation, a 15-ampere ordinary fuse or breaker might be required to allow the motor to start. This would provide the branch-circuit overcurrent protection, but it would not provide overload protection for the appliance.

If the motor does not have built-in overload protection, sizing the branch-circuit overcurrent protection as stated above would allow the motor to burn out if it were overloaded or stalled.

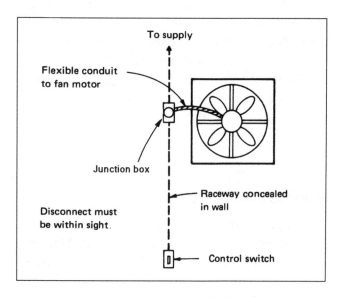

Fig. 12–2 Exhaust fan installation.

Abuse, lack of oil, bad bearings, and jammed V-belts are a few of the conditions that can cause a motor to draw more than normal current. Too much current results in the generation of too much heat within the windings of the motor. For every 10°C (18°F) above the maximum temperature rating of the motor, the expected life of the motor is reduced by 50%. This is sometimes referred to as the "half-life rule."

The solution is to install a time-delay fuse that permits the motor to start, yet opens on an overload before the motor is damaged.

Table 12–2 compares the load responses of ordinary fuses and time-delay fuses.

Table 12–3, Part I, shows how to size time-delay fuses for motors that are marked with a service factor of not less than 1.15 and for motors marked with a temperature rise of not over 40°C.

Table 12–3, Part II, shows how to size time-delay fuses for motors that have a service factor of less than 1.15 and for motors with a temperature rise of more than 40°C.

For abnormal installations, time-delay fuses larger than those shown in Parts I and II of Table 12–3 may be required. These larger-rated fuses (or circuit breakers) will provide short-circuit protection only for the motor branch circuit.

Abnormal conditions might include

- situations where the motor is started, stopped, jogged, inched, plugged, or reversed frequently
- high inertia loads such as large fans and centrifugal machines, extractors, separators, pulverizers, and machines having large flywheels

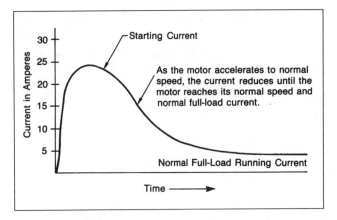

Fig. 12–3 This diagram shows how the starting current of a typical electric motor reaches a very high value for a short time, then decreases as the motor reaches its normal full-load speed. Time-delay cartridge fuses are required to carry five times their ampere rating for at least 10 seconds. This is why this type of fuse is able to handle the momentary starting inrush current of the motor.

- motors having a high code letter and full-voltage start as well as some older motors without code letters.

The higher the code letter, the higher is the starting current. For example, two motors have exactly the same full-load ampere rating. One motor is marked Code J. The other is marked Code D. The Code J motor will draw more momentary starting inrush current than the Code D. The code letter of a motor is found on the motor nameplate.

Refer to the *C.E.C., Part I* and to Bussmann's *Electrical Protection Handbook* for more data relating to motor overload and motor branch-circuit

Table 12–2 Load Response of Fuses
(Approximate Time in Seconds for Fuse to Blow)

Load (amperes)	4-Ampere Fuse (time-delay)	4-Ampere Ordinary Fuse (nontime-delay)	8-Ampere Ordinary Fuse (nontime-delay)	15-Ampere Ordinary Fuse (nontime-delay)
5	Over 300 s	Over 300 s	Won't blow	Won't blow
6	250 s	5 s	Won't blow	Won't blow
8	60 s	1 s	Won't blow	Won't blow
10	38 s	Less than 1 s	Over 300 s	Won't blow
15	17 s	Less than 1 s	5 s	Won't blow
20	9 s	Less than 1 s	Less than 1 s	300 s
25	5 s	Less than 1 s	Less than 1 s	10 s
30	2 s	Less than 1 s	Less than 1 s	4 s

184 Unit 12 Appliance Circuits (Bakery)

Table 12–3 Selection of Time-Delay Fuses for Motor Running Overload Protection Based on Motor Nameplate Ampere Rating

Part I Motors Marked With Not Less Than 1.15 Service Factor

Motor Ampere Rating	Time-Delay Fuse Ampere Rating** (Max. 125%)	Motor Ampere Rating	Time-Delay Fuse Ampere Rating ** (Max. 125%)
1.00 to 1.11	1 1/4	20.0 to 23.9	25
1.12 to 1.27	1 4/10	24.0 to 27.9	30
1.28 to 1.43	1 6/10	28.0 to 31.9	35
1.44 to 1.59	1 8/10	32.0 to 35.9	40
1.60 to 1.79	2	36.0 to 39.9	45
1.80 to 2.00	2 1/4	40.0 to 47.9	50
2.01 to 2.23	2 1/2	48.0 to 55.9	60
2.24 to 2.55	2 8/10	56.0 to 63.9	70
2.56 to 2.79	3 2/10	64.0 to 82.0	80
2.80 to 3.19	3 1/2	72.01 to 82.0	90
3.20 to 3.59	4	82.1 to 90	100
3.60 to 4.00	4 1/2	90.01 to 100	110
4.01 to 4.47	5	100.1 to 120	125
4.48 to 5.00	5 6/10	120.1 to 140	150
5.01 to 5.59	6 1/4	140.1 to 160	175
5.60 to 6.39	7	160.1 to 179	200
6.40 to 7.19	8	180 to 199	225
7.20 to 8.00	9	200 to 239	250
8.01 to 9.59	10	240 to 279	300
9.60 to 11.9	12	280 to 319	350
12.00 to 13.9	15	320 to 359*	400
14.00 to 16.0	17 1/2	360 to 399	450
16.01 to 19.9	20	400 to 479	500

Part II All Other Motors (i.e., Less than 1.15 Service Factor)

All Other Motors Ampere Rating	Time-Delay Fuse Ampere Rating** (Max. 115%)	All Other Motors Ampere Rating	Time-Delay Fuse Ampere Rating ** (Max. 115%)
1.00 to 1.08	1 1/8	17.4 to 20.0	20
1.09 to 1.21	1 1/4	21.8 to 25.0	25
1.22 to 1.39	1 4/10	26.1 to 30.0	30
1.40 to 1.56	1 6/10	30.5 to 34.7	35
1.57 to 1.73	1 8/10	34.8 to 39.1	40
1.74 to 1.95	2	39.2 to 43.4	45
1.96 to 2.17	2 1/4	43.5 to 50.0	50
2.18 to 2.43	2 1/2	52.2 to 60.0	60
2.44 to 2.78	2 8/10	60.9 to 69.5	70
2.79 to 3.04	3 2/10	69.6 to 78.2	80
3.05 to 3.47	3 1/2	78.3 to 86.9	90
3.48 to 3.91	4	87.0 to 95.6	100
3.92 to 4.34	4 1/2	95.7 to 108	110
4.35 to 4.86	5	109 to 125	125
4.87 to 5.43	5 6/10	131 to 150	150
5.44 to 6.08	6 1/4	153 to 173	175
6.09 to 6.95	7	174 to 195	200
6.96 to 7.82	8	196 to 217	225
7.83 to 8.69	9	218 to 250	250
8.70 to 10.0	10	261 to 300	300
10.5 to 12.0	12	305 to 347	350
13.1 to 15.0	15	348 to 391	400
15.3 to 17.3	17 1/2	392 to 434	450
		435 to 480	500

Note: Disconnect switches must comply with *Rule 28–602*.

protection, disconnecting means, controller size, conductor size, conduit size, and voltage drop. (Cooper Bussman, www.bussmann.com)

Time-delay fuses are designed to withstand motor-starting inrush currents and will not open needlessly. However, they will open if a sustained overload occurs. As illustrated in Unit 19, these fuses should be sized from 115% to 125% of the full-load ampere rating of the motor.

For the exhaust fan motor that has a 2.9-ampere full-load current draw, select a time-delay fuse in the range of:

$2.9 \times 1.15 = 3.34$ amperes
$2.9 \times 1.25 = 3.63$ amperes

Referring to Table 12–3, find that a 3.5-ampere size is available. If the motor has built-in or other overload protection, the 3.5 ampere fuse will provide backup overload protection for the motor.

Motor Branch-Circuit Overcurrent Protection

It is beyond the scope of this text to cover every aspect of motor circuit design, since an entire text could be devoted to this subject, however the basics of the typical motor installation will be presented.

- Motor branch-circuit conductors, controllers, and the motor must be provided with overcurrent protection, *Rules 28–200* and *14–012*.

- The overcurrent protection must have sufficient time-delay to permit the motor to be started, *Rule 28–200(d)*.

- *Rule 28–200(d)* and *Table 29* provide the maximum percentages permitted for fuses and circuit breakers, for different types of motors.

- When applying the percentages listed in *Table 29*, and the resulting ampere rating or setting does not correspond to standard equipment ratings, we must round down to the next lower standard rating.

- When applying the percentages listed in *Table 29*, and the resulting ampere rating or setting does not allow the motor to start, we are permitted to select a higher ampere rating. See Table 12–4 for maximum rating of overcurrent devices.

Table 12–4 Maximum Rating of Overcurrent Devices

Ratings as Percentage of FLA	Normal Maximum	Absolute Maximum
Nontime-delay fuses not over 600 amperes	300%	400%
Time-delay fuses	175%	225%
Inverse time circuit breakers 100 amperes or less	250%	400%
Inverse time circuit breakers over 100 amperes	250%	300%
Fuses over 600 amperes	300%	300%
Instant trip circuit breakers	1300%	1300%

The following example illustrates a typical squirrel cage motor, code J. Refer to *Table 29* for other types of motors.

- Always check the overload relay table on a motor controller to see if the manufacturer has indicated a maximum size or type of overcurrent device, such as "Maximum Size Fuse 25 amperes." Do not exceed this ampere rating, even if the percentage values listed above result in a higher ampere rating.

EXAMPLE: A motor, connected to a high inertia load, has a FLA rating of 27 amperes. Step one is to determine the maximum ampere rating for time-delay fuses. Refer to Table D12.

$27 \times 1.75 = 47.25$ amperes

The next lower standard rating fuse is 45 amperes, which is the maximum. If the 45-ampere fuse does not allow the motor to start, the fuse ampere rating may be increased, but cannot exceed 225%, *Rule 28–200(d)(ii)*.

$27 \times 2.25 = 60.75$ amperes

The next lower standard ampere rating fuse is 60 amperes, which would be the absolute maximum size permitted to be installed.

Unit 12 Appliance Circuits (Bakery)

SEVERAL MOTORS ON A SINGLE BRANCH CIRCUIT

Two or more motors may be grouped on a single branch circuit provided that certain conditions, as laid out in *Rule 28–206*, are met. Generally, such circuits are limited to 15-A overcurrent protection or are for a single co-ordinated drive system or a single machine that has been designed for the application.

CONDUCTORS SUPPLYING SEVERAL MOTORS

Rule 28–108(1)(a) requires that the conductors supplying several motors on one circuit shall have an ampacity of not less than 125% of the largest motor plus the full-load current ratings of all other motors on that circuit.

For example, consider the bakery's multi-mixer and dough divider, which are supplied by a single branch circuit and have full-load ratings of 2.2 and 3.96 amperes:

1.25 × 3.96 amperes =	4.95 amperes
plus	2.20 amperes
Total	7.15 amperes

The branch-circuit conductors must have a minimum ampacity of 7.15 amperes. Specifications for this commercial building call for No. 12 minimum. Checking *Table 2*, TW75/T90 Nylon copper conductor has an ampacity of 20 amperes, more than adequate to serve the two appliances. When derating ampacities for more than three conductors in one raceway or for high temperatures, we would begin by using the 90°C column ampacity to determine the TW75/T90 Nylon conductor's ampacity before derating.

SEVERAL MOTORS ON ONE FEEDER

Rules 28–204(1) and *28–108* set forth the requirements for this situation. Each motor branch-circuit has short-circuit and ground-fault overcurrent protection sized according to *Rule 28–200*. The individual motors have overload protection.

The feeder fuses or circuit breakers will be sized by:

1. determining the fuse or circuit breaker rating or setting per *Rule 28–200* for the largest motor in the group,
2. adding the value determined in step 1 to the full-load current ratings for all other motors in the group.

EXAMPLE: To determine the maximum overcurrent protection using dual-element time-delay fuses for a branch circuit serving three motors, refer to Table 12–4. The FLA ratings of the motors are 27, 14, and 11 amperes.

First determine the maximum time-delay fuse for the largest motor:

27 amperes × 1.75 = 47.25 amperes

Next add the remaining ampere ratings to this value:

(47.25 + 14 + 11) amperes = 72.25 amperes

For these three motors the maximum ampere rating of dual-element time-delay fuses is 72.25 amperes.

Sometimes there is confusion when one or more of the motors are protected with instant trip breakers, sized at 1300% of the motors full-load ampere rating. This would result in a very large (and possibly unsafe) feeder overcurrent device. *Rule 28–204(2)* limits this possibility by establishing 300% of feeder conductor ampacity for fuse sizing.

GROUNDING

Rule 10–408(1) states that exposed noncurrent-carrying metal parts of portable equipment must be bonded to ground.

Why? *Rule 10–002* explains why equipment must be grounded and bonded.

Where? *Rule 10–400* lists locations where equipment fastened in place or connected by permanent wiring methods (fixed) must be bonded to ground.

What? *Rule 10–402* explains what equipment that is fastened in place or connected by permanent wiring methods (fixed) must be bonded to ground. *Rule 10–408* explains what equipment that is cord-and-plug connected must be bonded to ground.

How? *Rule 10–510* explains how to ground equipment that is fastened in place and is connected by permanent wiring methods. *Rule 10–512* explains how to bond to ground equipment that is cord-and-plug connected.

Rule 10–408(3), *Appendix B* exempts tools and appliances that are double insulated and marked with the symbol ▣ from the bonding to ground requirement.

OVERCURRENT PROTECTION

Overcurrent protection for appliances is covered in *Sections 14* and *16*. If the appliance is motor driven, then *Section 28* applies. In most cases, the overload protection is built into the appliance. The appliance must comply with CSA standards. If the appliance is double insulated it does not need to be bonded to ground.

TYPICAL BAKERY EQUIPMENT

A bakery can have many types and sizes of food preparing equipment, such as blenders, choppers, cutters, disposals, mixers, dough dividers, grinders, molding and patting machines, peelers, slicers, wrappers, dishwashers, rack conveyors, water heaters, blower dryers, refrigerators, freezers, and hot-food cabinets; see Figs. 12–4 and 12–5. All of this equipment is designed and manufactured by companies that specialize in food preparation equipment. These manufacturers furnish specifications that clearly state any technical data that is required in order to connect the appliance in a safe manner, such as voltage, current, wattage, phase (single or three phase), minimum required branch-circuit rating, minimum supply circuit conductor ampacity, and maximum overcurrent protective device rating. If the appliance is furnished with a cord-and-plug set, the CSA size and type will be specified.

The data for the mixers and dough divider are shown in Table 12–5.

Fig. 12–4 Cake mixer. (Courtesy of Rondo.)

Fig. 12–5 Dough divider. (Courtesy of Rondo.)

Table 12–5 Ratings of Appliances

Type of Appliance	Voltage	Amperes	Phase	HP	How Connected
Multimixer	208	3.96	3	$\frac{3}{4}$	CSA 15–20P 4-wire cord and plug on appliance
Multimixer	208	7.48	3	$1\frac{1}{2}$	Same as above
Dough divider	208	2.2	3	$\frac{1}{2}$	Same as above

A CSA 15–20R receptacle outlet, Fig. 12–6, is provided at each appliance location. These are part of the electrical contract.

Each of these appliances is purchased as a complete unit. After the electrician provides the proper receptacle outlets, the appliances are ready for use as soon as they are moved into place and plugged in.

THE DOUGHNUT MACHINE

An individual branch circuit provides power to the doughnut machine. This machine consists of

1. a 2000-watt heating element that heats the liquid used in frying, and
2. a driving motor that has a full-load rating of 2.2 amperes.

As with most food preparation equipment, the appliance is purchased as a complete, prewired unit. This particular appliance is equipped with a four-wire cord to be plugged into a receptacle outlet of the proper configuration.

Fig. 12–7 is the control circuit diagram for the doughnut machine. The following components are indicated on the control circuit diagram:

S   A manual switch used to start and stop the machine

T1 A thermostat with its sensing element in the frying tank (This thermostat keeps the oil at the correct temperature.)

T2 Another thermostat with its sensing element in the drying tank (This thermostat controls the drive motor.)

Fig. 12–6 CSA 15–20R receptacle and plate.

A   A three-pole contactor controlling the heating element

B   A three-pole motor controller operating the drive motor

M   A three-phase motor

OL Overload units that provide overload protection for the motor (Note that there is one thermal overload unit in each phase.)

P   Pilot light to indicate when power is "on"

Since this appliance is supplied by an individual circuit, its current draw is limited to 80% of the branch-circuit rating. The branch circuit supplying the doughnut machine must have sufficient ampacity to meet the minimum load requirements as indicated on the appliance nameplate.

For example:

Heater load	= 2000 watts (VA)
Motor load	
2.2 amperes × 208 volts × 1.73 =	792 VA
Plus 25% of 792 VA	= 198 VA
Total	= 2990 VA

The maximum continuous load permitted on a 20-ampere, three-phase branch circuit is:

16 amperes × 208 volts × 1.73 = 5757.4 VA

The load of 2990 volt-amperes is well within the 5760 volt-amperes permitted loading of a 20-ampere branch circuit.

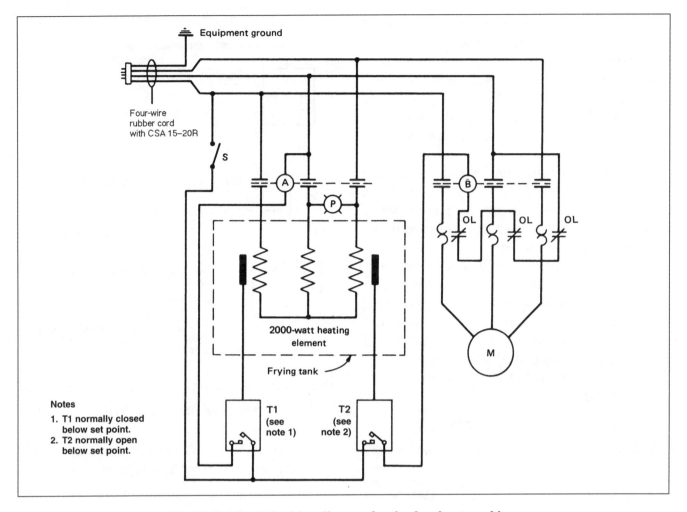

Fig. 12–7 Control wiring diagram for the doughnut machine.

THE BAKE OVEN

The bake oven installed in the bakery is an electrically heated commercial-type bake oven. See Fig. 12–8. The oven has a marked nameplate rating of 16 000 watts, 208 volts, three phase. This rated load includes all electrical heating elements, drive motors, timers, transformers, controls, operating coils, buzzers, lights, alarms, etc.

The instructions furnished with the bake oven, as well as the nameplate on the oven, specify that the supply conductors shall have a minimum ampacity of 60 amperes. The line-connecting terminals in the control panel and in the branch-circuit panelboard are suitable for 75°C. This would match the temperature rating of Type TWN75 wire. Type T90 Nylon wire has a temperature rating of 90°C, but for this application its ampacity is determined by referring to the 75°C column of *Table 2*.

Fig. 12–8 Bake oven. (Courtesy of Cutler Industries, Inc.)

According to the panel schedule, the bake oven is fed by a 60-ampere, three-phase, three-wire branch circuit using No. 6 T90 Nylon copper conductors. The metal raceway is considered acceptable for the equipment ground.

A three-pole, 60-ampere, 250-volt disconnect switch is mounted on the wall near the oven. From this disconnect switch, a conduit (flexible, rigid, or EMT as recommended by the oven manufacturer) is run to the control panel on the oven, Fig. 12–9.

The bake oven is furnished as a complete unit. Overcurrent protection is an integral part of the circuitry of the oven. The internal control circuit of the oven is 120 volts, which is supplied by an integral control transformer.

All of the previous discussion in this unit relating to circuit ampacity, conductor sizing, grounding, etc. also applies to the bake oven, dishwasher, and food-waste disposal. To avoid repetition, these *C.E.C., Part I* requirements will not be repeated. Additional information relating to appliances is found in Units 4 and 10.

For the bake oven:

1. Ampere rating = 44.5 amperes
2. Minimum conductor ampacity = 44.5 × 1.25 = 55.6 amperes
3. Conductor size No. 6 AWG T90 Nylon Copper

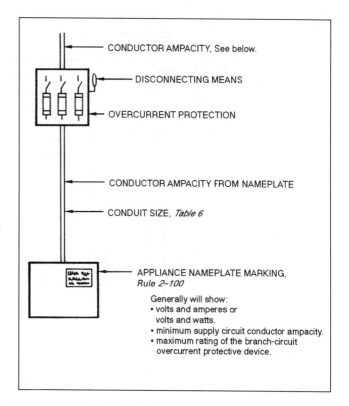

Fig. 12–9 Bake oven circuit.

4. Branch-circuit overcurrent protection = 44.5 amperes × 1.25 = 55.6 amperes

It is permitted to install 55- or 60-ampere fuses in the 60-ampere disconnect switch.

REVIEW

Note: Refer to the C.E.C., Part I or the plans as necessary.

1. What section of the *C.E.C., Part I* refers specifically to air-conditioning equipment?

   _____

2. What section of the *C.E.C., Part I* refers specifically to appliances?

   _____

3. A cord-and-plug connected appliance shall not exceed _____% of a branch-circuit ampere rating.

4. How many poles does a 15-30R receptacle have? _____

5. A disconnect switch for a motor must be _____ of the motor controller.

6. Where a branch circuit disconnect is used as the motor disconnect it must be located
   _____

7. What section of the *C.E.C., Part I* sets forth the requirements for the disconnecting means to be in sight of a motor and/or the controller for the motor?
   _____

8. The definition of "in sight" is that the disconnected means must be visible from the motor location (and) (or) be not more than 9 m (27.5 ft) from the motor location. Circle the correct answer.

9. Conductors that supply a motor shall be sized at not less than _____% of the motor full-load current rating.

10. Motor overload protection for a motor that is marked "S.F. 1.15" shall not exceed _____% of the motor full-load ampere rating.

11. What is the length of time that a time-delay fuse will carry five times its current rating?
    _____

12. When motors run too hot under overloaded conditions, the life expectancy of the motor is cut in half for every _____ °C that the motor is operating above the intended or marked temperature rating of the motor.

13. A 4-ampere, dual-element time-delay fuse protects a motor that momentarily draws 15 amperes when started. If the motor should fail to start, the fuse would protect the motor from burnout by opening the circuit in approximately _____ seconds. Refer to Table 12–2.

14. Motor #1 is marked Code B. Motor #2 is marked Code M. Both motors draw the same full-load current. Which motor has the larger momentary inrush current when started?
    _____

15. The recommended ampere rating for a dual-element time-delay fuse for an 8-ampere motor is _____ amperes.

16. The conductors that supply a group of motors must be sized at a minimum of _____% of the largest motor, plus the _____ of the full-load ampere ratings of the other motors.

17. A motor has a full-load ampere rating of 16 amperes. The preferred sizing for dual-element time-delay fuses is (a) _____% of the motor full-load ampere rating. If, for some reason, the characteristic of the installation is that the motor will be started and stopped repeatedly, *Table D12* of the *C.E.C., Part I* indicates that these dual-element time-delay fuses could be sized at (b) _____% of the motor full-load ampere rating. In fact, if the repeated starting and stopping nuisance blows the fuse size indicated in *Table D12*, *Rule 28–200* allows the electrician to size the fuses at a maximum of (c) _____%. Fill in the blanks a, b, and c and select the fuse and disconnect switch sizes for the above alternatives.

    a. _____
    b. _____
    c. _____

18. Three motors are connected to one feeder. Motor #1 draws 16 amperes. Motor #2 draws 11 amperes. Motor #3 draws 8 amperes. Calculate the minimum ampacity for the feeder conductors that supply these motors. The terminals in the switches and in the motor controllers are suitable for 75°C conductors. Type T90 Nylon conductors will be used.

_____

_____

_____

_____

19. What section of the *C.E.C., Part I* covers the subject of grounding?

_____

20. The doughnut machine has a _____-watt heating element plus a _____-ampere motor.

21. Calculate the current draw of an electric bake oven rated at 35 kilowatts, 208 volts, three phase. _____

_____

22. What size and type of conductors supply the electric bake oven in the bakery?

_____

23. When selecting overcurrent protection for a motor branch-circuit, if no standard ampere rating given in *Table 13* equals the value required for the circuit, the next (higher) (lower) standard rating should be chosen. Circle the correct answer.

24. When a motor controller's nameplate indicates that a 15-ampere maximum time-delay fuse is permitted for a given size overload relay, may a 20-ampere time-delay fuse be used? _____ May a 15-ampere circuit breaker be used? _____

UNIT 13

Luminaires and Lamps

Part I: Luminaires

OBJECTIVES

After completing the study of this part, the student will be able to
- install a lighting outlet
- install a luminaire
- identify different types of luminaires and state their application

DEFINITIONS

The terms "luminaire" and "lighting fixture" are used interchangeably. The Illuminating Engineering Society recommends the use of "luminaire" but the *C.E.C., Part I* uses either "lighting fixture" or just "fixture." There is a danger in using the term "fixture" alone, for in different contexts it may have different meanings to the reader. In a heating application, for example, it is common to use the term "fixture" to refer to a baseboard heater. For the sake of clarity, and to follow good practice, in this text the word "luminaire" will be used except when addressing *C.E.C., Part I* requirements, in which case "lighting fixture" will be used.

The following definition from the *IES Handbook* applies to either of the terms. A luminaire (lighting fixture) is a complete lighting unit consisting of a lamp or lamps together with the parts designed to distribute the light, to position and protect the lamps, and to connect the lamps to the power supply.

A lighting outlet can be defined as an outlet intended for the direct connection of a lampholder, a lighting fixture, or a pendant cord terminating in a lampholder.

INSTALLATION

The installation of luminaires is a frequent part of the work required for new building construction and for remodelling projects where customers are upgrading the illumination of their facilities. To execute work of this sort, the electrician must know how to install luminaires and, in some cases, select the luminaires.

The luminaires required for the commercial building are described in the specifications and

194 Unit 13 Luminaires and Lamps

indicated on the plans. The installation of luminaires, lighting outlets, and supports is covered in *Section 30*.

This section sets forth the basic requirements for the installation of the outlets and supports in what is commonly referred to as the rough-in.

The rough-in must be completed before the ceiling material can be installed. The exact location of the luminaires is rarely dimensioned on the electrical plans, and in some remodelling situations there are no plans. In either case, the electrician should be able to rough in the outlet boxes and supports so that the luminaires will be correctly spaced in the area.

If a single luminaire is to be installed in an area, the centre of the area can be found by drawing diagonals from each corner, Fig. 13–1. When more than one luminaire is required in an area, the procedure shown in Fig. 13–2 should be followed.

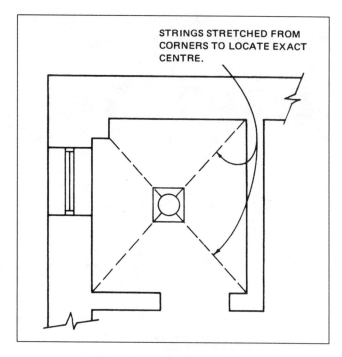

Fig. 13–1 Fixture location.

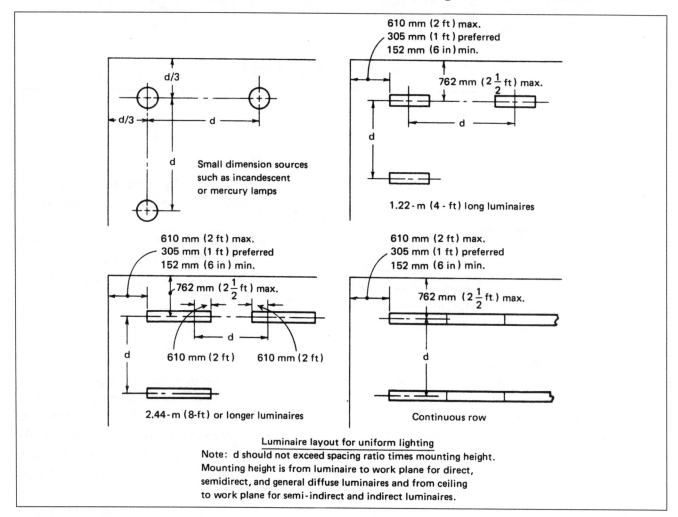

Fig. 13–2 Fixture spacing.

Uniform light distribution is achieved by spacing the luminaires so that the distances between the luminaires and between the luminaires and the walls follow these recommended spacing guides. The spacing ratios for specific luminaires are given in the data sheets and published by each manufacturer. This number, usually between 0.5 and 1.5, when multiplied by the mounting height, gives the maximum distance that the luminaires may be separated and provide uniform illuminance on the work surface.

Supports

Both the lighting outlet and the luminaire must be supported from a structural member of the building. To provide this support, a large variety of clamps and clips are available, Figs. 13–3 and 13–4. The selection of the type of support depends upon the way in which the building is constructed.

Surface-Mounted Luminaires

For surface-mounted and pendant-hung luminaires, the lighting outlets and supports must be

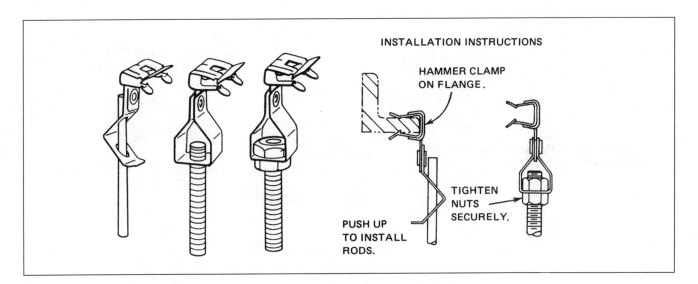

Fig. 13–3 Rod hangers for connection to flange.

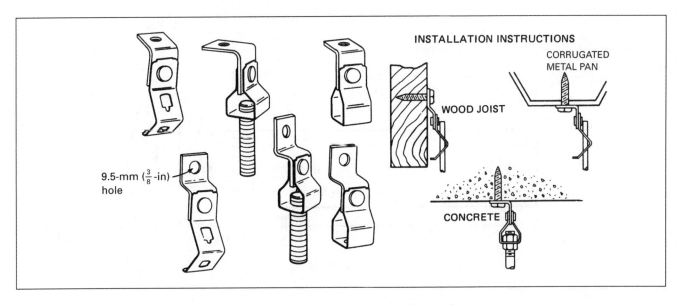

Fig. 13–4 Rod hanger supports for flat surfaces.

roughed in so that the luminaire can be installed after the ceiling is finished. Support rods should be placed so that they extend about 25 mm (1 in) below the finished ceiling. The support rod may be either a threaded rod or an unthreaded rod, Fig. 13–5. If the luminaires are not available when the rough-in is necessary, luminaire construction information should be requested from the manufacturer. The manufacturer can provide drawings that will indicate the exact dimensions of the mounting holes in the back of the luminaire, Fig. 13–6.

Recessed Luminaires

For recessed luminaires, the lighting outlet box will be located above the ceiling. This box must be accessible. It is connected to the luminaire by means of a metal raceway that is at least 450 mm (1.5 ft) long, but not more than 2 m (6.5 ft) long.

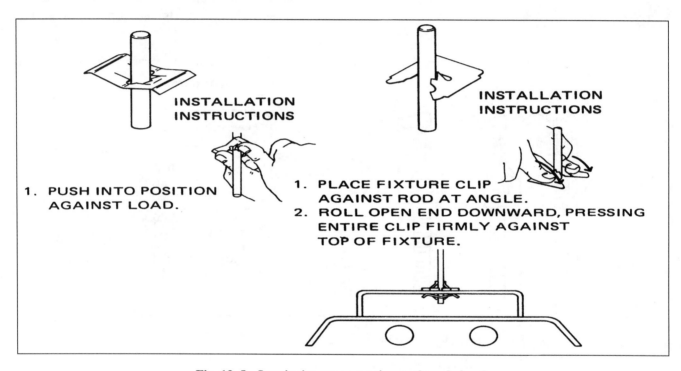

Fig. 13–5 Luminaire support using unthreaded rod.

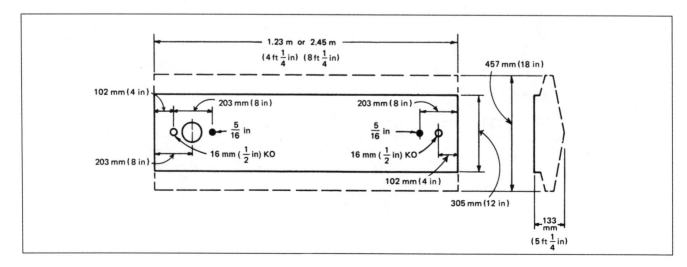

Fig. 13–6 Shop drawing indicating mounting holes for luminaire.

Conductors suitable for the temperatures encountered are necessary. This information is marked on the luminaire. Branch-circuit conductors may be run directly into the junction box on approved prewired luminaires (*Rule 30–910(2)*), Fig. 13–7.

Recessed luminaires are usually supported by rails installed on two sides of the rough-in opening. The rails can be heavy lather's channel or another substantial type of material.

LABELLING

Always carefully read the label(s) on the luminaires. The labels will list the types of lamps and the maximum wattage for the luminaires; a power factor for determining the volt-amperes from the lamp wattage may be listed as well. The labels may also indicate whether the luminaires are

- for wall mount only
- for ceiling mount only
- required to have access above ceiling
- suitable for air handling use
- for chain or hook suspension only
- suitable for operation in ambient not exceeding _____ °C (_____ °F)
- suitable for installation in poured concrete
- for installation in poured concrete only
- suitable for use in suspended ceilings
- suitable for use in uninsulated ceilings
- suitable for use insulated ceilings
- suitable for damp locations (such as bathrooms and under eaves)
- suitable for wet locations
- suitable for use as a raceway

The instructions on these labels and the bonding requirements of *Section 10* must be followed to ensure a safe installation.

Also the luminaire manufacturers' catalogues and literature are excellent sources of information on how to correctly install luminaires. The CSA

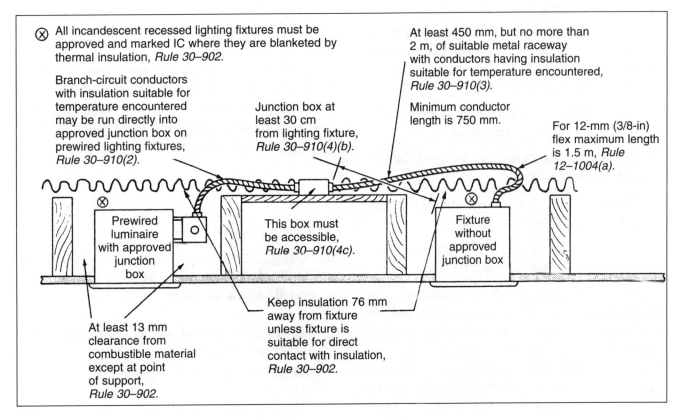

Fig. 13–7 **Clearance requirements for installing recessed lighting fixtures.**

lists, tests, and certifies luminaires and publishes relevant information.

Requirements for Installing Recessed Luminaires

The electrician must follow very carefully the requirements given in *Rules 30–900* and *30–910* for the installation and construction of recessed luminaires. Of particular importance are the restrictions on conductor temperature ratings, luminaire clearances from combustible materials, and maximum lamp wattage when luminaires are installed in thermal insulation.

Recessed luminaires generate a considerable amount of heat within the enclosure and are a **definite fire hazard** if not wired and installed properly, Figs. 13-7, 13-8, and 13-9. In addition, the excess heat will have an adverse effect on lamp and ballast life and performance.

If other than a Type IC luminaire is being installed, the electrician should work closely with the installer of the insulation to be sure that the clearances required by the *C.E.C., Part I* are followed.

According to *Rule 30–910(3)(4)*, a tap conductor must be run from the fixture terminal connection to an outlet box. For this installation, the following conditions must be met:

- The conductor insulation must be suitable for the temperatures encountered.
- The outlet box must be at least 30 cm (1 ft) from the fixture.
- The tap conductor must be installed in a suitable metal raceway or cable.
- The raceway shall be at least 450 mm (1.5 ft) but not more than 2 m (6.6 ft) long.

The branch-circuit conductors are run to the junction box. Here they are connected to conductors from the luminaire. These fixture wires have insulation suitable for the temperature encountered at the lampholder. Locating the junction box at least 30 cm (1 ft) from the luminaire ensures that the heat radiated from the luminaire cannot overheat the wires in the junction box. The conductors must run through at least 450 mm (1.5 ft) of the metal raceway (but not more than 2 m (6.6 ft)) between the luminaire and the junction box. Thus, any heat being conducted by the metal raceway will be dissipated considerably before reaching the junction box. Many recessed luminaires are factory equipped with a flexible metal raceway containing high-temperature wires that meet the requirements of *Rule 30–910(1)*.

Some recessed luminaires have a box mounted on the side of the luminaire so that the branch-circuit conductors can be run directly into the box and then connected to the conductors entering the luminaire.

Additional wiring is unnecessary with these prewired luminaires, Fig. 13-10. It is important to note that *Rule 30–910(8)* states that branch-circuit wiring shall not be passed through an outlet box that is an integral part of an incandescent luminaire unless the luminaire is identified for through wiring.

If a recessed luminaire is not prewired, the electrician must check the luminaire for a label indicating what conductor insulation temperature rating is required.

Recessed luminaires are inserted into the rough-in opening and fastened in place by various devices. One type of support and fastening method for recessed luminaires is shown in Fig. 13-11. The flag hanger remains against the luminaire until the screw is turned. The flag then swings into position and hooks over the support rail as the screw is tightened.

Thermal Protection

Recessed fixtures have an inherent heat problem. Accordingly, they must be suitable for the intended application, and must be properly installed.

To protect against overheating, the CSA requires that some types of recessed luminaires be equipped with an integral thermal protector, Fig. 13-12.

These devices will cycle *on–off–on–off* repeatedly until the heat problem is reduced. They

LUMINAIRE TYPE	COMMENT
Incandescent Recessed Luminaires Marked "Type IC" 	Usually of low wattage for installation in direct contact with insulation. May be blanketed with insulation. Recommended for homes, these luminaires operate under 90°C when covered with insulation. Separation from insulation is not required. Integral thermal protection is required in Type IC fixtures, which must be marked that thermal protection is provided. Check the manufacturer's markings for minimum clearances, maximum lamp wattage, and temperature rating for supply conductors. These fixtures may also be installed in noninsulated locations.
Incandescent Recessed Luminaires *Not* Marked "Type IC" 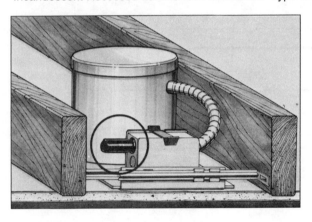	For recessed incandescent fixtures *not* marked "Type IC," or for those fixtures marked for installing directly in poured concrete. Insulation must *not* be placed over the top of the luminaires. Insulation must be kept back at least 75 mm (3 in) from the sides of the fixture. Combustible material must be kept at least 13 mm ($\frac{1}{2}$ in) from the fixture except at the point of support for the fixture. Watch for markings on the fixture for maximum clearances, maximum lamp wattages, and temperature ratings for the supply conductors.
Suspended Ceilings 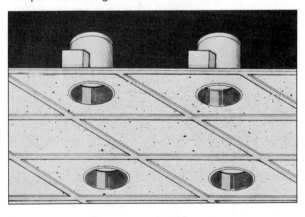	Presently, CSA lists luminaires that are marked "Suitable for use in suspended ceilings." Watch for markings on these luminaires showing spacing requirements between fixtures, walls, and ceiling above the fixture. Also check the markings for lamp types and maximum wattages as well as temperature ratings required for the supply conductors. Types IC and non-IC luminaires may be installed in suspended ceilings.

Fig. 13–8 Types of recessed fixtures.

200 Unit 13 Luminaires and Lamps

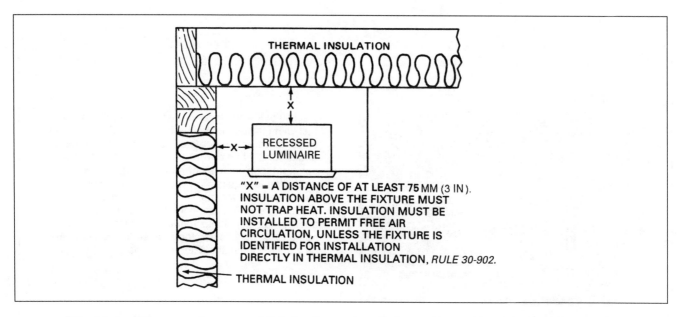

Fig. 13–9 Clearances for recessed lighting fixture installed near thermal insulation, *Rule 30–902*.

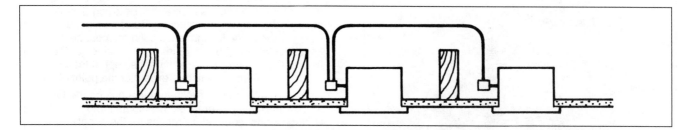

Fig. 13–10 Installation permissible only with prewired recessed lighting fixtures, *Rule 30–910(7)*.

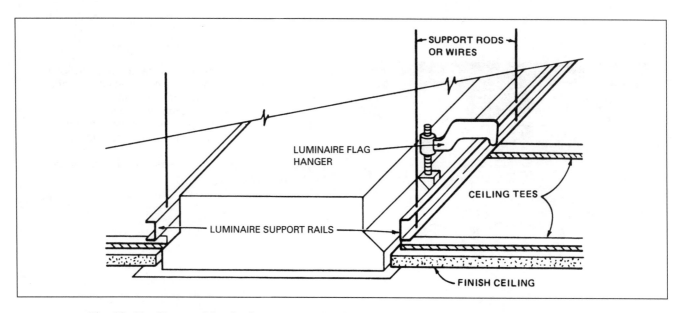

Fig. 13–11 Recessed luminaire support with flag hanger and support rails, *Rule 30–302(4)*.

are factory-installed by the manufacturer of the luminaire.

Both incandescent and fluorescent recessed luminaires are marked with the temperature ratings required for the supply conductors if over 60°C. In the case of fluorescent luminaires, branch-circuit conductors within 75 mm (3 in) of a ballast must have a temperature rating of at least 90°C, *Rule 30–308(3)*.

All fluorescent ballasts, including replacement ballasts, installed indoors must have integral thermal protection, per *CSA Standard C22.2 No. 9.0*. These thermally protected ballasts are called "Class P ballasts." This device will provide protection during normal operation, but it should not be expected to provide protection from the excessive heat that will be created by covering the luminaire with insulation. For this reason, fluorescent luminaires, just as incandescent luminaires, must have 13 mm (0.5 in) clearance from combustible materials and 75 mm (3 in) clearance from thermal insulation.

Because of the inherent risk of fires due to the heat problems associated with recessed luminaires, always read the markings on the luminaire and any instructions furnished with it.

Wiring

It was stated previously that it is very important to provide an exact rough-in for surface-mounted luminaires. *Rule 30–308* emphasizes this fact by requiring that the lighting outlet be accessible for inspection without removing the luminaire supports. The installation of the outlet meets the requirements of *Rule 30–308* if the lighting outlet is located so that the large opening in the back of the luminaire can be placed over it, Fig. 13–13.

To meet the requirements of *Rule 30–910(2)*, branch-circuit conductors with a rating of 90°C may be used to connect luminaires. However, these conductors must be from the single branch circuit supplying the luminaires. All of the conductors of multiwire branch circuits can be installed as long as these conductors are the grounded and ungrounded conductors of a single system, *Rule 30–310(1)*. For example, when a building has a three-phase, four-wire supply, the neutral and three hot wires, one on each phase, may be installed in a luminaire that has been approved as a raceway. This type of installation is suited to a long continuous row of luminaires, Fig. 13–14.

LOADING CALCULATIONS

The branch circuits are usually determined when the luminaire layout is completed. For incandescent luminaires, the VA allowance for each luminaire is based on the wattage rating of the luminaire. If an incandescent luminaire is rated at 300 watts, it must be included at 300 watts even though a smaller lamp is to be installed. For fluorescent and high-intensity discharge lamp luminaires, the VA allowance is based on the rating of the ballast. In the past it was a general practice to estimate this value, usually on the high side. With recent advances in ballast manufacture, and increased interest in reducing energy usage, the

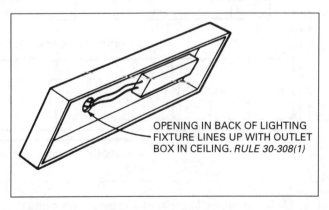

Fig. 13–13 Method of connecting fluorescent fixture to outlet box, *Rule 30–308(1)*.

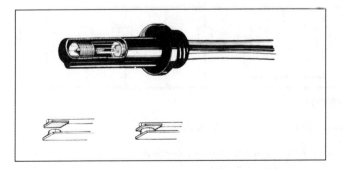

Fig. 13–12 A thermal protector.

Table 13–1   Luminaire Schedule for the Commercial Building

Style	Nominal Size	Lamp	Ballast	Lens-Louvre	Mounting	Description	VA	Watts
A	18 inches by 4 feet	Two F40/30U	Energy Saving	Wrap-Around	Surface		87	75
B	2 by 2 feet	Two FB40/30U	Energy Saving	Flat Opal	Surface		87	75
C	9 by 51 inches	Two F40/30U	Energy Saving	Clear Acrylic	Surface	Enclosed, Gasketed	87	75
D	Strip 4 feet long	One F48T12/CW	Standard	Luminous Ceiling	Surface		74	64
E	9 inches diameter	Two 26W Quad T4	Compatible	Gold Alzak Reflector	Recessed		144	74
F	2 by 4 feet	Four FO32/35K	Matching Electronic	24 Cell Lens—Louvre	Recessed		132	106
G	1 by 8 feet	Four FO32/35K	Matching Electronic	24 Cell Lens—Louvre	Recessed		132	106
H	9 inches by 8 feet	Two F40/CW	Energy Saving	Translucent Acrylic	Surface		87	75
I	12 inches by 4 feet	Two F40/CWX	Energy Saving	Translucent Acrylic	Troffer		87	75
J	8 inches diameter	One 150W, A21	N/A	Fresnel Lens	Recessed	IC Rated	150	150
K	16 inches square	One 70W, HPS	Standard	Vandal Resistant	Surface	With Photo Control	192	82
L	13 inches by 4 feet	Two F40/CW	Energy Saving	None	Surface or Hung		87	75
M	7 inches diameter	12V 50W NFL	Transformer	Coilex Baffle	Recessed	Adjustable Spot	50	50
N	20 inches long	One 60W, 120V	N/A	None	Surface	Exposed Lamp	60	60
O	4 feet by 20 inches	Three F40/CW	Two Ballasts	Low-brightness Lens	Recessed		143	129
P	2 feet by 20 inches	Two FB40/CW	One Ballast	Low-brightness Lens	Recessed		87	75
Q	4 feet by 20 inches	Three F40/CW	Two Ballasts	Small-cell Parabolic	Recessed		143	129

practice is to select a specific ballast type and base the allowance on the operation of that ballast. The contractor is required to install an "as good as" or "better than" ballast. Several types of ballasts are discussed later in the text. The design volt-amperes and watts for the luminaires selected for the commercial building are shown in Table 13–1. The following paragraphs discuss the styles of luminaires listed in this table.

The total wattage rating of all the luminaires on a branch circuit shall not exceed 80% of the branch overcurrent protection, *Rule 30–712(2)*.

Style A

The Style A luminaire is a popular fluorescent type that features a diffuser extending up the sides

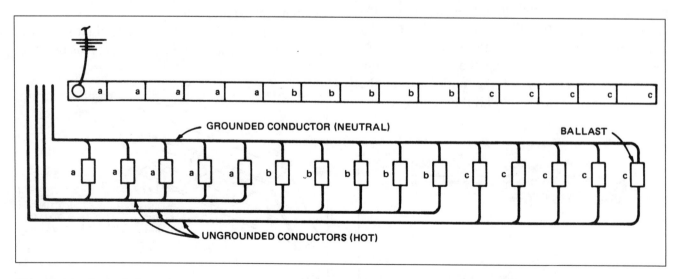

Fig. 13–14   A single branch, multiwire circuit consisting of one grounded and three ungrounded conductors of the same system supplying a continuous row of approved lighting fixtures, *Rule 30–310(1)*. Fixtures can also be connected alternately: a–b–c–a–b–c–a–b–c–a–b–c.

of the luminaire, Fig. 13–15. These are often called wrap-around lenses. This type of luminaire provides good ceiling lighting, which is particularly important for low ceilings. The diffuser is usually available in either an acrylic or polystyrene material. Although a polystyrene diffuser is less expensive, it will yellow as it ages, and quickly becomes unattractive. Diffusers can be specified to be made with an acrylic material that does not yellow with age.

The major disadvantage of the Style A luminaire is the difficulty of locating replacements for yellowed or broken diffusers. The ballast chosen for this luminaire has an A sound rating and is of an energy-efficient type that, when operated with two F40T12/RS lamps, has a line current of 0.725 ampere at 120 volts (87 volt-amperes) and a power rating of 75 watts. This type of luminaire is used in the beauty salon.

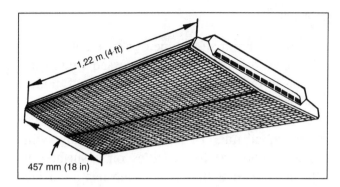

Fig. 13–15 Style A. A shallow, surface-mounted fluorescent luminaire, 1.22 m (4 ft) long and 457 mm (18 in) wide with a wrap-around acrylic diffuser and equipped for two F40CW rapid-start lamps.

Style B

The Style B luminaire, another popular style, has solid metal sides, Fig. 13–16. The bright sides of the Style A luminaire, when used on a low ceiling, may be objectionable to people who must look at them for long periods of time. The Style B reduces this problem. The opal glass of the Style B luminaire provides a soft diffusion of the light, but any flat diffuser may be used. Glass is easily cleaned and does not experience the same aging problems encountered by the plastic materials. This style is used in the sales area of the bakery.

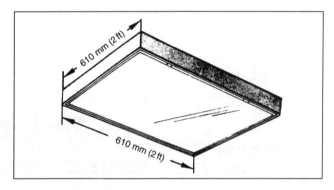

Fig. 13–16 Style B. A surface-mounted fluorescent luminaire, 610 mm (2 ft) square with solid sides, an opal glass diffuser, and equipped for two FB40 rapid-start lamps.

luminaire may enter from the top or from either end.

Style C

The Style C luminaire is used where the possible contamination of the area is an important consideration, such as in the bakery, where food is prepared. Style C luminaires are suitable for use in bakeries, kitchens, slaughterhouses, meat markets, and food-packaging plants. The clear acrylic diffuser of this luminaire protects the area in the event of a broken lamp. At the same time, the interior of the luminaire is kept dry and free from dirt or dust, Fig. 13–17. Raceways serving this

Style D

Luminous ceiling systems are used where a high level of diffuse light is required. The Style D system consists of fluorescent light strips (which may be ballasted for rapid-start, high-output, or very-high-output lamps) and a ceiling suspended 457 mm (18 in) or more below the lamps, Fig. 13–18. The ceiling may be of any translucent material, but usually consists of 0.6 m × 0.6 m (2 ft × 2 ft) or 0.6 m × 1.22 m (2 ft × 4 ft) panels that are easily removed for cleaning and lamp replacement. This lighting system is used in the drugstore.

Style E

The Style E luminaire is one of a type commonly referred to as downlights or recessed cans, Fig. 13–19. It requires a ceiling opening of under 230 mm (9 in) in diameter and has a height of 200 mm (8 in). This fluorescent version uses two 26-watt quad tube T4 lamps. A single unit will produce about 20 foot-candles on a surface at 2.74 m (9 ft) distance. Each lamp has a current rating of 0.6 ampere at 120 volts (72 volt-amperes) and uses 37 watts. These are used in several locations in the interior of the commercial building. This lamp is not recommended for exterior applications because of its poor starting characteristics during cold weather.

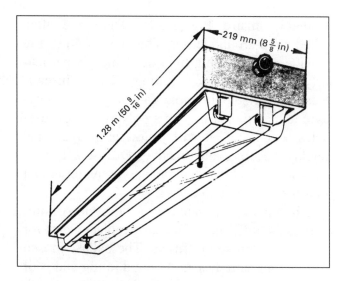

Fig. 13–17 Style C. A surface-mounted fluorescent luminaire 1.28 m (50 $\frac{9}{16}$ in) long and 219 mm (8 $\frac{5}{8}$ in) wide. Designed to prevent contamination of the area by having a totally enclosed and gasketed clear acrylic diffuser; equipped for two F40 rapid-start lamps.

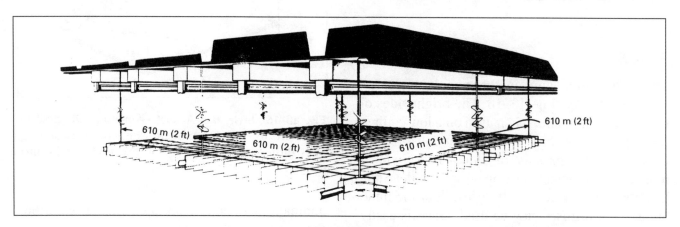

Fig. 13–18 Style D. A fluorescent light strip 1.22 m (4 ft) long, equipped with an F48 high-output lamp. Installed as part of a luminous ceiling system with 610 mm × 610 mm (2 ft × 2 ft) panels with 12.7 mm × 12.7 mm ($\frac{1}{2}$ in × $\frac{1}{2}$ in) cell size.

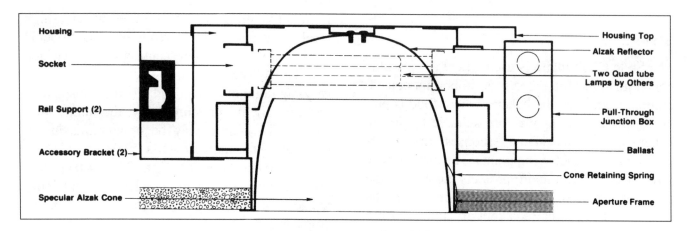

Fig. 13–19 Style E luminaire.

Style F

The Style F is a 0.6 m × 1.22 m (2 ft × 4 ft) recessed fluorescent luminaire, Fig. 13–20. This luminaire is equipped with four F032T8 lamps and an electronic ballast. The lamps and the ballast are a matched set or system. This system has been developed to maximize the ratio of light output to watts input. Compared to standard F40T12/RS lamps and a magnetic ballast, this combination provides about 160% as much light per watt. According to the manufacturer's data, each lamp in this luminaire will produce 2900 lumens initially, achieving a lamp–ballast efficacy of 110 lumens per watt.

This type of luminaire is available with many styles of lenses and louvers. For this installation, a lens has been chosen that has the features of a lens but the appearance of a louvre. A 51-mm (2-in) deep blade arrangement forms square light baffles on the surface of a lens. This gives a strong directionality to the light, concentrating the light downward to the work area. This luminaire, and the similar Style G, is used in the doctor's office. The lens-louvre is also used in the insurance office.

Rule 30–302 requires adequate support for the lighting fixtures in a T-bar ceiling. With T-bar ceiling lighting fixtures, chain supports attached to the construction above the T-bar ceiling should be provided. This is usually achieved by the use of S-hooks and jack chain.

Style G

The Style G luminaire is identical to the Style F luminaire except that it is 0.3 m × 2.44 m (1 ft × 8 ft). A single ballast is used with four F032T8 lamps. This is an identical luminaire to the Style F except for the dimensions.

Style H

The Style H luminaire shown in Fig. 13–22 is designed to light corridors, the narrow areas between storage shelves, and other long, narrow spaces. Style H luminaires are available in slightly different forms from various manufacturers and are relatively inexpensive. This luminaire has been chosen for use in the second-floor corridor. It is 178 mm (7 in) wide and 2.44 m (8 ft) long. It has two F40 lamps placed in tandem (end to end). A single two-lamp ballast serves 2.44 m (8 ft) of luminaire.

Style I

The Style I luminaire is similar to Style H but is 1.22-m (4-ft) long with the two lamps side by side. This is used in the drugstore where higher levels of illumination are desirable, Fig. 13–21.

Style I luminaires will be used in the merchandising areas of the drug store. They are 1 in (305 mm) × 4 in (1220 mm) fluorescent troffers c/w prismatic lens. Style I luminaires will be supplied with deluxe cool white lamps to improve the colour rendering in the display areas of the store. Style I luminaires are shown in Fig. 13–21.

Fig. 13–20 Style F luminaire.

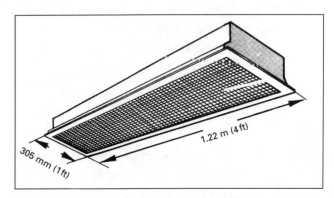

Fig. 13–21 Style I. This luminaire is 305 mm (1 ft) wide and 1.22 m (4 ft) long and equipped for two F40CW rapid-start lamps.

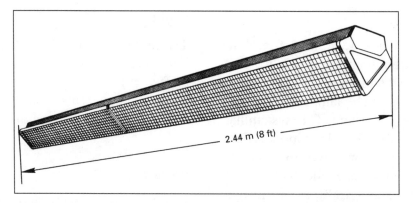

Fig. 13–22   Style H. A surface-mounted luminaire designed for corridor lighting with 9.5 mm ($\frac{3}{8}$ in) cube, V-shaped diffuser 2.44 m (8 ft) long; equipped for two F40CW rapid-start lamps in tandem.

Fig. 13–23   Style J luminaire.

Style J

The Style J luminaire, Fig. 13–23, is a recessed incandescent downlight luminaire that can be installed in the opening left by removing or omitting a single 0.3-m (1-ft) square ceiling tile. This luminaire is a Type IC, which has been approved to be covered with insulation. It has been specified to be equipped with a fresnel lens for wide distribution of the light and a 150-watt A21 lamp. This is used in protected exterior locations and in the doctor's office.

Style K

The Style K luminaire is used on vertical exterior walls, where it provides a light pattern that covers a large area. The Style K luminaire uses a high-intensity discharge source such as metal halide or high-pressure sodium, Fig. 13–24. This luminaire provides reliable security lighting around a building. The Style K luminaire is completely weatherproof and is equipped with a photoelectric cell to turn the light off during the day.

Style L

The Style L luminaire is of open construction, as shown in Fig. 13–25. This type is generally used in storage areas and other locations where it is not necessary to shield the lamps. It is often called an industrial fixture. The Style L luminaires are to be suspended from the ceiling on chains. In this type of installation armoured cable or flexible metal conduit will run from the luminaire to an outlet box on the ceiling. This type of fixture is used in the basement storage areas of both the bakery and the drugstore.

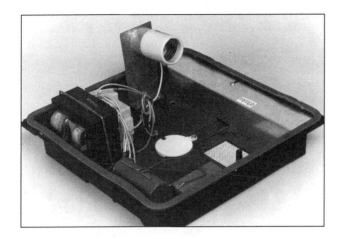

Fig. 13-24 Style K luminaire.

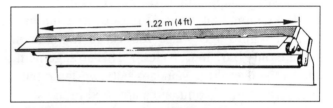

Fig. 13-25 Style L. An industrial-type fluorescent luminaire, 1.22 m (4 ft) long with a steel white enameled reflector; equipped for two F40 lamps.

Style M

The Style M luminaire is used to focus light on a specific object, Fig. 13-26. This luminaire has a diameter of 178 mm (7 in). The lamp can be swiveled through 358° laterally and 40° from the zenith. This luminaire utilizes an MR16 lamp in sizes up to 50 watts and in a variety of light distribution patterns. A low-voltage transformer is provided with the luminaire. At a distance of 1.22 m (4 ft), the 50-watt narrow flood lamp produces 26 footcandles. A dimmer switch used to operate these luminaires (such as that specified in the beauty salon) must be approved for use with low-voltage illumination systems. The dimmer is connected in the supply to the transformers. An ordinary dimmer will, at best, provide sporadic operation and is not approved for this application.

Fig. 13-26 Style M luminaire.

Style N

The Style N luminaire is especially designed for installations where the lamp will be exposed to viewing, Fig. 13-27. This luminaire is installed on both sides of a mirror to provide excellent illumination of the face but avoid the glare that can be a problem when conventional incandescent lamps are installed in a similar fashion. The lamp is a 0.51-m (20-in) linear incandescent with a 60-watt rating.

Style O

The Style O luminaire is 1.22 m (4 ft) long and 0.51 m (20 in) wide. It is equipped with three F40T12/RS lamps and two ballasts. One of the ballasts supplies two lamps and the other one lamp. This arrangement provides for three levels of illumination in the room by using the single lamp in each luminaire, by using a pair of lamps in each luminaire, or by using all three lamps. In each of these arrangements, the illumination is uniformly distributed throughout the room. The luminaire is fitted with a white louvre and a lens, which provides a high-quality light in addition to having a very pleasing effect on the appearance of the room. This luminaire is used in the staff and the reception areas of the insurance office.

Style P

The Style P luminaire is similar to the Style O luminaire but is only 0.61 m (2 ft) long and has a single ballast supplying two FB40/RS lamps. The FB40 is a lamp that was 1.22 m (4 ft) long but has been bent into a U-shape for use in luminaires that are only 0.61 m (2 ft) long. This arrangement allows the use of the standard F40 two-lamp ballast. Two of these luminaires are used in the staff area of the insurance office to fill in areas of the room where the Style O was too large.

Style Q

The Style Q luminaire is similar to the Style O except for the louvres. The lens specified in this room is a low-brightness lens especially designed for use in computer rooms. A high percentage of the light is directed downward to the horizontal surfaces. A minimum amount of light is produced on the computer monitor's vertical surface.

Fig. 13–27 Style N luminaire.

LOCATION OF LUMINAIRES IN CLOTHES CLOSETS

Clothing, boxes, and other material normally stored in clothes closets are a potential fire hazard. These items may ignite on contact with the hot surface of an exposed light bulb. The bulb, in turn, may shatter and spray hot sparks and hot glass onto other combustible materials.

Rule 30–204 covers the special requirements for installing lighting fixtures in clothes closets. It is significant to note that these rules cover *all* clothes closets—residential, commercial, and industrial.

Part II: Lamps for Lighting

OBJECTIVES

After completing the study of this unit, the student will be able to
- list the types of lamps used in the commercial building
- define the technical terms relating to lamp selection and installation
- list the parts of three types of lamps
- list the operating characteristics of lamp types
- recognize the significance of lamp designations

For most construction projects, the electrical contractor is required to purchase and install lamps in the luminaires (lighting fixtures). Thomas Edison provided the talent and perseverance that led to the development of the incandescent lamp in 1879 and the fluorescent lamp in 1896. Peter Cooper Hewitt produced the mercury lamp in 1901. All three of these lamp types have been refined and greatly improved since they were first developed.

Section 30 contains the provisions for the wiring and installation of lighting fixtures, lampholders, lamps, receptacles, and rosettes.

Several types of lamps are used in the commercial building. In the *C.E.C., Part I* these lamps are referred to as either incandescent or electric discharge lamps, *Rule 30–100*. In the industry, the incandescent lamp is also referred to as a filament lamp because the light is produced by a heated wire filament. The electric discharge lamps include a variety of types but all require a ballast. The most common types of electric discharge lamps are fluorescent, mercury, metal halide, high-pressure sodium, and low-pressure sodium. Mercury, metal halide, and high-pressure sodium lamps are also classed as high-intensity discharge (HID) lamps.

LIGHTING TERMINOLOGY

Candela (cd): The luminous intensity of a source, when expressed in candelas, is the candlepower (cp) rating of the source.

Lumen (lm): The amount of light received in a unit of time on a unit area at a unit distance from a point source of one candela, Fig. 13–28. The surface area of a sphere is 12.57 times the square of its radius; therefore, a one-candela source produces 12.57 lumens. When the measurement is in imperial units, the unit area is 1 square foot and the unit distance is 1 foot. If the units are in SI, then the unit area is 1 square metre and the unit distance is 1 metre.

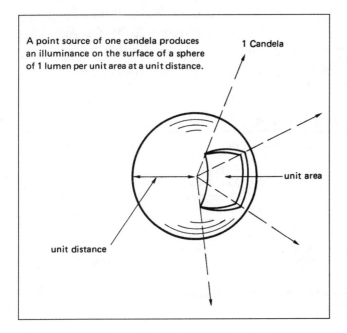

Fig. 13–28 Diagram showing definition of candela (cd).

Illuminance: The measure of illuminance on a surface is the lumen-per-unit area expressed in footcandles (fc) or lux (lx). The recommended illuminance levels vary greatly, depending on the task to be performed and the ambient lighting conditions. For example, while 54 lux (5 footcandles) is often accepted as adequate illumination for a dance hall, 2152 lx (200 fc) may be necessary on a drafting table for detailed work.

Lumen per watt (lm/W): This is defined as the measure of the effectiveness (efficacy) of a light source in producing light from electrical energy. A 100-watt incandescent lamp producing 1670 lumens has an effective value of 16.7 lumens per watt.

Kelvin (K): The kelvin (sometimes incorrectly called degree Kelvin) is measured from absolute zero; it is equivalent to a degree value in the Celsius scale plus 273.16. The colour temperature of lamps is given in Kelvin. The lower the number, the warmer the light (more red content); the higher the number, the cooler the light (more blue content).

Colour Rendering Index (CRI): This value is often given for lamps so the user can have an idea of the colour rendering probability. The CRI uses filament light as a base for 100 and the warm white fluorescent for 50. The CRI can be used only to compare lamps that have the same colour temperature. The only sure way to determine if a lamp will provide good colour rendition is to see the material in the lamplight.

Refer to Table 13–2 for a comparison of the various types of electric lamps.

LAMP ENERGY EFFICIENCY

In 1992, the Federal Government put in place *The Energy Efficiency Act* (EEACT). The purpose of this legislation is to ensure the use of more energy efficient lamps and thereby reduce Canada's contribution to global greenhouse gas emissions. The current amendment to the Energy Efficiency Regulations establishes energy efficiency standards (lumens per watt) for certain types of lamps imported into Canada or traded interprovincially. Failing to abide by these regulations could result in very large fines.

Products restricted under this legislation include:

- all R and PAR shaped medium base, 115 to 130V incandescent reflector lamps of 40 to 205 watts.
- full wattage 4-foot medium bi-pin (T8, T10, T12) fluorescent lamps
- full wattage 2-foot U-shaped (T12, T8) fluorescent lamps
- full wattage 8-foot high-output fluorescent lamps
- full wattage 8-foot slimline fluorescent lamps

Certain types of R and PAR incandescent lamps such as halogen, coloured, and special purpose lamps are exempt. The ER and BR shaped lamps are also acceptable because they meet the minimum lumens/watt required by the Act. The restricted full wattage fluorescent lamps have been replaced with reduced wattage products of

Characteristics of Electric Lamps

	Filament	Fluorescent	Mercury	Metal Halide	HPS	LPS
Lumen per watt	6 to 23	25 to 100	30 to 65	65 to 120	75 to 140	130 to 180
Wattage range	40 to 33k	4 to 215	40 to 1000	175 to 1500	35 to 1000	35 to 180
Life (hours)	750 to 8k	9k to 20k	16k to 24k	5k to 15k	20k to 24k	18k
Colour temperature	2400 to 3100	2700 to 7500	3000 to 6000	3000 to 5000	2000	1700
Colour rendition index	90 to 100	50 to 110	25 to 55	60 to 70	20 to 25	0
Potential for good colour rendition	high	highest	fair	good	colour discrimination	no colour discrimination
Lamp cost	low	moderate	moderate	high	high	moderate
Operational cost	high	good	moderate	moderate	low	low

The values gives above are generic for general service lamps. A survey of lamp manufacturers' catalogues should be made before specifying or purchasing any lamp.

Table 13–2 Characteristics of electric lamps.

the same physical size that will function with existing ballasts. General service fluorescent lamps that are exempted include lamps with a colour rendering index of 82 or greater, lamps for use in cold temperature applications, grow lamps, reprographic and UV radiation lamps, as well as coloured, impact resistant, and reflectorized types.

INCANDESCENT LAMPS

The incandescent lamp has the lowest efficacy of the types listed in Table 13–2. However, incandescent lamps are very popular and account for more than 50% of the lamps sold in North America. This popularity is due largely to the low cost of incandescent lamps and luminaires.

Construction

The light-producing element in the incandescent lamp is a tungsten wire called the *filament*, Fig. 13–29. This filament is supported in a glass envelope or bulb. The air is evacuated from the bulb and is replaced with an inert gas such as argon. The filament is connected to the base by the lead-in wires. The base of the incandescent lamp supports the lamp and provides the connection means to the power source. The lamp's base may be any one of the styles shown in Fig. 13–30.

Characteristics

Incandescent lamps are classified according to the following characteristics.

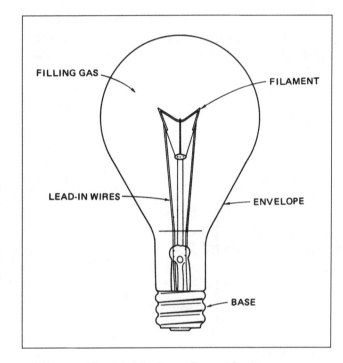

Fig. 13–29 Incandescent lamp.

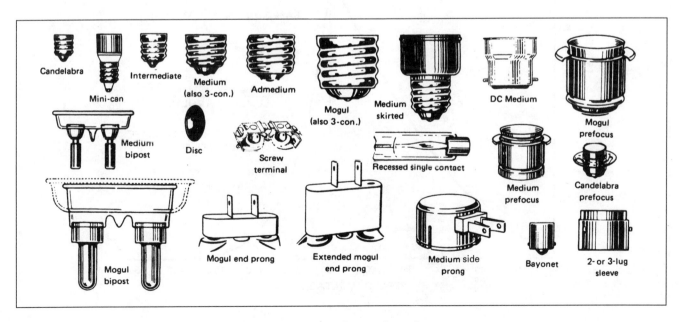

Fig. 13–30 Incandescent lamp bases.

212   Unit 13   Luminaires and Lamps

Voltage Rating. Incandescent lamps are available with many different voltage ratings. When installing lamps the electrician should be sure that a lamp with the correct rating is selected, since a small difference between the rating and the actual voltage has a great effect on the lamp life and lumen output, Fig. 13–31.

Wattage. Lamps are usually selected according to their wattage rating. This rating is an indication of the consumption of electrical energy but is not a true measure of light output. For example, at the rated voltage, a 60-watt lamp produces 840 lumens and a 300-watt lamp produces 6000 lumens; therefore, one 300-watt lamp produces more light than seven 60-watt lamps.

Size. Fig. 13–32 illustrates the common lamp configurations and their letter designations.

Shapes. Lamp size is usually indicated in eighths of an inch and is the diameter of the lamp at the widest place. Thus, the lamp designation A19 has an arbitrary shape and is 60.3 mm ($\frac{19}{8}$ or $2\frac{3}{8}$ in) in diameter, Fig. 13–33.

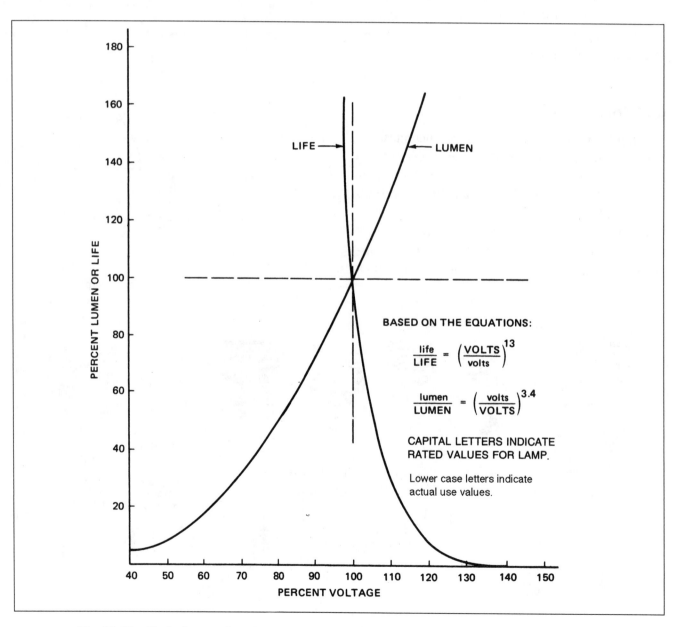

Fig. 13–31   Typical operating characteristics of an incandescent lamp as a function of voltage.

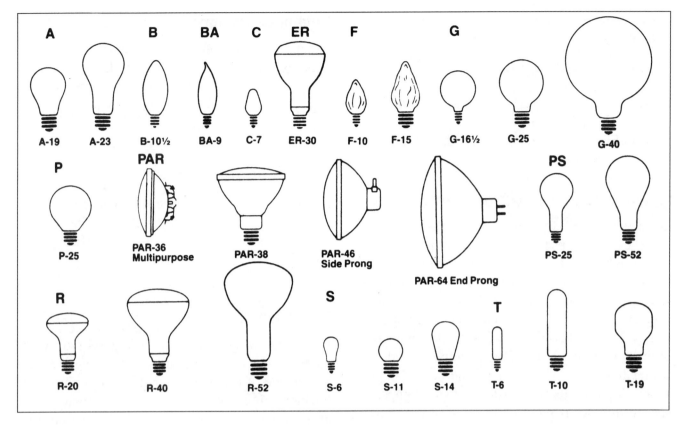

Fig. 13–32 Incandescent lamps.

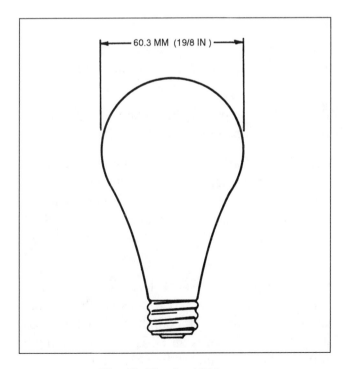

Fig. 13–33 An A19 lamp.

Operation

The light-producing filament in an incandescent lamp is a resistance load that is heated to a high temperature by the electric current going through the lamp. This filament is usually made of tungsten, which has a melting point of 3655 K. At this temperature, the tungsten filament produces light with an efficacy of 53 lumens per watt. However, to increase the life of the lamp, the operating temperature is lowered, which also means a lower efficacy. For example, if a 500-watt lamp filament is heated to a temperature of 3000 K, the resulting efficacy is 21 lumens per watt.

Catalogue Designations

Catalogue designations for incandescent lamps usually consist of the lamp wattage followed by the shape and ending with the diameter and other special designations as appropriate. Common examples are:

60A19	60 watt, arbitrary shape, 19/8 inches diameter
75PAR38	75 watt, parabolic reflector, 38/8 inches diameter
100R40/FL	100 watt, reflector flood, 40/8 inches diameter

LOW-VOLTAGE INCANDESCENT LAMPS

In recent years, low-voltage (usually 12-volt) incandescent lamps have become very popular for accent lighting. Many of these lamps are tungsten halogen lamps. They have a very small source size, as shown in Fig. 13–34. This feature allows very precise control of the light beam. A popular size is the MR16, which has a reflector diameter of just 51 mm (2 in). These lamps provide a whiter light than do regular incandescent lamps. When dimming these tungsten halogen lamps, a special dimmer is required because of the transformer that is installed to reduce the voltage. The dimmer is installed in the line voltage circuit supplying the transformer. Dimmed lamps will darken if they are not occasionally operated at full voltage.

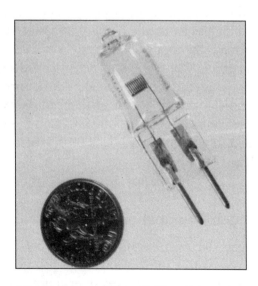

Fig. 13–34 A 100-watt, 12-volt tungsten halogen lamp as compared to a dime.

Catalogue Designations

Except for some special cases, catalogue designations for low-voltage incandescent lamps are similar to other incandescent lamps, for example:

MR16 mirrored reflector, $\frac{16}{8}$ inches diameter

FLUORESCENT LAMPS

Luminaires using fluorescent lamps are considered to be electric-discharge lighting. Fluorescent lighting has the advantages of a high efficacy and long life.

Construction

A fluorescent lamp consists of a glass bulb with an electrode and a base at each end, Fig. 13–35. The inside of the bulb is coated with a phosphor (a fluorescing material), the air is evacuated, and an inert gas plus a small quantity of mercury is released into the bulb. The base styles for fluorescent lamps are shown in Fig. 13–36.

Characteristics

Fluorescent lamps are classified according to type, length or wattage, shape, and colour. See Fig. 13–37.

Type. The lamps may be preheat, rapid start, or instant start depending upon the ballast circuit.

Length or Wattage. Depending on the lamp type, either the length or the wattage is designated. For example, both the F40 preheat and the F48 instant start are 40-watt lamps, 1.22 m (48 in) long. The bases of these two lamps are different, however.

Shapes. The fluorescent lamp usually has a straight tubular shape. Exceptions are the circline lamp, which forms a complete circle; the U-shaped lamp, which is an F40T12 lamp having a 180° bend in the centre to fit a 610-mm (2-ft) long luminaire; and the PL lamp, which has two parallel tubes with a short connecting bridge at the ends opposite the base.

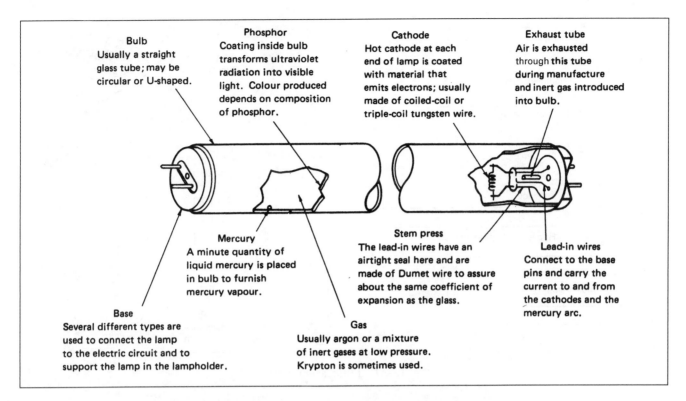

Fig. 13–35 Basic parts of a typical hot cathode fluorescent lamp.

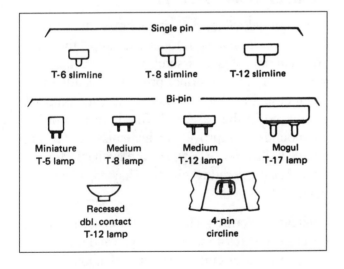

Fig. 13–36 Bases for fluorescent lamps.

Colour. The colour of a fluorescent lamp depends upon the phosphor mixture used to coat the inside of the lamp. The colour is indicated in the lamp designation. For example, an F40T12CW is a cool white lamp and accents blue colours wherever it is used. Additional colour designations are warm white (WW), daylight (D), white (W), cool white deluxe (CWX), and warm white deluxe (WWX). Decorative colours such as blue (B), pink (PK), and green (G) are also available.

Catalogue Designations

Catalogue designations for fluorescent lamps begin with an "F," except for compact fluorescent and a few other cases. The following numbers indicate the wattage for rapid start lamps and the length for most of the other types. This is followed by shape, diameter, colour, and special designations.

F40T12/CW	fluorescent, 40 watts, tubular shape, $\frac{12}{8}$ inches diameter, cool white
F48T12/CW/HO	fluorescent, 48 inches in length, tubular shape, $\frac{12}{8}$ inches diameter, cool white, high output
PL18/27	compact fluorescent, 18 watts, 2700 K colour temperature

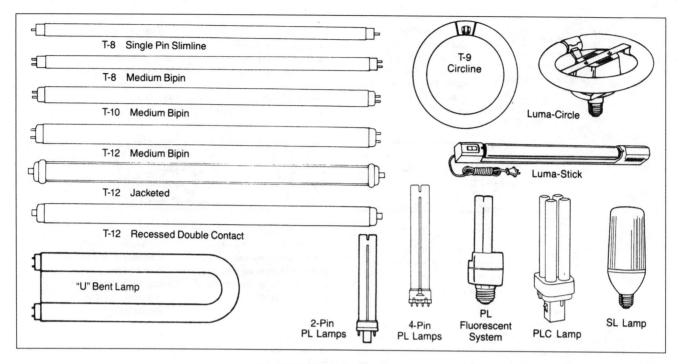

Fig. 13-37 Fluorescent lamps.

Operation

If a substance is exposed to ultraviolet light or X-rays and emits visible light as a result, then the substance is said to be fluorescing. The inside of the fluorescent lamp is coated with a phosphor material that serves as the light-emitting substance. When sufficient voltage is applied to the lamp electrodes, electrons are released. Some of these electrons travel between the electrodes to establish an electric discharge or arc through the mercury vapour in the lamp. As the electrons strike the mercury atoms, ultraviolet radiation is emitted by the atoms. This radiation is converted into visible light by the phosphor coating on the tube, Fig. 13-38.

As the mercury atoms are ionized, the resistance of the gas is lowered. The resulting increase in current ionizes more atoms. If allowed to continue, this process will cause the lamp to destroy itself. As a result, the arc current must be limited. The standard method of limiting the arc current is to connect a reactance (ballast) in series with the lamp.

Ballasts (*Rule 30-700*)

Although inductive, capacitive, or resistive means can be used to ballast fluorescent lamps, the most practical ballast is an assembly of a core and coil, a capacitor, and a thermal protector installed in a metal case, Fig. 13-39. Once the assembled parts are placed in the case, it is filled with a potting compound to improve the heat dissipation and reduce ballast noise. Ballasts are available for the three basic operating circuits that are discussed next.

Preheat Circuit. The first fluorescent lamps developed were of the preheat type and required a starter in the circuit. This type of lamp is now obsolete and is seldom found except in smaller sizes that may be used for items such as desk lamps. The starter serves as a switch and closes the circuit until the cathodes are hot enough. The starter then opens and the lamp lights. The cathode temperature is maintained by the heat of the arc after the starter opens. Note in Fig. 13-40 that the ballast is in series with the lamp and acts as a choke to limit the current through the lamp.

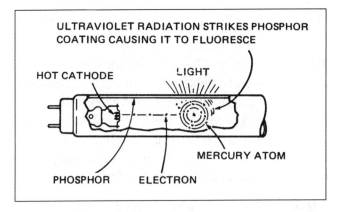

Fig. 13–38 How light is produced in a typical hot cathode flourescent lamp.

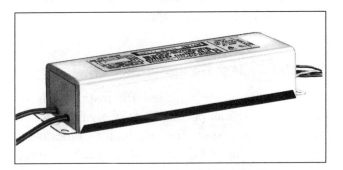

Fig. 13–39 Fluorescent ballast.

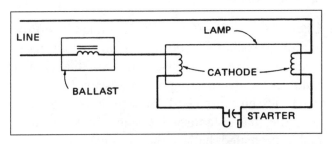

Fig. 13–40 Basic preheat circuit.

Rapid-Start Circuit. In the rapid-start circuit, the cathodes are heated continuously by a separate winding in the ballast, Fig. 13–41, with the result that almost instantaneous starting is possible. This type of fluorescent lamp requires the installation of a continuous grounded metal strip within 25 mm (1 in) of the lamp. The metal wiring channel or the reflector of the luminaire can serve as this grounded strip. The standard rapid-start circuit operates with a lamp current of 430 mA. Two variations of the basic circuit are available: the high-output (HO) circuit operates with a lamp current of 800 mA and the very high-output circuit has 1500 mA of current. Although high-current lamps are not as efficacious as the standard lamp, they do provide a greater concentration of light, thus reducing the required number of luminaires.

Instant-Start Circuit. The lamp cathodes in the instant-start circuit are not preheated. Sufficient voltage is applied across the cathodes to create an instantaneous arc, Fig. 13–42. As in the preheat circuit, the cathodes are heated during lamp operation by the arc. The instant-start lamps

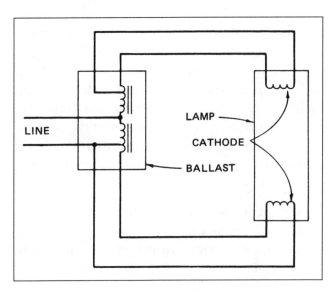

Fig. 13–41 Basic rapid-start circuit.

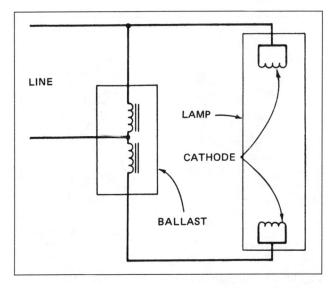

Fig. 13–42 Basic instant-start (cold cathode) circuit.

require single-pin bases, Fig. 13–43, and are generally called *slimline lamps*. Some bi-pin base instant-start fluorescent lamps are available, such as the 40-watt F40T12/CW/IS lamp. For this style of lamp, the pins are shorted together so that the lamp will not operate if it is mistakenly installed in a rapid-start circuit.

Special Circuits

Most fluorescent lamps are operated by one of the circuits just covered: the preheat, rapid-start, or instant-start circuits. Variations of these circuits, however, are available for special applications.

Dimming Circuit. The light output of a fluorescent lamp can be adjusted by maintaining a constant voltage on the cathodes and controlling the current passing through the lamp. Devices such as thyratrons, silicon-controlled rectifiers, and autotransformers can provide this type of control. The manufacturer of the ballast should be consulted about the installation instructions for dimming circuits.

Flashing Circuit. The burning life of a fluorescent lamp is greatly reduced if the lamp is turned on and off frequently. Special ballasts are available that maintain a constant voltage on the cathodes and interrupt the arc current to provide flashing.

High-Frequency Circuit. Fluorescent lamps operate more efficiently at frequencies above 60 hertz. The gain in efficacy varies according to the lamp size and type. However, the gain in efficacy and the lower ballast cost generally are offset by the initial cost and maintenance of the equipment necessary to generate the higher frequency.

Direct-Current Circuit. Fluorescent lamps can be operated on a dc power system if the proper ballasts are used. A ballast for this type of system contains a current-limited resistor that provides an inductive kick to start the lamp.

Special Ballast Designation

Special ballasts are required for installations in cold areas such as out of doors. Generally, these ballasts are necessary for installations in temperatures lower than 10°C (50°F). These ballasts have a higher open-circuit voltage and are marked with the minimum temperature at which they will operate properly.

Class P (*Rule 30–708*). The National Fire Protection Association reports that the second most frequent cause of electrical fires in the United States is the overheating of fluorescent ballasts.

To lessen this hazard, Underwriters Laboratories, Inc. has established a standard for a thermally protected ballast that is designated as a Class P ballast. This type of ballast has an internal protective device that is sensitive to the ballast temperature. This device opens the circuit to the ballast if the average ballast case temperature exceeds 90°C when operated in a 25°C ambient temperature. After the ballast cools, the protective device is automatically reset. As a result, a fluorescent lamp with a Class P ballast is subject to intermittent off-on cycling when the ballast overheats.

It is possible for the internal thermal protector of a Class P ballast to fail. The failure can be in the "welded shut" mode or it can be in the "open" mode. Since the welded shut mode can result in overheating and possible fire, it is recommended

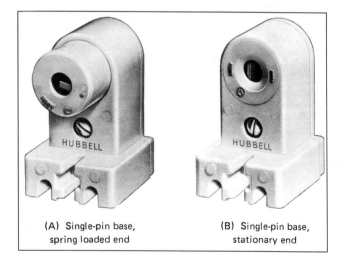

Fig. 13–43 **Single-pin base for instant-start fluorescent lamp.**

that the ballast be protected with in-line fuses as indicated in Fig. 13–44.

External fuses can be added for each ballast so that a faulty ballast can be isolated to prevent the shutdown of the entire circuit because of a single failure, Fig. 13–44. The ballast manufacturer normally provides information on the fuse type and its ampere rating. The specifications for the commercial building require that all ballasts shall be individually fused; the fuse size and type are selected according to the ballast manufacturer's recommendations, Fig. 13–45.

CSA standard C22.2 No. 9.0 requires that all fluorescent ballasts installed indoors, both for new installations and replacements, must have built-in thermal protection.

Sound Rating. All ballasts emit a hum that is caused by magnetic vibrations in the ballast core. Ballasts are given a sound rating (from A to F) to indicate the severity of the hum. The quietest ballast has a rating of A. The need for a quiet ballast is determined by the ambient noise level of the location where the ballast is to be installed. For example, the additional cost of the A ballast is justified when it is to be installed in a doctor's waiting room. In the bakery work area, however, a ballast with a C rating is acceptable; in a factory, the noise of an F ballast probably will not be noticed.

Power Factor. The ballast limits the current through the lamp by providing a coil with a high

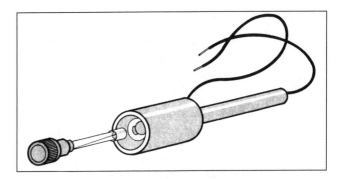

Fig. 13–45   In-line fuseholder for ballast protection.

reactance in series with the lamp. An inductive circuit of this type has an uncorrected power factor of from 40% to 60%; however, the power factor can be corrected to within 5% of unity by the addition of a capacitor. In any installation where there are to be a large number of ballasts, it is advisable to install ballasts with a high power factor.

Compact Fluorescent Lamps

Compact fluorescent lamps usually consist of a twin tube arrangement that has a connecting bridge at the end of the tubes, Fig. 13–46. Lamps are available that have two sets of tubes; these are called double twin tube or quad tube lamps, Fig. 13–47.

This type of lamp has a rated life ten times that of an incandescent lamp and provides about three times the light per watt of power. A typical

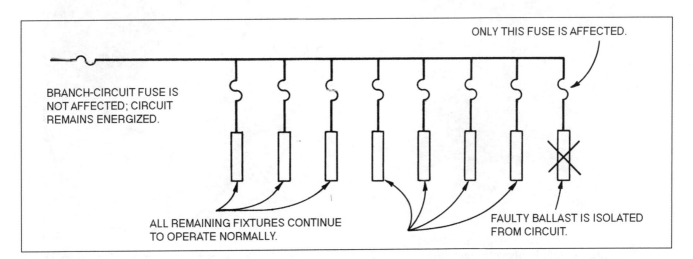

Fig. 13–44   Each ballast is individually protected by a fuse. In this system, a faulty ballast is isolated from the circuit.

220 Unit 13 Luminaires and Lamps

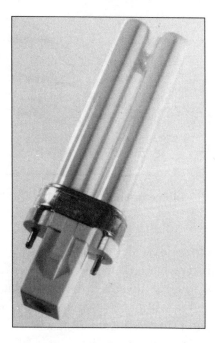

Fig. 13–46 Five-watt, twin tube, compact fluorescent lamp.

socket, along with a low-power factor ballast, is shown in Fig. 13–48. Some lamps have a medium screw base, with the ballast in the base, Fig. 13–49. This type of lamp can directly replace an incandescent lamp. The bases for these lamps often have a retractable pin for the base connection, Fig. 13–50. This allows the positioning of the lamp.

Class E Ballast

Manufacturers of certain fluorescent ballasts are required to meet a Ballast Efficiency Factor (BEF). Initially, this applied only to F40T12 one- and two-lamp ballasts and to two-lamp ballasts for F96T12 and F96T12HO lamps. The BEF for the two-lamp F40T12, as used in the commercial building, requires that the lamp operate at a minimum 84.8 lumens per watt efficacy.

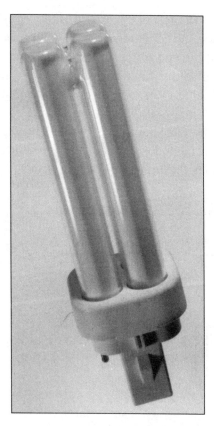

Fig. 13–47 Ten-watt, double twin tube or quad tube compact fluorescent lamp.

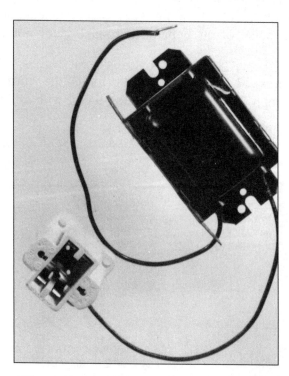

Fig. 13–48 Socket and ballast for compact fluorescent lamp.

HIGH-INTENSITY DISCHARGE (HID) LAMPS

Two of the lamps in this category, mercury and metal halide, are similar in that they use mercury as an element in the light-producing process. The other HID lamp, high-pressure sodium, uses sodium in the light-producing process. It produces a golden-white light in the 1900–2700 terahertz range. Low-pressure sodium produces a yellow–orange light around 1700 THz. Mercury vapour produces a blue–green light in the 3000–7000 THz range.

In all three lamps, the light is produced in an arc tube that is enclosed in an outer glass bulb. This bulb serves to protect the arc tube from the elements and to block ultraviolet produced by the arc. An HID lamp will continue to give light after the bulb is broken, but it should be promptly removed from service. When the outer bulb is broken, people can be exposed to harmful ultraviolet radiation.

Fig. 13–49 Medium base socket with ballast for compact fluorescent lamp.

Mercury Lamps

Many people consider the mercury lamps to be obsolete. They have the lowest efficacy of the HID family, which ranges from 30 lumens per watt for the smaller-wattage lamps to 65 lumens per watt for the larger-wattage lamps. Some of the positive features of mercury lamps are that they have a long life, with many lamps still functioning at 24 000 hours. With a clear bulb, they give a greenish light that makes them popular for landscape lighting.

Catalogue Designations

Catalogue designations vary considerably for HID lamps, depending on the manufacturer. In general, the designation for a mercury lamp will begin with an "H," for metal halide lamps it will begin with an "M," and either an "L" or "C" for high-pressure sodium lamps. Designations for two common types are:

MH250/C/U	metal halide, 250 watts, phosphor coated, base up (Philips)
MVR250/C/U	same as above (General Electric)
M250/C/U	same as above (Sylvania)

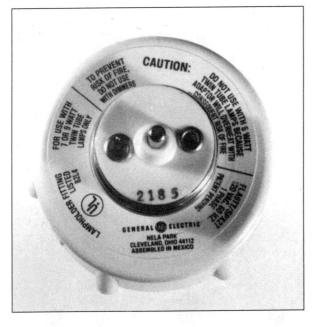

Fig. 13–50 Medium base screw-in socket with retractable pin for positioning twin tube compact fluorescent lamp.

H38MP-100/DX mercury, Type 38 ballast, 100 watts, deluxe white (Philips)

HR100DX38/A23 same as above (General Electric)

H38AV-100/DX same as above (Sylvania)

HID lamp shapes are shown in Fig. 13-51.

Metal Halide Lamps

The metal halide lamp has the disadvantages of a relatively short life and a rapid drop-off in light output as the lamp ages. Rated lamp life varies from 5000 to 15 000 hours. During this period the light output can be expected to drop by 30% or more. These lamps are considered to have good colour-rendering characteristics and are often used in retail clothing and furniture stores. The lamp has a high efficacy rating, which ranges from 65 to 120 lumens per watt. Only a few styles are available with ratings below 175 watts. Operating position (horizontal or vertical) is critical with many of these lamps and should be checked before a lamp is installed.

High-Pressure Sodium (HPS) Lamps

This type of lamp is ideal for applications in warehouses, parking lots, and similar places where colour recognition is necessary but high-quality colour rendition is not required. The light output is rather orange in colour. The lamp has a life rating equal to or better than that of any other HID lamp and has very stable light output over the life of the lamp. The efficacy is very good, ranging as high as 130 lumens per watt.

Low-Pressure Sodium Lamps

This lamp has the highest efficacy of any of the lamps, ranging from 140 to above 180 lumens per watt. The light is monochromatic, containing

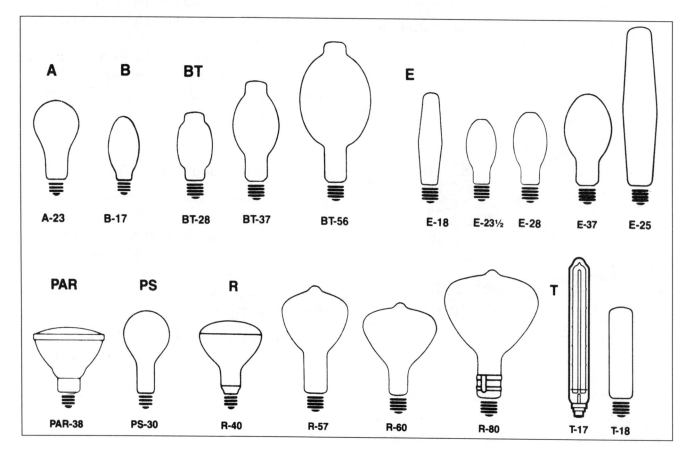

Fig. 13–51 HID lamps.

energy in only a very narrow band of yellow. This lamp is usually used only in parking and storage areas where no color recognition is required. The lamp has a good life rating of 18 000 hours. It maintains a very constant light output throughout its life. The lamp is physically longer than HID lamps but generally is shorter than fluorescent lamps.

REVIEW

PART I: LUMINAIRES

Note: Refer to the *C.E.C., Part I* or the plans as necessary.

1. A luminaire weighing 3.2 kg (7 lb) shall not be supported by the screw-shell of the lampshade. True or false? Explain.

   _____

   _____

2. Flexible raceway or cable installed to a recessed luminaire must be not more than _____ m (_____ ft) or less than _____ m (_____ ft) long.

3. The neutral and phase wires of a three-phase, four-wire system may be installed in a continuous row of luminaires, provided the conductors have a _____ °C rating if they are within 76 mm (3 in) of the ballast.

4. A 2.44-m (8-ft) and a 1.22-m (4-ft) luminaire are to be installed end to end. Using the dimensions given in Fig. 13–6, provide the following measurements from a wall 0.61 m (2 ft) from the end of the 2.44-m (8-ft) luminaires.

   a. The first support rod hole is _____ m from the wall.

   b. The first outlet box opening is _____ m from the wall.

   c. The second support rod hole is _____ m from the wall.

   d. The third support rod hole is _____ m from the wall.

   e. The final support rod hole is _____ m from the wall.

5. The basic loading allowance for the beauty salon is _____ watts.

6. Book shelving, such as in a library, is best illuminated by using a luminaire similar to Style _____.

7. What are the requirements for installing a recessed luminaire that does not have an approved box furnished with the luminaire? _____

   _____

   _____

   _____

PART II: LAMPS FOR LIGHTING

Note: Refer to the *C.E.C., Part I* or the plans as necessary.

1. Describe the lamps specified by the following designations:
   a. 150 R _____
   b. 200 A _____
   c. 150 PAR _____
   d. F48T12/CW _____
   e. H37-5KC/DX _____

2. What is the size and shape of an A23 lamp?
   _____

3. The operating current of a 40-watt standard rapid-start lamp is _____ mA.

4. Match the following items:

   Lumen per square metre   _____   a. Intensity
   Lumen per watt   _____   b. Footcandle
   One candela   _____   c. Efficacy
   Starter   _____   d. Instant start
   Cold cathode   _____   e. Rapid start
   Constant heated cathode   _____   f. Preheat

5. The _____ is the light-producing element in the incandescent lamp.
6. The _____ is the light-producing element in the mercury lamp.
7. The _____ is the light-producing element in the fluorescent lamp.
8. The main purposes of a fluorescent ballast are to _____
   _____
   _____

9. When calculating the lighting load for a circuit, _____ volt-amperes are allowed for a Style A fluorescent luminaire.

10. Describe a Class P ballast. _____
    _____
    _____

11. Explain a method used to isolate a faulty ballast to prevent the shutdown of the entire circuit because of a single failure. _____
    _____
    _____
    _____

UNIT 14

Reading Electrical Drawings—Insurance Office/Beauty Salon

OBJECTIVES

After completing the study of this unit, the student will be able to

- list (tabulate) the materials required to install an electrical system
- make branch-circuit calculations
- lay out the raceway system
- cite *C.E.C., Part I* references

PRINTS

A journeyman electrician should be able to look at an electrical drawing (Fig. 14–1) and prepare a list of the materials required to install the wiring system. A great deal of time can be lost if the proper materials are not on the job when needed. It is essential that the electrician prepare in advance for the installation so that the correct variety of material is available in sufficient quantities to complete the job.

LOADING SCHEDULE

The loading schedules for the insurance office and the beauty salon are included at this point to assist the student in reviewing the project. See Tables 14–1 and 14–2. In the insurance office there are 56 receptacles in the multioutlet assembly and 15 elsewhere in the occupancy that qualify.

APPLIANCE CIRCUITS
Water Heater

A beauty salon (Fig. 14–2) uses a large amount of hot water. To accommodate this need, the specifications indicate that a circuit in the beauty salon is to supply an electric water heater. The water heater is not furnished, but is to be connected, by the electrical contractor, Fig. 14–3. The water heater is a full demand heater and is rated for 3800 watts at 208 volts single phase. In this case, it is assumed that the water heater is a full demand heater and is not marked with the maximum overcurrent protection. The water heater is connected to an individual branch circuit.

The *branch-circuit rating* must be at least 125% of the marked rating. The water heater in the beauty salon is considered to be "continuously loaded." See *Rules 8–104(3)* and *8–302(2)*.

Current rating: 3800/208 = 18.3 amperes
Minimum BC rating: 18.3 × 1.25 = 22.9 amperes

226 Unit 14 Reading Electrical Drawings—Insurance Office/Beauty Salon

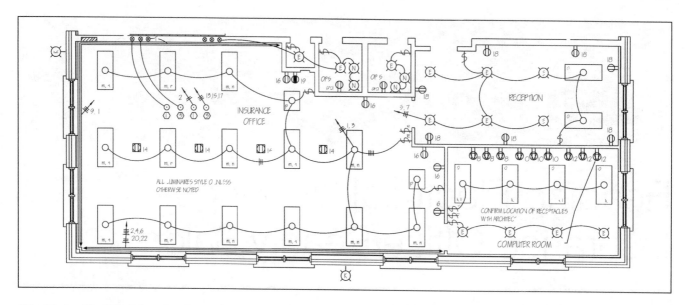

Fig. 14–1 Electrical drawing for an insurance office. *Note*: For complete blueprint, refer to blueprint E4 in back of text.

Table 14–1 Insurance Office Service Calculation

Item	Count	VA/unit	Demand Load	Installed Load	Demand Factor Feeder	Service	Use Feeder	Service
MINIMUM BASIC LOAD								
(C.E.C. Rule 8-210)	122 m²	50 W/m²	6100 W		1	0.9	6100	5490
BASIC INSTALLED LOADS								
Style E luminaires	10	144		1440			1440	1440
Style O luminaires	16	143		2288			2288	2288
Style P luminaires	2	87		174			174	174
Style Q luminaires	4	143		572			572	572
General receptacles	14	120		1680			1680	1680
Basic installed loads							6154	6154
BASIC LOAD		(Use the greater of the demand load or installed loads, C.E.C. Rule 8–106(2).)					6159	6159
SPECIAL LOADS								
Copier outlet	1	1500		1 500			1 500	1 500
Computer outlets	9	500		4 500			4 500	4 500
Roof receptacle	1	1440		1 440			1 440	1 440
Multioutlet assembly	56	180		10 080			10 080	10 080
Floor receptacles	4	120		480			480	480
Motors	**Volts**	**FLA**	**Phase**					
Cooling unit								
- compressor	208	20.2	3	7269	1.25		9 086	7 269
- evaporator motor	208	3.2	1	666			666	666
- condenser motor	208	3.2	1	666			666	666
Total load							34 572	32 755

$$I = \frac{P}{E \times 1.73} = \frac{34\,572}{208 \times 1.73} = 96.1 \text{ A}$$

Assuming all loads are continuous, using standard-rated equipment and RW90-XLPE copper conductors from *Table 2, C.E.C.*, the ampere rating of the circuit would be:

$$\frac{96.1}{0.8} = 120.1 \text{ A}$$

The minimum size feeder would be 4 No. 1 RW90-XLPE copper conductors in a 35-mm (1¼-in) conduit, supplied by a 200 A-disconnect c/w 125-A fuses.

Table 14–2 Beauty Salon Service Calculation

Item	Count	VA/unit	Demand Load	Installed Load	Demand Factor Feeder	Demand Factor Service	Use Feeder	Use Service
MINIMUM BASIC LOAD (C.E.C., Rule 8-210)	45 m²	30 W/m²	1350 W		1	0.9	1350	1215
BASIC INSTALLED LOADS								
Style A luminaires	5	87		435			435	435
Style E luminaires	5	144		720			720	720
Style M luminaires	9	50		450			450	450
General receptacles	6	120		720			720	720
Basic installed loads							2325	2325
BASIC LOAD		(Use the greater of the demand load or the installed load. C.E.C., Rule 8–106(2).)					2325	2325
SPECIAL LOADS								
Receptacle outlets	3	1500		4500			4500	4500
Roof receptacle	1	1440		1440			1440	1440
Washer/dryer	1	4000		4000			4000	4000
Water heater	1	3800		3800			3800	3800
Motors	Volts	FLA	Phase					
Cooling unit								
- compressor	208	14.1	3	5074	1.25		6342	5074
- condenser/evaporator	208	3.3	1	686			686	686
Total load							23 093	21 825

$$I = \frac{P}{E \times 1.73} = \frac{23\,093}{208 \times 1.73} = 64.2 \text{ A}$$

Assuming all loads are continuous, using standard-rated equipment and RW90 copper conductors from *Table 2, C.E.C.*, the ampere rating of the circuit would be:

$$\frac{64.2}{0.8} = 80.25 \text{ A}$$

The minimum size feeder would be 4 No. 4 RW90-XLPE copper conductors in a 27-mm (1-in) conduit, supplied by a 100-A disconnect c/w 90-A fuses.

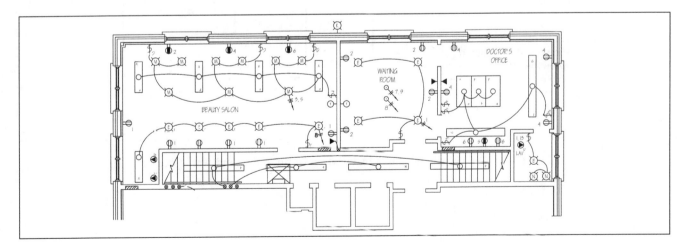

Fig. 14–2 Electrical drawing for a beauty salon and doctor's office. *Note:* For complete blueprint, refer to blueprint E4 in back of text.

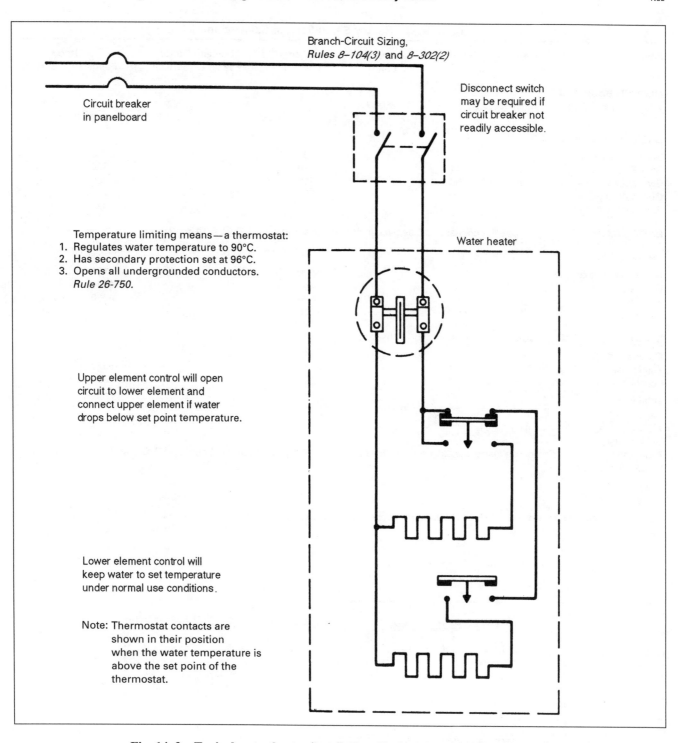

Fig. 14–3 Typical water heater installation (limited demand water heater).

Conductor size is selected from *Table 2*. No. 10 Type TWN75/T90 copper conductors have an ampacity of 30 amperes.

The *branch-circuit overcurrent protection* for single, nonmotor operated appliances might be marked on the nameplate of the appliance. If not marked, then refer to *Table 13*. In this case, a 30-ampere conductor requires a 30-ampere fuse as the maximum overcurrent protection.

Washer–Dryer Combination

One of the uses of the hot water is to supply a washer–dryer combination used to launder the many towels and other items commonly used in a beauty salon. The unit is rated for 4000 volt-amperes at 120/208 volts single phase. The branch circuit supplying this unit must have a rating of at least 125% of the appliance load.

- The branch-circuit demand ampacity is 20 amperes (4000 / 208 = 19.23).
- The overcurrent protective device must be rated at least 25 amperes (load in amperes multiplied by 1.25, then raised to the next standard size overcurrent device).
- The conductor ampacity must be at least 25 amperes to allow the use of an overcurrent device with a rating of 25 amperes, *Rule 8–104(5)(a)*.

For appliances that contain one or more motors and other loads, such as heating elements, reference should be made to the manufacturer's nameplate, which must state the minimum supply circuit conductor ampacity and the maximum rating of the circuit overcurrent protective device.

When stackable (combination) washer-dryers are installed, all of the power required for the unit comes from a single receptacle. For the combination washer-dryer in the beauty salon, a standard dryer receptacle (14-30R) is used and will be protected by a 30 amp breaker. The minimum size conductor permitted to supply the receptacle would be No. 10 awg copper. If the conductors for the receptacle were run in the same raceway as conductors to other loads, they will have to be derated in accordance with *Table 5C*.

REVIEW

Note: Refer to the *C.E.C., Part I* or the plans as necessary.

1. How many of each style of luminaire are located in the insurance office?
    a. Style E luminaire _____
    b. Style O luminaire _____
    c. Style P luminaire _____
    d. Style Q luminaire _____
2. How many lamps are required?
    a. F40/CW _____
    b. FB40/CW _____
    c. 26-watt quad T4 _____
3. Indicate the number of switches used in the insurance office.
    a. Single-pole switches _____
    b. Three-way switches _____

4. Three types of receptacles are specified for use in the insurance office. Identify the three types and the required number of each type.

   a. Type _____ Count _____

   b. Type _____ Count _____

   c. Type _____ Count _____

5. Branch circuit No. 14 serves the receptacle outlet for the copy machine.

   a. How many 3-m (10-ft) lengths of conduit will be needed to install this circuit if the conduit can be placed in the slab? _____ lengths

   b. How many 3-m (10–ft) lengths of conduit will be needed to install this circuit if the conduit is run overhead? _____ lengths

   Note: Although colour coding is not required by the *C.E.C., Part I*, the use of coloured conductors is highly recommended when it is necessary to keep track of specific circuits.

   c. How many metres (feet) and which colours of conductors should be provided for this run if it is installed in the slab?

   Colours _____

   Length _____

Complete the following loading schedule by referring to the *C.E.C., Part I*, the plans, and the specifications.

6. The water heater (can) (cannot) be connected to the electrical power by a receptacle outlet. (Circle the correct answer.) Explain why and cite the *C.E.C., Part I* reference.

   _____

   _____

7. The rating of the circuit breaker or fuse that should be installed for the water heater circuit is _____ amperes. Explain and cite the *C.E.C., Part I* reference.

   _____

   _____

8. A separate disconnect switch (is) (is not) required for the water heater. (Circle the correct answer.) Explain and cite the *C.E.C., Part I* reference.

   _____

   _____

9. Make a list of the materials required to install the raceway system for the beauty salon. In actual practice many of the luminaires will be equipped with a pull-through junction box. For this exercise, assume that all connections will require a junction box. Indicate in the blanks provided the quantities, sizes, and styles.

   Wall boxes for switches and receptacles: _____

   _____

   _____

   _____

Ceiling boxes for lighting and junctions: _____

_____

_____

_____

Receptacles: _____

_____

_____

_____

_____

Switches: _____

_____

_____

Conduit: _____

_____

_____

10. Draw the configurations for the different styles of receptacles used in the beauty salon.

11. If an electric water heater is not marked with the maximum overcurrent protection rating, the water heater's ampere rating is to be sized not greater than _____% of the maximum overcurrent protection.

12. When an appliance is marked with a maximum branch-circuit rating that is not equal to a standard rating given in *Table 13*, then a standard rating that is next (lower) (higher) than the marked rating should be selected. Circle the correct answer.

UNIT 15

Special Systems

OBJECTIVES

After completing the study of this unit, the student will be able to

- install surface metal raceway
- install multioutlet assemblies
- install communication circuits
- install floor outlets

A number of special electrical systems are found in almost every commercial building. Although these systems usually are a minor part of the total electrical work to be done, they are essential systems and it is recommended that the electrician be familiar with the installation requirements of these special systems.

SURFACE RACEWAYS

Surface raceways, either metal or nonmetallic, are generally installed as extensions to an existing electrical raceway system or used where it is impossible to conceal conduits, such as in desks, counters, cabinets, and modular partitions. This installation of surface metal raceways is governed by *Rules 12–1600* through *12–1614*. The number and size of the conductors to be installed in surface raceways is limited by the design of the raceway. Catalogue data from the raceway manufacturer will specify the permitted number and size of the conductors for specific raceways. Conductors to be installed in raceways may be spliced at junction boxes or within the raceway if the cover of the raceway is removable. See *Rule 12–1610*.

Surface raceways are available in various sizes, Fig. 15–1, and a wide variety of special fittings, such as the supports in Fig. 15–2, make it possible to use surface metal raceway in almost any dry location. Two examples of the use of surface raceways are shown in Figs. 15–3 and 15–4.

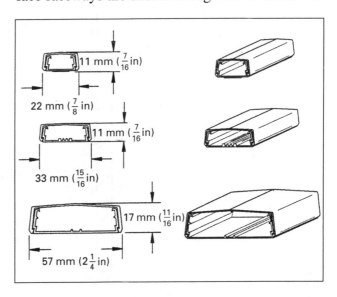

Fig. 15–1 Surface nonmetallic raceways.

MULTIOUTLET ASSEMBLIES

The definition of a multioutlet assembly is given in *Section 0* of the *C.E.C., Part I*. Multioutlet assemblies, Fig. 15–5, are similar to surface raceways and are designed to hold both conductors and devices. These assemblies offer a high degree of flexibility in an installation and are particularly suited to heavy-use areas where many outlets are required or where there is a likelihood of changes in the installation requirements.

The plans for the insurance office specify the use of a multioutlet assembly that will accommodate both power and communication cables, Fig. 15–6. This installation will allow the tenant in the insurance office to alter and expand the office facilities as the need arises.

Loading Allowance

The load allowance for a multioutlet assembly is specified by *Rule 12–3000(4)*:

- each 1.5 m of assembly counts as one outlet when normal loading conditions exist, or
- each 300 mm of assembly counts as one outlet when heavy loading conditions exist.

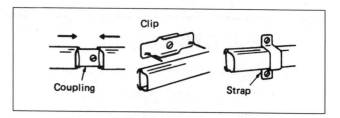

Fig. 15–2 Surface metal raceway supports.

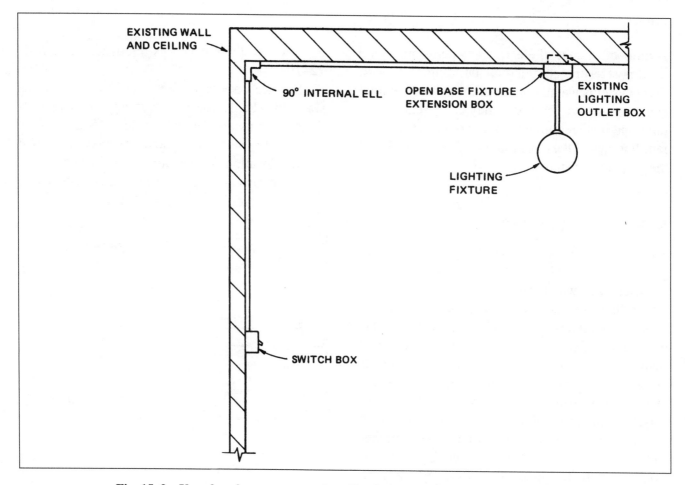

Fig. 15–3 Use of surface raceway to install switch on existing lighting installation.

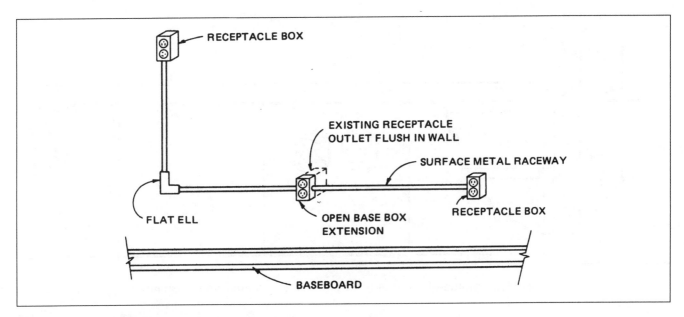

Fig. 15-4 Use of surface metal raceway to install additional receptacle outlets.

The usage in the insurance office is expected to be intermittent and made up of only small appliances; thus it would qualify for the allowable minimum of one outlet for each 1.5 metres. However, the contractor is required to install a duplex receptacle every 450 mm for a total of 56 receptacles in the office space. The number of receptacles is limited to 12 outlets per circuit, *Rule 12–3000(1)*. This load is connected to five branch circuits, which satisfies the continuous loading maximum as set out in *Rule 12–3000*. This allows for some growth, and additional circuits can be easily installed if required.

When determining the feeder size, the requirement for the receptacle load is calculated by the application of the demand factors set out in *Table 14*.

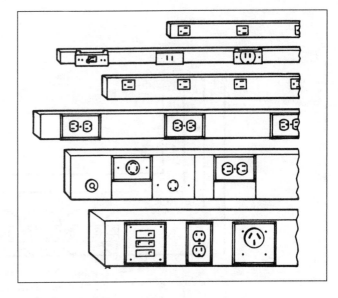

Fig. 15-5 Multioutlet assemblies.

Receptacle Wiring

The plans indicate that the receptacles to be mounted in the multioutlet assembly must be spaced 450 mm apart. The receptacles may be connected in either of the arrangements shown in Fig. 15-7. That is, all of the receptacles on a phase can be connected in a continuous row, or the receptacles can be connected on alternate phases.

The connection of the identified conductor or the neutral (if fed from a three-phase, four-wire system) to the receptacle must utilize tails and wire connectors. This permits a receptacle to be removed without the neutral or identified conductor being broken. The neutral or identified conductor provides the return for the other two phases.

An electrician could be working on circuit A receptacle, which is off (dead), while circuits B

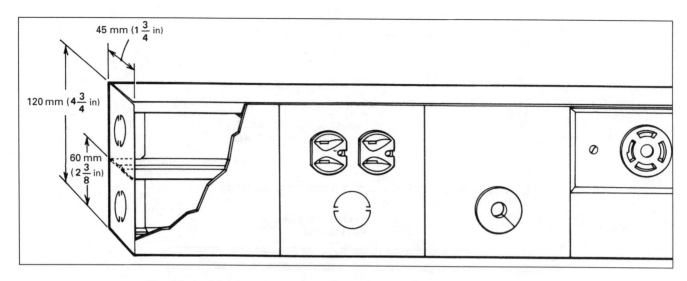

Fig. 15–6 Multioutlet assembly for power and communication systems.

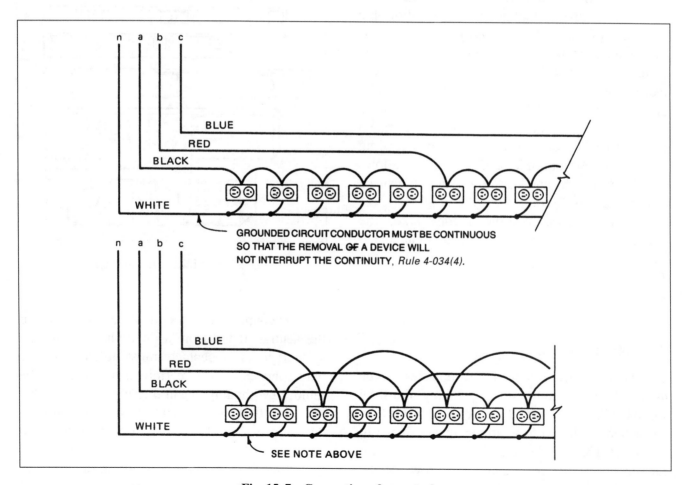

Fig. 15–7 Connection of receptacles.

and C remain on (alive). These circuits are fed from a multiwire branch circuit. Since the neutral has not been broken there is less possibility of the electrician becoming the return path to ground. **Any shock current greater than 6 milliamperes is dangerous.**

COMMUNICATIONS SYSTEMS

The installation of the telephone system in the commercial building will consist of two separate installations. The *electrical contractor* will install an empty conduit system according to the specifications for the commercial building and in the locations indicated on the plans. In addition, the installation will meet the rules, regulations, and requirements of the communications company that will serve the building. Once the conduit system is complete, the electrical contractor, telephone installing company, or the telephone company will install a complete telephone system.

Power Requirements

An allowance of 2500 volt-amperes is made for the installation of the telephone equipment to provide the power required to operate the special switching equipment for a large number of telephones. Because of the importance of the telephone as a means of communication, the receptacle outlets for this equipment are connected to the emergency power system.

The Telephone System

Telephone service requirements vary widely according to the type of business and the extent of the communication convenience desired. In many situations, the telephone lines may be installed in exposed locations. However, improvements in building construction techniques have made it more important to provide facilities for installing concealed telephone wiring. The use of new wall materials and wall insulation, the reduction or the complete omission of trim around windows and doors, the omission of baseboards, and the increased use of metal trim make it more difficult to attach exposed wires. For these reasons (and many others), the wiring is more conspicuous if the installation is exposed. In addition, the unprotected wiring is more subject to damage and interference.

To solve the problems of exposed wiring, conduits are installed and the proper outlet boxes and junction boxes are placed in position during the construction. The material and construction costs are low when compared to the benefits gained from this type of installation.

Since it is generally more difficult to conceal the telephone wiring to the floors above the first floor after the building construction is completed, it is recommended that telephone conduits be provided to these floors. The materials used in the installation and the method of installing the conduits for the telephone lines are the same as those used for light and power wiring, with the following modifications:

- To avoid small openings and limited space, junction boxes are used rather than the standard conduit fittings such as ells and tees.

- Since multipair conductors are used extensively in telephone installations, the size of the conduit should be at least 21-mm ($\frac{3}{4}$-in) trade size. The current and potential ratings for an installation of this type are not governed by the same rules as conventional wiring for light and power.

- The number of bends and offsets is kept to a minimum; when possible, bends and offsets are made using greater minimum radii or larger sweeps. The allowable minimum radius is 152 mm (6 in).

- A fishwire is installed in each conduit for use in pulling in the cables.

- When basement telephone wiring is to be exposed, the conduits are dropped into the basement and terminated with a bushing (junction boxes are not required). No more than 50 mm (2 in) of conduit should project beyond the joists or ceiling level in the basement.

- The inside conduit drops *must* be grounded in accordance with *Section 10* and *Rule 60–302*.
- The service conduit carrying telephone cables from the exterior of a building to the interior must be permanently and effectively grounded, *Rule 60–302*.
- If the cable has a metal shield or sheath, the covering shall be bonded to ground, *Rules 60–306* and *60–700*.

Installing the Telephone Outlets

The plans for the commercial building indicate that each of the occupancies requires telephone service. As shown previously, the insurance office uses a multioutlet assembly. Telephone outlets are installed in the balance of the occupancies. These outlets are supplied by EMT, which runs to the basement, Fig. 15–8, where the cable then runs exposed to the main terminal connections. Telephone company personnel will install the cable and connect the equipment.

FLOOR OUTLETS

In an area the size of the insurance office, it may be necessary to place equipment and desks where wall outlets are not available. Floor outlets may be installed to provide the necessary electrical supply to such equipment. Two methods can be used to provide floor outlets:

1. installing underfloor raceway, or
2. installing floor boxes.

Underfloor Raceway (*Subsection 12–1700*)

In general, underfloor raceway is installed to provide both power and communication outlets using a dual duct system similar to the one shown in Fig. 15–9. The junction box is constructed so that the power and communication systems are always separated from each other. Service fittings are available for the outlets, Fig.15–10.

Floor Boxes

Floor boxes, either metallic or nonmetallic, can be installed using any approved raceway such as rigid conduit, rigid nonmetallic conduit, or EMT. Some boxes must be installed to the correct height by adjusting the levelling screws before the concrete is poured, Fig. 15–11. One type of nonmetallic box can be installed without adjustment and cut to the desired height after the concrete is poured, Figs. 15–12 and 15–13.

COMPUTER ROOM CIRCUITS

Although a room in the insurance office is especially designed to be a computer room, there are no special requirements applicable. Three actions were taken to comply with good practice: special receptacles are specified, a special grounding system is provided, and a separate neutral is provided with each of the 120-volt, single-phase circuits.

The special receptacles were discussed in Unit 11; they have an isolated grounding connection and provide surge protection. The grounding terminal of each receptacle is connected to an insulated grounding conductor. This grounding conductor is installed in the conduit with the circuit conductors. The conduit is grounded according to *Section 10*. Since there are three conduits involved, the three grounding conductors are connected to a special isolated grounding terminal in the panelboard; see *Rule 10–906(8)*.

From here a grounding conductor is installed with the feeder circuit and is connected to a grounding electrode at the switchgear. This grounding system reduces, if not eliminates, the electromagnetic interference (EMI) that often is present in the conventional grounding system. The surge protection feature of the receptacle provides protection from lightning and other severe electrical surges that may occur in the electrical system. These surges are often severe enough to cause a loss of computer memory and in some cases damage to the computer. Surge protectors are available in separate units, which can be used to provide protection on any type of sensitive equipment, such as radios and televisions.

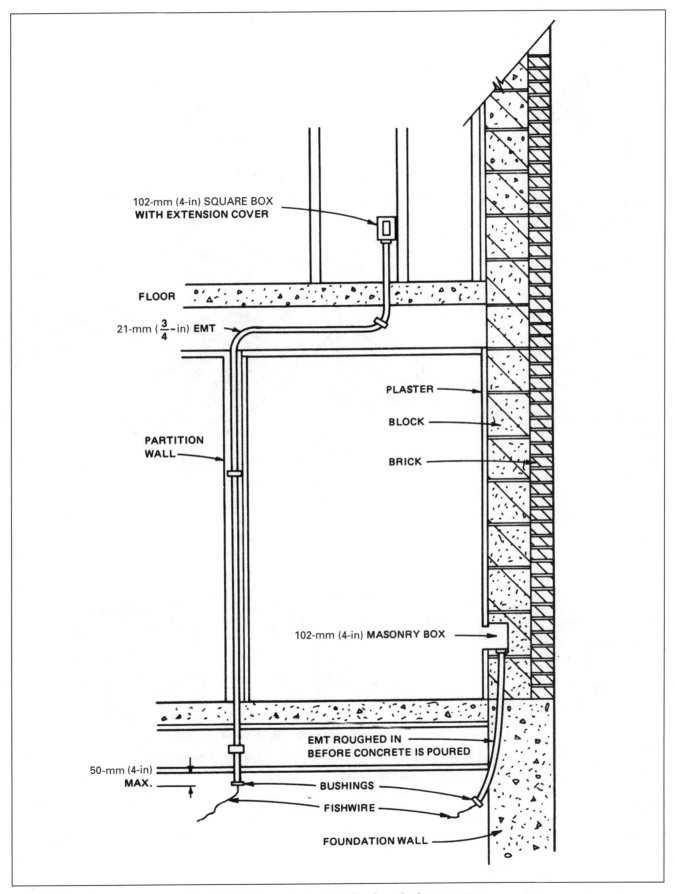

Fig. 15–8 Raceway installation for telephone system.

240　Unit 15　Special Systems

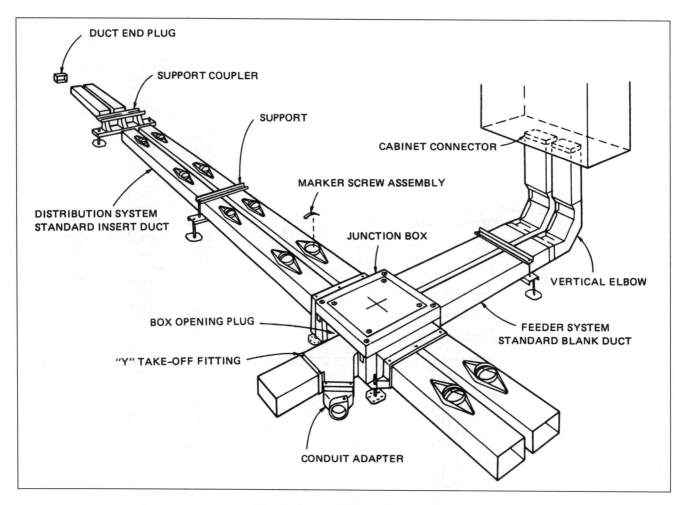

Fig. 15–9　Underfloor raceway.

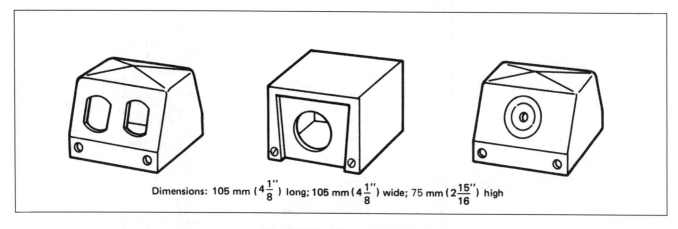

Dimensions: 105 mm ($4\frac{1}{8}''$) long; 105 mm ($4\frac{1}{8}''$) wide; 75 mm ($2\frac{15}{16}''$) high

Fig. 15–10　Service fittings.

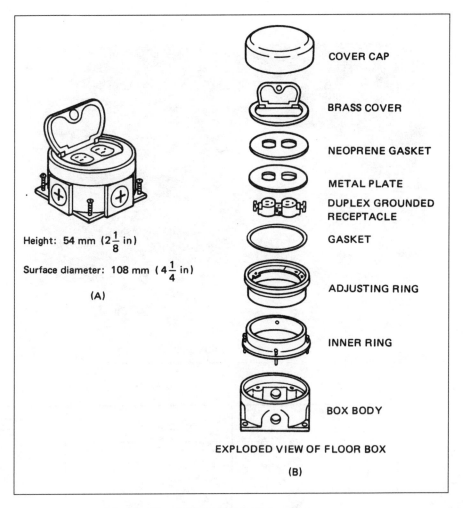

Fig. 15–11 A floor box with levelling screws.

The separate neutrals or identified conductors are installed to greatly reduce the possibility of conductor overheating. The common practice is to use one neutral to serve three single-phase circuits of a four-wire wye system. Loads such as computers produce odd harmonics—the third, the fifth, the seventh, and so on. For example, the third harmonic is $3 \times 60 = 180$ Hz. Since the loads are not all resistance, the harmonics from different phases do not cancel out as ordinary current does. Under these conditions, the common neutral current could be the sum of the individual circuits and damage could occur to the wiring system. Doubling the size of the neutral or installing zero-sequence filters, *Rule 26–266*, will help to alleviate the effects of the additive harmonic RMS current on the neutral of a three-phase, four-wire system.

242 Unit 15 Special Systems

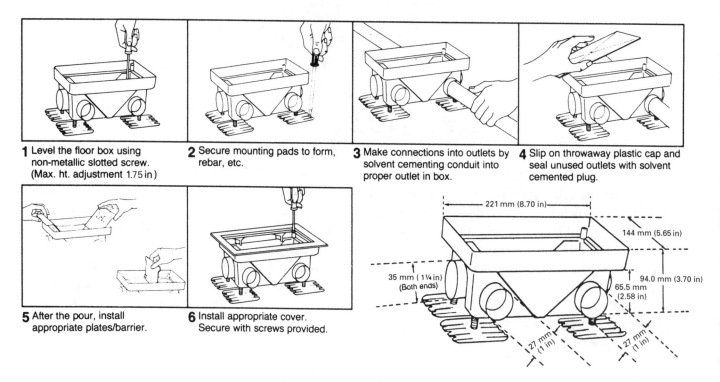

Fig. 15–12 Installation of multifunction nonmetallic floor box.

Unit 15 Special Systems 243

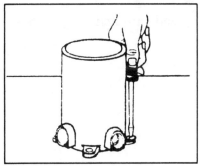

1 Fasten the box to the form or set on level surface.

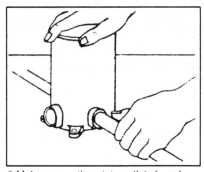

2 Make connections into outlets by solvent cementing conduit into the proper outlet in box.

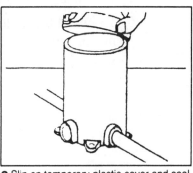

3 Slip on temporary plastic cover and seal unused outlets with solvent cemented plugs.

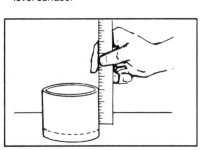

4 Remove temporary plastic cover and determine thickness of flooring to be used. Scribe a line around box at this distance from the floor.

5 Using a handsaw, cut off box at scribed line.

6 Install leveling ring to underside of cover, so that four circular posts extend through slots in outer ring.

7 Apply PVC cement to leveling ring and to top inside edge of box.

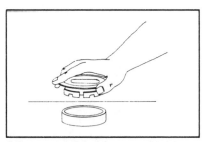

8 Press outer ring cover assembly into floor box for perfect flush fit.

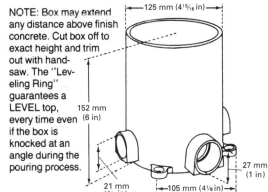

NOTE: Box may extend any distance above finish concrete. Cut box off to exact height and trim out with handsaw. The "Leveling Ring" guarantees a LEVEL top, every time even if the box is knocked at an angle during the pouring process.

Fig. 15–13 Installation and trim-out of nonmetallic floor box.

REVIEW

Note: Refer to the *C.E.C., Part I* or the plans as necessary.

1. Surface raceway may be extended through dry partitions if the _____
   _____

2. In surface raceway, the conductors (including taps and splices) shall fit not more than _____ % of the area.

3. Where power and communications circuits are run in combination raceway, the different systems shall be run in _____
   _____

4. In areas of light usage, _____ outlet(s) is/are allowed for each _____ metres of multioutlet assembly.

5. How many floor outlets are to be installed in the insurance office? _____

6. To install the multioutlet assembly in the insurance office, _____ end caps and _____ internal ells are required.

7. How many telephone conduits, for all occupants, are stubbed into the basement? _____

UNIT 16

Reading Electrical and Architectural Drawings (Prints)—Drugstore

OBJECTIVES

After completing the study of this unit, the student will be able to

- identify symbols associated with the reading of electrical drawings
- determine the requirements of the electrical contract
- calculate plan dimensions

THE DRUGSTORE PRINTS

Branch circuits are an important part of the electrical system in a building. Unit 4 presents the essential information on branch circuits. This information includes how to determine, according to the *C.E.C., Part I*, the number of branch circuits required and their correct size. The student should give close attention to the drugstore branch-circuit panelboard schedule on sheet E3. This schedule gives, in addition to other information, the branch-circuit numbers and a brief description of the loads served. These items provide important links between the panelboard schedule and the electrical drawings. See Fig. 16–1.

A feature of the drugstore wiring is the low-voltage remote-control system. See Unit 20 for a complete discussion. This system offers flexibility of control that is not available in the traditional control system. The switches used in this system operate on 24 volts, and the power wiring, at 120 volts, goes directly to the electrical load. This reduces branch-circuit length and voltage drop. A switching schedule gives details on the system operation, and a wiring diagram provides valuable information to the installer.

The illumination system in each area of the drugstore is somewhat different. In the merchandise area, nine luminaires are installed in a continuous row. It is necessary to install electrical power to only one point of a continuous row of luminaires. From this point the conductors are installed in the wiring channel of the luminaire. In the pharmacy area, a luminous ceiling is shown. This illumination system consists of rows of strip fluorescents and a ceiling that will transmit light. The installation of the ceiling, in many jurisdictions, is the responsibility of the electrician. For this system to be efficient, the surfaces above the ceiling must be highly reflective (white).

Drugstore Feeder Loading Schedule

The various fixtures and other loads are calculated on Table 16–1 to show the total demand watts on the feeders and the service. This will require a 100-ampere switch with a 90-ampere fuse in the main switchgear. The No. 4 RW90-XLPE copper conductor will feed the 100-ampere panel located in the drugstore.

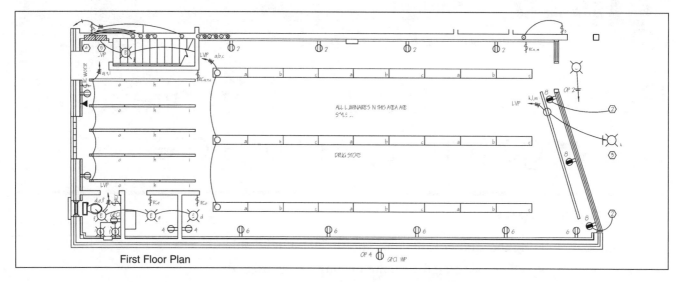

Fig. 16–1 Electrical drawing for a drugstore. *Note*: For complete blueprint, refer to blueprint E3 in back of text.

Table 16–1 Drug Store Service Calculation

Item	Count	VA/unit	Demand Load	Installed Load	Demand Factor Feeder	Demand Factor Service	Use Feeder	Use Service
MINIMUM BASIC LOAD								
(C.E.C. Rule 8-210)	225 m²	30 W/m²	6750		1	1	6750	6750
BASIC INSTALLED LOADS								
Style D luminaires	15	74		1110			1110	1110
Style E luminaires	4	144		576			576	576
Style I luminaires	27	87		2349			2349	2349
Style N luminaires	2	60		120			120	120
Style L luminaires	9	87		783			783	783
General receptacles	21	120		2520			2520	2520
Basic installed loads							7458	7458
BASIC LOAD	(Use the greater of the demand load or installed loads, *Rule 8–106(2)*.)						7458	7458
SPECIAL LOADS								
Show window rec.	3	500		1500			1500	1500
Track lighting	15	110		1650			1650	1650
Sign	1	1200		1200			1200	1200
Roof receptacle	1	120		120			120	120
Motors	Volts	FLA	Phase					
Cooling unit								
• compressor	208	20.2	3	7269	1.25		9086	7269
• evaporator motor	208	3.2	1	666			666	666
• condenser motor	208	3.2	1	666			666	666
Total load							22 346	20 529

$$I = \frac{P}{E \times 1.73} = \frac{22\,346}{208 \times 1.73} = 62.1 \text{ A.}$$

Assuming all loads are continuous, standard-rated equipment and using RW90 copper conductors from *Table 2, C.E.C.*, the ampere rating of the circuit would be:

$$\frac{62.1}{0.8} = 77.6 \text{ A.}$$

The minimum size feeder would be 4 No. 4 RW90-XLPE copper conductors in a 27-mm (1-in) conduit, supplied by a 100-A disconnect c/w 90-A fuses.

REVIEW

1. Identify the following symbols. Refer to Electrical Symbol Schedule Sheet E1.

   a. _____

   b. _____

   c. _____

   d. _____

   e. _____

   f. _____

   g. _____

In the following questions, fill in the blanks.

2. The elevation of the finished first floor is _____ m.
3. The drugstore ceiling material is _____.
4. The north wall of the drugstore is constructed of _____ and is _____ mm (_____ inches) plus finish material thick.
5. The south wall of the drugstore is constructed of _____ and is _____ mm (_____ inches) plus finish material thick.
6. The ceiling height of the drugstore is _____.
7. The outside dimensions of the building are _____ by _____.
8. The drugstore basement has an area of _____ m² (_____ ft²).
9. How many luminaires are controlled by switch RC_a? _____.
10. How many receptacles are installed on branch circuit #2? _____.

UNIT 17

Overcurrent Protection: Fuses and Circuit Breakers

OBJECTIVES

After completing the study of this unit, the student will be able to

- list and identify the types, classes, and ratings of fuses and circuit breakers
- describe the operation of fuses and circuit breakers
- develop an understanding of switch sizes, ratings, and requirements
- define interrupting rating, short-circuit currents, I^2t, I_p, RMS, and current limitation
- apply the *C.E.C., Part I* to the selection and installation of overcurrent protective devices
- use time–current characteristics curves and peak let-through charts

---

Overcurrent protection is one of the most important components of an electrical system. The overcurrent device opens an electrical circuit whenever an overload or short circuit occurs. Overcurrent devices in an electrical circuit may be compared to pressure relief valves on a boiler. If dangerously high pressures develop within a boiler, the pressure relief valve opens to relieve the high pressure. The overcurrent device in an electrical system also acts as a "safety valve."

The *C.E.C., Part I* requirements for overcurrent protection are contained in *Section 14*. *Rule 14–010* states, in part, that devices must be provided to automatically open the electrical circuit if the current reaches a value that will produce a dangerous temperature in the equipment or conductor. Additionally, in the event of a ground fault all normally ungrounded conductors must be de-energized, subject to *Rule 14–102*. Also refer to *Rule 14–012* for voltage and interrupting current rating requirements.

Two types of overcurrent protective devices are commonly used: fuses and circuit breakers. The Canadian Standards Association (CSA) and the National Electrical Manufacturers Association (NEMA) establish standards for the ratings, types, classifications, and testing procedures for fuses and circuit breakers.

The standard ampere ratings for fuses and/or nonadjustable circuit breakers are listed in *Table 13* of the *C.E.C.* for ampacities up to and including 600 amperes. Standard ampacities in excess of 600 amperes are 700, 800, 1000, 1200, 1600, 2000, 2500, 3000, 4000, 5000, and 6000. Although fuses are readily available in all of these ampacities, circuit breakers may not be available in all sizes, depending on the manufacturer. In addition, there are standard ratings for fuses of 1, 3, 6, and 10 amperes.

The *C.E.C., Part I* describes circumstances in which it is permissible to increase the rating or setting of the overcurrent device. For example, *Rule 14–104(a)* states that if the ampacity of a conductor is less than 600 amperes and does not match a standard ampere rating fuse or breaker, then it is allowable to use the rating found in *Table 13*. Therefore, for a conductor that has an ampacity of 115 amperes, the *C.E.C., Part I* permits the overcurrent protection to be sized at 125 amperes.

When protecting conductors where the overcurrent protection is above 600 amperes, it is vital to ensure that the overcurrent device is equal to or lower than the ampacity of the conductor. For example, three 500 kcmil copper conductors are installed in parallel. Each has an ampacity of 380 amperes. Therefore, the total ampacity is $380 \times 3 = 1140$ amperes.

It would be a *C.E.C., Part I* violation to protect these conductors with a 1200-ampere fuse. The correct thing to do is to round down to an ampere rating less than 1140 amperes. Fuse manufacturers provide 1100 ampere rated fuses.

Fused switches are available in ratings of 30, 60, 100, 200, 400, 600, 800, 1200, 1600, 2000, 2500, 3000, 5000, and 6000 amperes in both 250- and 600-volt versions. They are for use with copper conductors only unless marked that the terminals are suitable for use with aluminum conductors. The ratings of the switches are based on 60°C wire (No. 14 through No. 1) and 75°C for wires No. 1/0 and larger, unless otherwise marked. Switches also may be equipped with ground-fault sensing devices. These will be clearly marked as such.

Switches may have labels that indicate their intended application, such as "Continuous load current not to exceed 80% of the rating of the fuses employed in other than motor circuits."

- Switches intended for isolating use only are marked "For isolation use only—Do not open under load."
- Switches that are suitable for use as service switches are marked "Suitable for use as service equipment."
- When a switch is marked "Motor-circuit switch," it is only for use in motor circuits.
- Switches of higher quality may have markings such as "Suitable for use on a circuit capable of delivering not more than 100 000 RMS symmetrical amperes, 600 volts maximum: Use Class T fuses having an interrupting rating of no less than the maximum available short-circuit current of the circuit."
- Enclosed switches with horsepower ratings in addition to ampere ratings are suitable for use in motor circuits as well as for general use.
- Some switches have a dual-horsepower rating. The larger horsepower rating is applicable when using dual-element time-delay fuses. Dual-element time-delay fuses are discussed later on in this unit.
- Fusible bolted pressure contact switches are tested for use at 100% of their current rating at 600 volts ac, and are marked for use on systems having available fault currents of 100 000, 150 000, and 200 000 RMS symmetrical amperes.

In the case of externally adjustable trip circuit breakers, the rating is considered to be the breaker's maximum trip setting. The exceptions to this are:

1. if the breaker has a removable and sealable cover over the adjusting screws, or
2. if it is located behind locked doors accessible only to qualified personnel, or
3. if it is located behind bolted equipment enclosure doors.

In these cases, the adjustable setting is considered to be the breaker's ampere rating. This is an important consideration when selecting proper size phase and neutral conductors, equipment grounding conductors, overload relays in motor controllers, and other situations where sizing is based upon the rating or setting of the overcurrent protective device.

FUSES AND CIRCUIT BREAKERS

For general applications, the voltage rating, the continuous current rating, the interrupting rating, and the speed of response are factors that must be considered when selecting the proper fuses and circuit breakers.

Voltage Rating

According to *Rule 14–012*, the voltage rating of a fuse or circuit breaker must be sufficient for the voltage of the circuit in which the fuse or circuit breaker is to be used.

Continuous Current Rating

The continuous current rating of a fuse or circuit breaker should be equal to the ampacity of the conductor being protected. For example, referencing *Table 2*, we find that a No. 8 type T90 Nylon conductor has an ampacity of 45 amperes. The proper overcurrent protection for this No. 8 type T90 Nylon conductor would be a 45-ampere fuse or 50-ampere circuit breaker. See *Table 13*.

There are several situations that permit the overcurrent protective device rating to exceed the ampacity of the conductor. For example, the rating of a fuse or circuit breaker may be greater than the current rating of the conductors that supply an electric motor or a refrigeration compressor.

Protection of Conductors

- *When the overcurrent device is rated 600 amperes or less.* When the ampacity of a conductor does not match the ampere rating of a standard fuse, or does not match the ampere rating of a standard nonadjustable circuit breaker, the use of the next higher standard ampere-rated fuse or circuit breaker is permitted, *Rule 14–104*. Refer to Fig. 17–1.

- *When the overcurrent device is rated above 600 amperes.* When the ampere rating of a fuse or of a circuit breaker that does not have an overload trip adjustment exceeds 600 amperes, the conductor ampacity must be equal to or greater than the rating of the fuse or circuit breaker. The *C.E.C., Part I* allows the use of fuses and circuit breakers that have ratings less than the standard sizes. Adjustable trip circuit breakers were discussed above. See *Rule 14–104* and refer to Fig. 17–2.

Interrupting Rating (*Rule 14–012(a)*)

The interrupting rating of an overcurrent device is a measure of the ability of a fuse or circuit breaker to safely open an electrical circuit under fault conditions, such as overload currents, short-circuit currents, and ground-fault currents. In other words, the interrupting rating of an overcurrent device is the maximum short-circuit current that it can interrupt safely at its rated voltage.

CAUTION: The interrupting rating of a circuit breaker or fuse is *that* circuit breaker's or fuse's ability to safely interrupt fault currents not exceeding its interrupting rating. This marked interrupting rating has nothing to do with the *withstand ratings* of the downstream components, such as other circuit breakers, panelboards, switchboards, bus bars and their bracing, conduc-

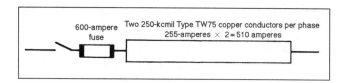

Fig. 17–1 *Rule 14–104* allows the use of a 600-ampere fuse or circuit breaker.

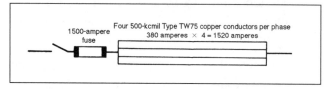

Fig. 17–2 *Rule 14–104* requires the use of a 1500-ampere fuse or circuit breaker.

tors, contactors, controllers, relays, ground-fault circuit interrupters, and so on.

It can be very hazardous to consider the interrupting ratings of a circuit breaker while ignoring the withstand ratings of the downstream connected components.

The withstand rating of all electrical components is based upon **how much current will flow** and **how long the current will flow**. Units 18 and 19 discuss the important issues of "current limitation" and "peak let-through current."

Overload currents have the following characteristics:

- They are greater than the normal current flow.
- They are contained within the normal conducting current path.
- If allowed to continue, they will cause overheating of the equipment, conductors, and the insulation of the conductors.

Short-circuit and ground-fault currents have the following characteristics:

- They flow outside of the normal current path.
- They may be less than, equal to, or greater than the normal current flow.

Short-circuit and ground-fault currents, which flow outside of the normal current paths, can cause conductors to overheat. In addition, mechanical damage to equipment can occur as a result of arcing and the magnetic forces of the large current flow. Some short-circuit and ground-fault currents may be no larger than the normal load current, but they can be thousands of times larger.

The terms *interrupting rating* and *interrupting capacity* are interchangeable in the electrical industry, but there is a significant difference. Interrupting rating is the value of the test circuit capability. Interrupting capacity is the actual current that the contacts of a circuit breaker "see" when opening under fault conditions.

For example, the test circuit in Fig. 17–3 is calibrated to deliver 14 000 amperes of fault current. The standard test circuit allows 1.22 m (4 ft) of conductor to connect between the test bus and the breaker's terminals. The standard test further allows 254 mm (10 in) of conductor per pole to be used as the shorting wire. Therefore, when the impedance of the connecting conductors is taken into consideration, the actual fault current that the contacts of the breaker "see" as they open is approximately 9900 amperes. The label on this circuit breaker is marked "14 000 amperes interrupting rating," yet its true interrupting capacity is 9900 amperes.

As described above, it is important to ensure that the ratings on equipment conform to the requirements of *Rule 14–012*.

For those interested in the complete details of how test circuits for breakers are connected, reference should be made to *CSA Standards C22.2 No. 4 M89* and *C22.2 No. 5 M91*.

Speed of Response

The time required for the fusible element of a fuse to open varies inversely with the magnitude of the current that flows through the fuse. In other words, as the current increases, the time required

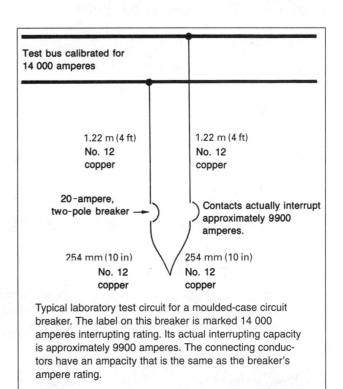

Typical laboratory test circuit for a moulded-case circuit breaker. The label on this breaker is marked 14 000 amperes interrupting rating. Its actual interrupting capacity is approximately 9900 amperes. The connecting conductors have an ampacity that is the same as the breaker's ampere rating.

Fig. 17–3 Test circuit for a moulded case circuit breaker.

for the fuse to open decreases. The time–current characteristic of a fuse depends upon its rating and type. A circuit breaker also has a time–current characteristic. For the circuit breaker, however, there is a point at which the opening time cannot be reduced further due to the inertia of the moving parts within the breaker. The time–current characteristic of a fuse or circuit breaker should be selected to match the connected load of the circuit to be protected. Time–current characteristic curves are available from the manufacturers of fuses and circuit breakers.

TYPES OF FUSES

Dual-Element Time-Delay Fuse

The dual-element time-delay fuse, Fig. 17–4, provides a time delay in the low-overload range to eliminate unnecessary opening of the circuit because of harmless overloads. However, this type of fuse is extremely responsive in opening on short circuits. This fuse has two fusible elements connected in series. Depending upon the magnitude of the current flow, one element or the other will open. The thermal cutout element is designed to have a delayed opening for currents up to approximately 500% of the fuse rating. The short-circuit element opens immediately when a large short circuit or overload occurs. That is, this the element opens at current values of approximately 500% or more of the fuse rating.

The thermal element is designed to open at approximately 140°C (284°F) as well as on damaging overloads. Thus, the thermal element will open whenever a loose connection or a poor contact in the fuseholder causes heat to develop. As a result, a true dual-element fuse also offers thermal protection to the equipment in which it is installed.

Dual-element fuses are suitable for use on motor circuits and other circuits having high-inrush characteristics. This type of fuse can be used as well for mains, feeders, subfeeders, and branch circuits. Dual-element fuses may be used to provide back-up protection for circuit breakers, bus ducts, and other circuit components that lack an adequate interrupting rating, bracing, or withstand rating (covered later in this unit).

Using Fuses for Motor Overload Protection

Dual-element, time-delay fuses used on single-motor branch circuits are generally sized not to exceed 125% of the full-load running current of the motor, in accordance with *Rule 28–302(2)* and *28–306* for motor overload protection. Sizing these fuses slightly higher than the motor's over-

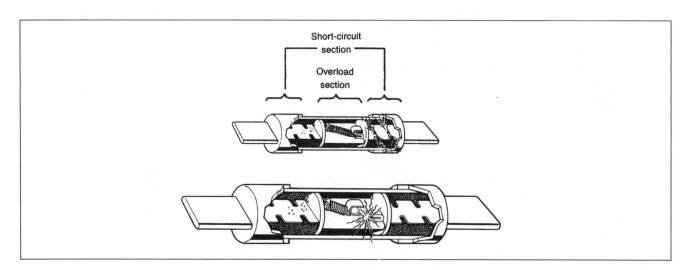

Fig. 17–4 Cutaway view of a Fusetron dual-element time-delay fuse. On overloads, the spring-loaded trigger assembly opens. On short-circuits or heavy ground faults, the fuse elements in the short-circuit section open. The fuse elements are generally made of copper. (Courtesy of Bussmann, Cooper Industries.)

load setting will provide "back-up" overload protection for the motor. The motor now has double overload protection.

Experience has shown that sizing dual-element time-delay fuses in the range of 115% to 125% of the motor's full-load current rating meets the *C.E.C., Part I* requirements for branch-circuit, short-circuit, and ground-fault protection, as well as providing "back-up" motor overload protection.

Rule 28–300 to *28–318* set forth motor overload protection requirements. There are many questions to ask when searching for the proper level of motor overload protection. These include: What are the motor's starting characteristics? Is the motor manually or automatically started? Is the motor continuous duty or intermittent duty? Does the motor have integral built-in protection? Is the motor impedance protected? Is the motor larger than one horsepower? Is the motor fed by a general-purpose branch circuit? Is the motor cord-and-plug connected? Will the motor automatically restart if the overload trips? When these questions are answered, refer to *Rules 28–300* to *28–318* for the level of overload protection needed.

The procedure for selecting the overload protection is shown in Fig. 17–5. The motor type is selected in the first column. The motor nameplate full-load current rating is increased according to the value in the second column.

EXAMPLE: What is the ampere rating of a dual-element time-delay fuse that is to be installed to provide branch-circuit protection as well as overload protection for a motor with a full-load current rating of 16 amperes?

16 × 1.25 = 20 amperes

	Maximum Overload Size, *Rule 28–306*
Service factor not less than 1.15	125%
All other motors	115%

Fig. 17–5 Maximum overload size for motors.

Using Fuses for Motor Branch-Circuit Overcurrent Protection

Table 29 shows that for typical motor branch-circuit, overcurrent protection, the maximum size permitted for dual-element fuses is 175% of the full-load current of the motor. An exception to the 175% value is given in *Rule 28–200(d)(ii)*, where permission is given to go as high as 225% of the motor full-load current if the lower value is not sufficient to allow the motor to start.

EXAMPLE: Overcurrent protection for the motor branch circuit in the previous example:

16 × 1.75 = 28 amperes (maximum)

The next lower standard rating is 25 amperes.

If for some reason the 25-ampere fuse cannot handle the starting current, *Rule 28–200(d)(ii)* allows the selection of a dual-element time-delay fuse not to exceed:

16 × 2.25 = 36 amperes (maximum)

The next lower standard rating is 35 amperes.

Dual-Element, Time-Delay, Current-Limiting Fuses

The dual-element, time-delay, current-limiting fuse, Fig. 17–6, operates in the same manner as the standard dual-element time-delay fuse, but has a faster response in the short-circuit range and thus is more current limiting. Sizing this type of fuse for motor circuits, where time-delay is necessary for starting the motor, is done according to the procedure discussed in the previous paragraph. The short-circuit element in the current-limiting fuse can be silver or copper surrounded by a quartz sand arc-quenching filler. Silver-link fuses are more current-limiting than copper-link fuses.

Some of the common terms referring to fuses are described in Fig. 17–7b.

Current-Limiting Fuses (Nontime Delay)

The straight current-limiting fuse, Fig. 17–7c, has an extremely fast response to both overloads and short circuits. When compared to other types of fuses, this one has the lowest energy let-through

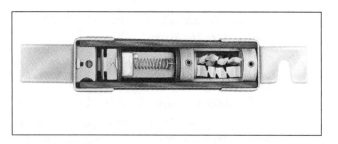

Fig. 17–6 Cutaway view of a low-peak, dual-element, time-delay, current-limiting fuse. On overloads, the spring-loaded trigger assembly opens. On large short circuits or ground faults, the fuse elements in the short-circuit section open. The fuse elements are generally made of silver. (Courtesy of Bussmann, Cooper Industries.)

values. Current-limiting fuses are used to provide better protection to mains, feeders, subfeeders, circuit breakers, bus duct, switchboards, and other circuit components that lack adequate interrupting rating, bracing, or withstand rating.

Current-limiting fuse elements can be made of silver or copper surrounded by a quartz sand arc-quenching filler. Silver-link fuses are more current limiting than copper-link fuses.

Current-limiting fuses may be selected and loaded at 100% of their rating for noninductive circuit loads. However, it is a good design practice not to exceed 80% of the rating for feeder and circuit loading.

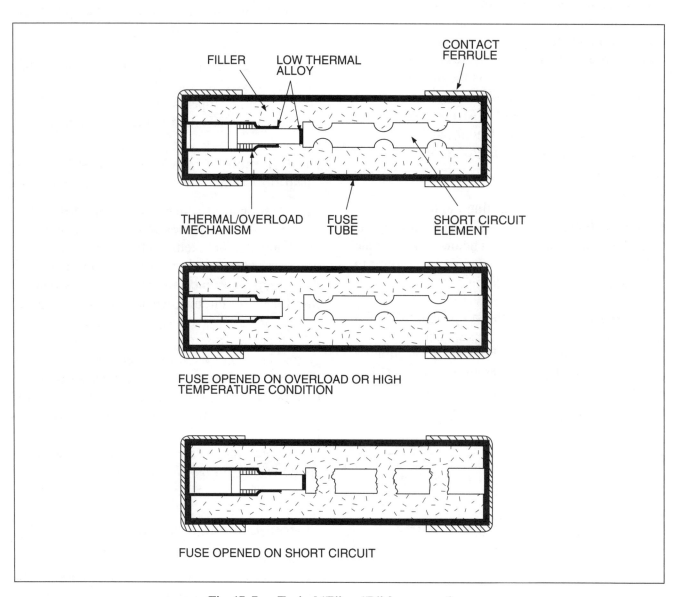

Fig. 17–7a Typical "P" or "D" fuse operation.

A standard current-limiting fuse does not have the spring-loaded overload assembly found in dual-element fuses. To use straight current-limiting fuses for motor circuits, refer to *Table 29* (Fig. 17–8) under the column "Non-Time Delay Fuse."

To be classified as current limiting, the overcurrent device must reduce heavy (high magnitude) fault currents, to a value less than the fault current that would have flowed into the circuit had there been no fuse or breaker.

When used on motor circuits or other circuits having high current-inrush characteristics, the current-limiting nontime-delay fuses must be sized at a much higher rating than the actual load. That is, for a motor with a full-load current rating of 10 amperes, a 30- or 40-ampere current-limiting fuse may be required to start the motor. In this case, the fuse is considered to be the motor branch-circuit short-circuit protection only.

CARTRIDGE FUSES

A cartridge fuse must be marked to show its

- ampere rating
- voltage rating
- interrupting rating when over 10 000 amperes
- current-limiting type, if applicable
- trade name or name of manufacturer

Fig. 17–7c Cutaway view of a Limitron, current-limiting, fast-acting, single-element fuse. (Courtesy of Bussmann, Cooper Industries.)

Some Popular Fuse Terms

Interrupting Rating: The maximum, short-circuit current a fuse can safely interrupt.

HRC: A low voltage fuse with a minimum interrupting rating of 100 kA. Most HRC fuses have an interrupting rating of 200 kA.

HRCI or Form I: Denotes that the fuse provides both overload and short-circuit protection. Can also be used for short-circuit protection only, e.g., in motor circuits. Usually a Class J or Class R fuse.

HRCII or Form II: Denotes that the fuse provides short-circuit protection only. Usually a Class C fuse.

Time Delay: A CSA-specified time–current characteristic, which generally means that the fuse will carry:
HRC and Code   –500% ampere rating for 10 seconds minimum
Plug Fuses   –200% ampere rating for 12 seconds minimum
Supplemental Fuses (a) 3 amperes or less: 200% rating for 5 seconds minimum.
                    (b) more than 3 amperes: 200% rating for 12 seconds minimum.

Code Fuse: The "Standard" fuse that is required to have a minimum interrupting rating of 10 kA.

Supplemental Fuse: A midget or small dimension fuse intended for use in electrical or electronic equipment; i.e., not a branch-circuit fuse.

Dual Element: Not a CSA-recognized term. Often confused with Time Delay, the term Dual Element is simply a manufacturer's term to describe the construction of the fuse element. Dual Element fuses can be either Time Delay or Nontime Delay.

Fast Acting: The popular term for a Nontime Delay HRC fuse.

Type "D": A Time Delay fuse with a low melting point characteristic—Gould Shawmut types TD (Plug), CRN/CRS (Code), and TRNR/TRSR (HRCI).

Type "P": A Nontime Delay fuse with a low melting point characteristic. Available in Plug (Type GP, 15–30A) and 250-volt one-time cartridge fuses. Gould Shawmut Types NRN and OTN, 15–60 amperes.

Note: Manufacturer's terms such as "Time Lag," "Delay Action," etc. are not, in fact, CSA-certified and usually indicate that the fuse is not, in fact, a CSA-certified "Time Delay" fuse. A CSA-certified Time Delay fuse has to be marked with the words "Time Delay" or "Type D."

Fig. 17–7b Some popular fuse terms. (Courtesy of Gould Shawmut.)

Table 29
(See *Rules 28–200, 28–204, 28–208,* and *28–308.*)
Rating or Setting of Overcurrent Devices for the Protection of Motor Branch Circuits

Type of Motor	Percent of Full-Load Current		
	Maximum Fuse Rating		Maximum Setting Time-Limit Type Circuit Breaker
	Time-Delay* Fuses	Nontime Delay	
Alternating Current			
Single Phase, All Types	175	300	250
Squirrel-Cage and Synchronous:			
Full-Voltage, Resistor and Reactor Starting	175	300	250
Auto-Transformer and Star Delta Starting:			
Not more than 30 A	175	250	200
More than 30 A	175	200	200
Wound Rotor	150	150	150
Direct Current	150	150	150

*Includes time-delay "D" fuses referred to in *Rule 14–200*.
Notes: (1) Synchronous motors of the low-torque low-speed type (usually 450 rpm, or lower) such as are used to drive reciprocating compressors, pumps, etc., and which start up unloaded, do not require a fuse rating or circuit-breaker setting in excess of 200% of full-load current.
(2) For the use of instantaneous trip (magnetic only, circuit interrupters in motor branch circuits see *Rule 28–210*.

Fig. 17–8   *Table 29* of the *C.E.C., Part I.*

The CSA requires that the fuse class be indicated on the fuse (for example, Class R or Class C1).

Plug fuses and all fuses carrying the CSA Class listings are tested on ac circuits and are marked for ac use. When fuses are to be used on dc systems, the electrician should consult the fuse manufacturer since it may be necessary to reduce the fuse voltage rating and the interrupting rating to ensure safe operation.

The variables in the physical appearance of fuses include length, ferrule diameter, and blade size (length, width, and thickness), as well as other distinctive features. For these reasons, it is difficult to insert a fuse of a given ampere rating into a fuseholder rated for less amperage than the fuse. The differences in fuse construction also make it difficult to insert a fuse of a given voltage into a fuseholder with a higher voltage rating. Fig. 17–9 indicates several examples of the method of ensuring that fuses and fuseholders are not mismatched.

Fuseholders for current-limiting fuses should be designed so that they cannot accept fuses that are noncurrent limiting, Fig. 17–10.

In general, fuses may be used at system voltages that are less than the voltage rating of the fuse. For example, a 600-volt fuse combined with a 600-volt switch may be used on 575-volt, 277/480-volt, 120/208-volt, 240-volt, 50-volt, and 32-volt systems. Fuses that have silver links perform the best, followed by fuses with copper links, followed by fuses with zinc links.

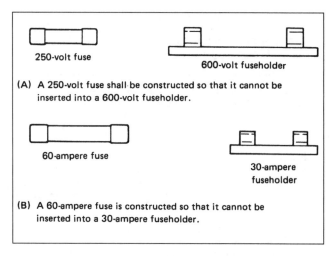

Fig. 17–9   Examples of fuse construction ensuring that fuses and fuseholders are not mismatched.

P- and D-Type Fuses

P- and D-type fuses are standard fuses that have low melting point characteristics. P-type fuses are nontime-delay; D-type fuses are time-delay. They have been designed primarily to offer better protection for residential applications. This type of fuse provides overload and short-circuit protection and has low melting point characteristics. This means that the fuse opens in an ambient temperature of 200°C within 24 hours with no current flowing. These fuses are not current limiting.

Fuse Standards

Fuses that are made and sold in Canada are required to meet CSA standards. As a result of the North American free trade agreement fuse standards in Canada have been harmonized with those of the USA and Mexico. The table below identifies current Canadian and American standards for fuses and their old CSA designations.

North American Low Voltage Fuse Standard

New Standard Number			
U.S.A.	Canada	Subject	Old Canadian Designation
UL 248-1	C22.2 No. 248.1	General Requirements[1]	
UL 248-2	C22.2 No. 248.2	Class C	HRCII-C
UL 248-3	C22.2 No. 248.3	Class CA and CB	HRCI-CA and CB
UL 248-4	C22.2 No. 248.4	Class CC	HRCI- CC
UL 248-5	C22.2 No. 248.5	Class G	SUPPLEMENTAL
UL 248-6	C22.2 No. 248.6	Class H Non-renewable	STANDARD
UL 248-7	C22.2 No. 248.7	Class H Renewable	STANDARD
UL 248-8	C22.2 No. 248.8	Class J	HRCI-J
UL 248-9	C22.2 No. 248.9	Class K	STANDARD
UL 248-10	C22.2 No. 248.10	Class L	HRC-L
UL 248-11	C22.2 No. 248.11	Plug	STANDARD
UL 248-12	C22.2 No. 248.12	Class RK1 and RK5	HRCI-R
UL 248-13	C22.2 No. 248.13	Semiconductor	SEMICONDUCTOR
UL 248-14	C22.2 No. 248.14	Supplemental	HRCI-MISC
UL 248-15	C22.2 No. 248.15	Class T	HRCI-T
UL 248-16	C22.2 No. 248.16	Test Limiters	TEST LIMITERS

Notes:
1. Part No. 1 of the standard covers the general requirements for all fuses such as ratings, markings, and testing. Additional parts describe the dimensions and characteristics of each fuse class.
2. Class H fuses were previously known as standard or code fuses.

(Courtesy of Gould Shawmut)

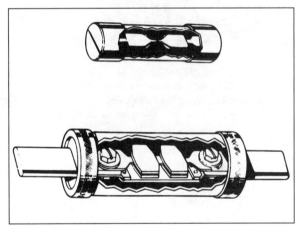

Fig. 17–10 Examples of *Rule 14–210* requirements.

Fig. 17–11 Class H cartridge fuse. Illustration shows renewable-type fuse in which the blown link may be replaced. (Courtesy of Bussmann, Cooper Industries.)

Standard Code Fuses

In Canada, standard (code) fuses have included the current Class H and Class K. Most low-cost renewable and non-renewable fuses are Class H. If you have a fusible disconnect in your house, it will probably contain Class H fuses. Class H fuses have two voltage ratings (250 V and 600 V) and are available in current ratings from 1/10 A to 600 A. They have an interrupting rating of 10 000 A. If a fuse does not have its class marked on the label, it will be Class H fuse.

Class K fuses

Class K fuses have the same physical dimensions as Class H fuses. They have higher interrupting ratings than Class H (K5 has a 50 000 A interrupting rating) and are available in the same

(A) One-time, single-element, noncurrent-limiting fuses. 1/8–60 amperes Class K5—50 000 amperes interrupting rating. 250- and 600-volt ratings. 70–600 amperes Class H—10 000 amperes interrupting rating. 250- and 600-volt ratings. (Courtesy Bussmann, Cooper Industries.)

(B) Limitron, single-element, current-limiting, fast-acting fuses. 1–600 amperes Class RK1—200 000 amperes interrupting rating. 250- and 600-volt ratings. (Courtesy Bussmann, Cooper Industries.)

(C) Fusetron dual-element, time-delay, current-limiting fuses. 1/10–600 amperes Class RK5—200 000 amperes interrupting rating. 250- and 600-volt ratings. (Courtesy Bussmann, Cooper Industries.)

(D) Low-peak, dual-element, time-delay, current-limiting fuses. 1/10–600 amperes Class RK1—300 000 amperes interrupting rating. 250- and 600-volt ratings. (Courtesy Bussmann, Cooper Industries.)

Fig. 17–12 Class H, K5, RK1, and RK5 fuses.

voltage and current rating as Class H fuses. Class K is an older UL class fuse that has been supplanted by the newer Class R. Both Class H and Class K fuses are available in time delay and non-time delay versions.

HRC Fuses (High Rupturing Capacity)

These fuses were developed to protect against fault currents higher than 10 000 amperes. These fuses provide protection for a minimum of 200 000 amperes RMS symmetrical and are able to clear a high short-circuit fault current before it reaches its maximum available peak value. The fault current is reduced to zero in a specified time period. This gives these fuses current-limiting characteristics. Within the HRC type of fuse there are various classes and types of fuses, each with specific characteristics that make the fuse suitable for specific applications. The two large categories are HRCI (formerly called HRC-I or Form I) and HRCII (formerly called HRC-II or Form II).

HRCI fuses provide both overload and overcurrent (short-circuit) protection.

HRCII fuses *only* provide overcurrent (short-circuit) protection.

Another term often referred to on fuses is *time delay*. If a fuse is marked "time delay" it must carry 500% of its ampere rating for 10 seconds minimum. In view of the fact that some manufacturers have had difficulty in meeting the 10-second delay requirement, the CSA standard has been revised to allow for an 8-second delay on 250-volt 15- to 30-ampere rated fuses only. Some manufacturers mark their fuses with phrases such as "dual element," "delay fuse," "time lag," or "super lag," but these manufacturers' terms do not mean that the fuses are a time-delay type.

HRCI-R Fuses (Class R)

A higher quality nonrenewable one-time fuse is also available, called an HRCI-R fuse. This class of fuses is divided into Class RK1 or Class RK5 in the United States. CSA only recognizes the higher Class RK5 level of protection, therefore caution should be exercised when replacing HRCI-R fuses. They are available in two types of overload performance characteristics; time-delay and nontime-delay. They have a 200 000-ampere interrupting rating. The peak let-through current (I_p) and the total clearing energy (I^2t) values are specified for this class of fuse. It is easy to identify this better grade of fuse because the HRCI-R mark and its interrupting rating are marked on the label.

The ferrule-type HRCI-R fuse has a rating range of $\frac{1}{10}$ to 60 amperes and can be distinguished by the annular ring on one end of the case, Fig. 17–13(A). The knife blade-type HRCI-R fuse has a rating range of 70 to 600 amperes and has a slot in the blade on one end, Fig. 17–13(B). When a fuseholder is designed to accept an HRCI-R fuse, it will be impossible to install a standard Class H or Code fuse (these fuses do not have the annular ring or slot of the HRCI-R fuse). However, the HRCI-R fuse can be installed in older-style fuse clips on existing installations. As a result, the Class R fuse may be called a *one-way rejection fuse*.

Electrical equipment manufacturers will provide the necessary rejection-type fuseholders in their equipment, which is then tested with an HRCI-R fuse at short-circuit current values of 200 000 amperes.

A Class R fuse will have a 200 000 ampere interrupting rating only if installed in a similarly rated fuseholder. Such fuseholders have the rejection feature so that only HRCI-R fuses can be installed. If Class R fuses are installed in fuseholders without the rejection feature, the assembly will have an interrupting rating of only 10 000 amperes.

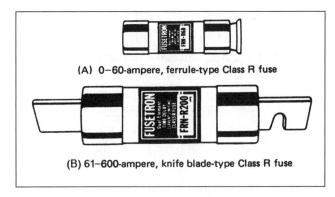

(A) 0–60-ampere, ferrule-type Class R fuse

(B) 61–600-ampere, knife blade-type Class R fuse

Fig. 17–13 Class R cartridge fuses (may be RK1 or RK5).

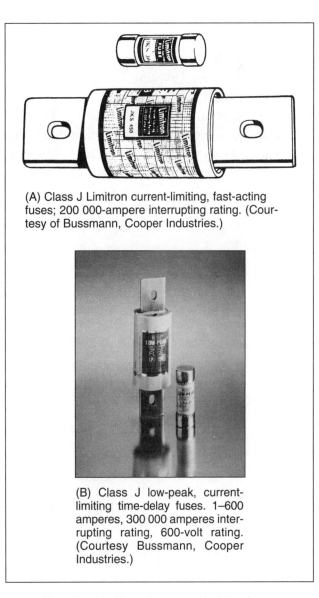

(A) Class J Limitron current-limiting, fast-acting fuses; 200 000-ampere interrupting rating. (Courtesy of Bussmann, Cooper Industries.)

(B) Class J low-peak, current-limiting time-delay fuses. 1–600 amperes, 300 000 amperes interrupting rating, 600-volt rating. (Courtesy Bussmann, Cooper Industries.)

Fig. 17–14 Class J current-limiting fuses.

HRCI-J (Class J). HRCI-J fuses are current limiting and are so marked, Fig. 17–14. They are approved by CSA with an interrupting rating of 200 000 RMS symmetrical amperes. HRCI-J fuses are physically smaller than standard Code fuses. Therefore, when a fuseholder is installed to accept a HRCI-J fuse, it will be impossible to install a Code fuse (Class H) in the fuseholder, *Rule 14–210*.

Both fast-acting, current-limiting HRCI-J fuses and time-delay, current-limiting HRCI-J fuses are available in ratings ranging from 1 to 600 amperes at 600 volts ac.

HRCII-C (Class C). HRCII-C fuses are designed to meet British Standard dimensions. They are different from HRCI-J and HRCI-R fuses in that they are intended to be bolted directly to bus bars or installed in dead-front fuseholders. The current-limiting levels are the same as those of HRCI-R fuses, with limits not to exceed maximum values of I^2t and I_p. Form II overload characteristic is neither time-delay nor nontime-delay, but is specified to provide short-circuit protection for motor circuits that already have overload protection, *Rule 14–212(c)*.

These fuses have ratings from 2 to 600 amperes at 600 volts or less, with a 200 000 A.I.R.. See Fig. 17–14A.

HRCI-L (Class L). HRCI-L fuses, Fig. 17–15 parts A, B, and C, are listed by the CSA in sizes ranging from 601 to 6000 amperes at 600 volts ac. These fuses have specified maximum values of I^2t and I_p. They are current-limiting fuses and have an interrupting rating of 200 000 RMS symmetrical amperes (A.I.R.). These bolt-type fuses are used in bolted pressure contact switches and are not interchangeable with any other fuse type. HRCI-L fuses are available in both a fast-acting, current-limiting type and a time-delay, current-limiting type. Both types of HRCI-L fuses meet CSA requirements.

HRCI-T (Class T). HRCI-T fuses, Fig. 17–16, are current-limiting fuses and are so marked. These fuses are the same as the UL listed HRCI-T fuse used in the United States. They have an interrupting capacity of 200 000 RMS symmetrical amperes. HRCI-T fuses are physically smaller than HRCI-J fuses. The configuration of this type of fuse limits its use to fuseholders and switches that will reject all other types of fuses.

HRCI-T fuses rated 600 volts have electrical characteristics similar to those of HRCI-J fuses and are tested in a similar manner by CSA and UL. HRCI-T fuses rated at 300 volts have electri-

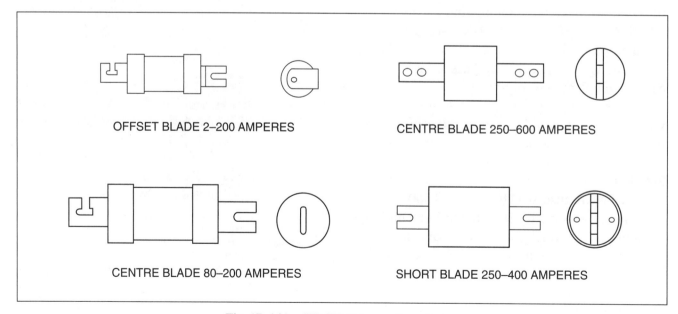

Fig. 17–14A HRCII-C low-voltage fuses.

(A) Limitron current-limiting, fast-acting Class L fuse; 200 000-ampere interrupting rating. Links are made of silver. Has very little time delay. Good for protection of circuit breakers and on circuits that *do not* have high inrush loads (such as motors, transformers). Size at 300% on motor circuits and other higher inrush loads. (Courtesy of Bussmann, Cooper Industries.)

(B) Low-peak, current-limiting, time-delay multi-element fuse. 601–6000 amperes Class L—300 000 amperes interrupting rating. 600-volt rating. Links made of silver. Will hold 500% of rated current for a minimum of 4 seconds. Good for use on high inrush circuits (motors, transformers, and other inductive loads). The best choice for overall protection. Generally sized at 150–225% for motors and other high inrush loads. (Courtesy Bussmann, Cooper Industries.)

(C) Limitron current-limiting, time-delay Class L fuse; 200 000-ampere interrupting rating. Links are generally made of copper. Will hold 500% of rated current for a minimum of 10 seconds. Good for use on high inrush circuits (such as motors, transformers), but is the least current-limiting of Class L fuses. (Courtesy of Bussmann, Cooper Industries.)

Fig. 17–15 Class L fuses. All Class L fuses are rated 600 volts. Listed is 601–6000 ampere ratings. The smallest switch for Class L fuses is 800 amperes. Class L fuses that have fuse elements rated at 600 amperes and less are available. These special ampere-rated fuses are physically the same size as the 800-ampere size.

cal characteristics superior to those of Class J fuses in that their peak let-through currents, as well as the I^2t values, are much lower than those of Class J fuses. The CSA standard C22.2 No. 106–M1985 is similar to the UL standard for HRCI-T (Class T) fuses.

CSA presently lists the 600-volt HRCI-T fuses in sizes from 0 to 800 amperes. CSA lists the 300-volt HRCI-T fuses in sizes from 1 to 800 amperes. The 300-volt HRCI-T fuses may be installed on single-phase line to neutral circuits derived from a three-phase, four-wire solidly grounded system when the line-to-neutral voltage does not to exceed 300 volts. On a 277/480-volt system, the line-to-line voltage is 480 volts, and the line-to-neutral voltage is 277 volts. On a 347/600-volt system, the 300-volt HRCI-T fuse would not be permitted; a 600-volt HRCI-T-rated fuse would be required.

HRCI-T fuses may be used to protect mains, feeders, branch circuits, and circuit breakers when the equipment is designed to accept this type of fuse.

Supplemental Fuses. These fuses are referred to as midget fuses and are cartridge fuses with small physical dimensions of 10 mm × 38 mm ($\frac{13}{32}$ in × $1\frac{1}{2}$ in). They are used on circuits up to

Fig. 17–16 Class T current-limiting, fast-acting fuse; 200 000-ampere interrupting rating. Links are made of silver. Has very little time delay. Good for the protection of circuit breakers and on circuits that *do not* have high inrush loads. Size at 300% for motors and other high inrush loads.

HRCI-MISC (Class CC). HRCI-MISC fuses are primarily used for the protection of motor control circuits, ballasts, and small transformers in industrial applications. HRCI-MISC fuses are also referred to as Class CC fuses and are rated at 600 volts or less and have a 200 000-ampere interrupting rating in sizes from $\frac{1}{10}$ ampere through 30 amperes. These fuses measure 10 mm × 38 mm ($\frac{13}{32}$ in × $1\frac{1}{2}$ in) and can be recognized by a "button" on one end of the fuse, Fig. 17–18. This button is unique to HRCI-MISC Class CC fuses. When a fuseblock or fuseholder that has the matching Class CC rejection feature is installed, it is impossible to insert any other fuse type. Only a Class CC fuse will fit into these special fuseblocks and fuseholders. A Class CC fuse can be installed in a standard fuseholder but would have to be derated to an interrupting capacity of 10 000 amperes.

Ampere Rating	Dimensions
0–15	$\frac{13}{32}$ in × $1\frac{5}{16}$ in
16–20	$\frac{13}{32}$ in × $1\frac{13}{32}$ in
21–30	$\frac{13}{32}$ in × $1\frac{5}{8}$ in
31–60	$\frac{13}{32}$ in × $2\frac{1}{4}$ in

Fig. 17–17 Class G fuses.

600 volts. These fuses are similar to UL-listed Class G fuses, Fig. 17–17, and are available in sizes ranging from 0 to 60 amperes. They have an interrupting capacity of 100 000 RMS symmetrical amperes.

These fuses are current limiting. They may be used for the protection of ballasts, electric heaters, small motors, and motor control transformers. They are also available in a fast-acting version used for semiconductor protection.

Plug Fuses (*Rules 14–202* to *14–208*)

Rules 14–202 through *14–208* cover the requirements for plug fuses. The electrician will be most concerned with the following requirements for plug fuses, fuseholders, and adapters:

- They shall not be used in circuits exceeding 125 volts between conductors, except on systems having a grounded neutral with no conductor having more than 150 volts to ground. This situation is found in the 120/208-volt system in the commercial

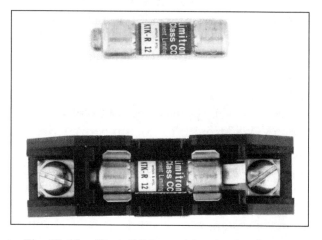

Fig. 17–18 Class CC fuse with rejection feature.

building covered in this text, or in the case of a 120/240-volt, single-phase system.

- They shall have ampere ratings of 0 to 30 amperes.
- The screw shell must be connected to the load side of the circuit.
- All plug fuses must be installed such that they cannot be interchanged with a fuse of a higher ampacity.
- Type S plug fuses are classified 0 to 15 amperes, 16 to 20 amperes, and 21 to 30 amperes.

Prior to the installation of Type S fuses, the electrician must determine the ampere rating of the various circuits. An adapter of the proper size is then inserted into the Edison-base fuseholder; only the proper size of Type S fuse can be inserted into the fuseholder. The adapter makes the fuseholder nontamperable and noninterchangeable. For example, after a 15-ampere adapter is inserted into a fuseholder for a 15-ampere circuit, it is impossible to insert a Type S or an Edison-base fuse having a larger ampere rating into this fuseholder without removing the 15-ampere adapter. The adapters are designed so that they are extremely difficult to remove.

Type S fuses and suitable adapters, Fig. 17–19, are available in a large selection of ampere ratings ranging from 3/10 ampere to 30 amperes. When a Type S fuse has dual elements and a time-delay feature, it does not blow unnecessarily under momentary overloads, such as the current surge caused by the startup of an electric motor. On heavy overloads or short circuits, this type of fuse opens very rapidly.

Standard Edison base plug fuses may also be used along with a fuse rejector. A fuse rejector is a plastic disc with a hole in the middle which corresponds to the diameter of the nipple on the bottom of the plug fuse. This disc is screwed into the bottom of the fuse holder. Because the nipple on the bottom of the plug fuse increases in diameter with an increase in ampacity, it is impossible to put a higher ampacity fuse into a fuseholder with a lower ampacity fuse rejector.

Tables 17–1, 17–2, and 17–3 provide ratings and application information for the fuses supplied by various manufacturers. The tables indicate the Class, voltage rating, current range, interrupting ratings, and typical applications for the various fuses. Tables of this type may be used to select the proper fuse to meet the needs of a given situation.

Special Interrupting Ratings

Some HRCI-R, HRCI-J, and HRCI-L fuses have passed special tests that qualify them to be marked "Interrupting rating 300 000 amperes." This information is found on the label of the fuse. In addition, these fuses will have the letter designation "SP." However, CSA presently only recognizes fuseholders rated up to 200 000 amperes interrupting capacity.

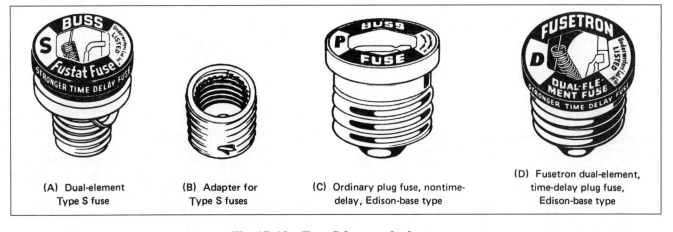

(A) Dual-element Type S fuse
(B) Adapter for Type S fuses
(C) Ordinary plug fuse, nontime-delay, Edison-base type
(D) Fusetron dual-element, time-delay plug fuse, Edison-base type

Fig. 17–19 Type S fuses and adapter.

Gould Shawmut Fuse
By manufacturers' type reference or series number.

Cross Reference Guide
Ampere ratings must be added for ordering purposes.

Fuse		Volt	Gould Shawmut OLD	Gould Shawmut CANADA	Gould Shawmut U.S.A.	Buss	Littelfuse	Reliance	Brush	Dorman Smith	English Electric	Aeroflex
HRC (Current Limiting)												
Class R (UL Class RK5)	Time Delay	250	CRN-R	TRNR	TR-R	FRN-R	FLN-R	ECNR				HB*
		600	CRS-R	TRSR	TRS-R	FRS-R	FLS-R	ESNR				HA*
	Non Time Delay	250		FT-R								
		600		FTS-R								
Class R (UL Class RK1)	Time Delay	250		A2D-R		LPN-RK	LLN-RK					
		600		A6D-R		LPS-RK	LLS-RK					
	Non Time Delay	250	HNR	A2K-R		KTN-R	KLN-R	NCLR	KN	RHN	C-HG	HB
		600	HSR	A6K-R		KTS-R	KLS-R	SCLR	KS	RHS	C-HR	HA
Class J	Time Delay	600		AJT		JHC						JA*
	Non Time Delay	600		CJ	A4J	JKS	JLS	JCL	CJ	SJ	C-J	
Class L	Time Delay	600		A4BT		KLU	KLLU	LCU				
	Non Time Delay	600		CL	A4BY	KTU / KRP-C	KLP-C	LCL	L		C-L	L8,L12, L16,L20
Class C	Offset Blade	600	ES	FES		CGL			H07C	AAO	CIA	932
			ES	FES		CGL			K07C	BAO	CIS	933
			ES	FES		CGL			L14C	CEO	CCP	944
			ES	FES		—			M14C	DEO	CFP	945
	Centre Blade		ESC	FESC		—			L09C	CD	CC	964
			ESC	FESC		CGL			M09C	DD	CF	965
			ES	FESC		CGL			N11C/P11C	EF	CM	976
			ES	FESC		CGL			R11C	FF	CLM	977
			ESF	FESF		—			N09C/P09C	ED	CMF	966
UL Class CC				ATMR		KTK-R	KLK-R	MCL-R				
Type K		600		ESK						CIH/CIK/CIL	C-K	
Miniature				MS						CIF21	C-N	

*Note a true Time Delay fuse — for motor applications (full voltage start) substitute a Time Delay fuse rated 150-175% motor full load current.

Standard (Code)

Fuse	Volt	Gould Shawmut OLD	Gould Shawmut CANADA	Gould Shawmut U.S.A.	Buss	Littelfuse	Reliance
Super One Time (50 K.A.I.R.) (UL Class K5)	250	SPC-N	OTN*	OT			
	600	SPC-S	OTS	OTS			
One Time	250	OT-/250	NRN*		NON	NLN	EON
	600	OT-/600	NRS		NOS	NLS	EOS
Renewable	250	R-/250	RFN	RF	REN	RLN	ERN
	600	R-/600	RFS	RFS	RES	RLS	ERS
Renewable Links	250	RL-/250	RLN	RL	LKN	LKN	ELN
	600	RL-/600	RLS	RLS	LKS	LKS	ELS
Time Delay	250		CRN				
	600		CRS				

*OTN15-60 & NRN 15-60 are Type CSA-"P"

Semiconductor (Rectifier)

Gould Shawmut	Volt	Buss	Reliance	Brush/Int. Rect.
A13X	130	FWA/KAA	RFA	SF13X
A25X	250	FWX/KAX	REN/RFN	SF25X
A50P	500	FWH/KBH	RFV	SF50P XL50P
A60X	600	KBC	RFS*	SF60X
A70P	700	FWP KBP KAC*	—	SF70P XL70P
A70Q	700	KAC*	—	SF60C*
A100P	1000	—	RFK	SF100P

*Mounting dimensions should be verified

Supplemental (Midget) $\frac{13}{32}$ in × $1\frac{1}{2}$ in

Gould Shawmut	Buss	Littelfuse	Brush/Reliance
ATM	KTK	KLK	MCL
ATQ	FNQ	FLQ	MEQ
OTM	BAF/BAN	BLF/BLN	MOF/MOL
TRM	FNM	FLM	MEN
GGU	AGU	511/512	—
GFN	FNA	FLA	MID

Supplemental (Miniature)

Gould	Buss	Littelfuse	Gould	Buss	Littelfuse
GAB	ABC	314	GGL	GLH	312***
GDL	MDL	313	GGM	GMA	212
GDV	MDV	315	GGX	AGX	361/362
GGC	AGC	312*	GTH	MTH	312**
GGJ	GJV	318	CT	HBO	312**

NOTE: Fuse characteristics can vary between manufacturers and should be checked for proper co-ordiantion as required.

*Up to 3A, **4-6A, ***7-10A

Table 17–1 Cross-reference fuse guide. (Courtesy of Gould Shawmut.)

WHICH FUSE? — A "quick guide" for Low Voltage distribution applications 600A and less.

Fuse Clips or Contacts	Fuse Type		Feeders	Motors	Trans-formers	Breakers & Panelboards	Electric Heating		Capacitors	Welders
							Industrial	Residential		
Class J 600V or less	HRCI-J	Time Delay	AJT	AJT	AJT	AJT	AJT		AJT	
		Non Time Delay	CJ	*		CJ	CJ		CJ	
Class R 250 & 600V	HRCI-R (UL Class RK1)	Time Delay	A2D-R A6D-R	A2D-R A6D-R	A2D-R A6D-R		A2D-R A6D-R			
		Non Time Delay	HNR HSR	*			HNR HSR			
	HRCI-R (UL Class RK5)	Time Delay	TRNR TRSR	TRNR TRSR	TRNR TRSR		TRNR TRSR			TRNR TRSR
		Non Time Delay	FT-R FTS-R	*			FT-R FTS-R			
Class C 600V or less	HRCII-C	Form II		FES FESC FESF					FES FESC FESF	
Standard (Code)	10-kA Time Delay Type D 250 & 600V		CRN CRS	CRN CRS	CRN CRS		CRN CRS	CRN —		CRN CRS
NOTE: Class RK1 and RK5 fuses may also be used in Standard Clips.	50 kA Super One Time 250 & 600V		OTN OTS†	*			OTN	OTN		
	10kA One Time 250 & 600V		NRN NRS†				Type P Ratings only 250V, 15-60A NRN	NRN		
	10 kA Renewable 250 & 600V		RFN RFS†							

* Time Delay usually preferred for easier application and lower ratings.
† For Non-Cycling loads only. Refer also to *C.E.C., Rule 26-000* re cable ampacities.
Voltage Ratings — where two catalogue numbers are shown above, the upper is 250 Volts A.C. and the lower is 600 Volts A.C.
NOTE: When upgrading a Standard (Code) fuse installation where One Time or Renewable fuses have been rated at 300% of the motor FLC, it is sometimes more practical to substitute Non Time Delay HRCI-R fuses, rating for rating, rather than downsizing the clips to accommodate Time Delay fuses at 175% maximum of the motor FLC.

Table 17–2 A quick guide for low-voltage distribution applications. (Courtesy of Gould Shawmut.)

Some MAXIMUM permitted ratings for Overcurrent Protection (*C.E.C.*, 2002)

Application Low Voltage Distribution	Feeders	Motors Full Voltage Start		Transformers and secondary protection				Electric Heating	Capacitors	Welders	
		Time Delay	Non T-Delay	Dry Type	Liquid Filled—primary rated at					Trans Arc	Resistance
					9A and over	2A up to 9A	less than 2A				
HRCI	100%	175%	300%	125%	*150%	*167%	*300%	125%	250%	200%	300%
Standard Code	100%	175%	†300%	125%	*150%	*167%	*300%	125% (P&D types only)	—	200% (D type only)	300% (D type only)
HRCII-C Rule 14-212(c)	—	—	255%	—	†127%	†142%	†255%	†106%	212%	†170%	†255%
Fuse Rating as a percent of	Cable Amps	Motor Full Load Amps		Rated Primary Amps				Cable Amps	Capacitor Amps	Rated Primary Amps	
C.E.C., Rule	14-104	28-200		26-256	26-254			62-114	26-210	42-008	42-016

*Usually requires Time Delay fuse to prevent nuisance blowing on transformer switching current. Small transformers with very high inrush currents may need HRCI-R or Standard Code Time Delay fuses.
**Total connected load must not exceed 80% of the fuse rating.
†Not recommended for these applications.

Table 17–3 Maximum permitted ratings for overcurrent protection. (Courtesy of Gould Shawmut.)

TESTING FUSES

WHEN POWER IS TURNED ON. On live circuits, extreme caution must be exercised when checking fuses. There are many different voltage readings that can be taken, such as line-to-line, line-to-ground, line-to-neutral, etc.

Using a voltmeter, the first step is to be sure to set the scale to its highest voltage setting, then change to a lower scale after you are sure the voltage is within the range of the voltmeter. For example, when testing what you believe to be a 120-volt circuit, it is wise to first use the 600-volt scale, then try the 300-volt scale, and then use the 150-volt scale—**just to be sure!**

Taking a voltage reading across the bottom (load side) of fuses—either fuse-to-fuse, fuse-to-neutral, or fuse-to-ground—can show full voltage even though a fuse might have opened, because there can be "feedback" through the load. You could come to a wrong conclusion. Taking a voltage reading from the line side of a fuse to the load side of a fuse will show open-circuit voltage if the fuse has "blown" and the load is still connected. This also can result in a wrong conclusion.

Reading from the line side to the load side of a good fuse should show zero voltage or else an extremely small voltage.

Always carefully read the instructions furnished with electrical test equipment such as voltmeters, ohmmeters, etc.

WHEN POWER IS TURNED OFF. This is the safest way to test fuses. Remove the fuse from the switch, then take a resistance reading across the fuse using an ohmmeter. A good fuse will show zero to very minimal resistance. An open (blown) fuse will generally show a very high resistance reading.

Cable Protectors

Cable protectors, sometimes referred to as cable limiters, are used quite often in commercial and industrial installations where parallel cables are used on service entrances and feeders.

Cable protectors differ in purpose from fuses, which are used for overload protection. Cable protectors are short-circuit devices that can *isolate* a faulted cable, rather than having the fault open the entire phase. They are selected on the basis of conductor size; for example, a 500-kcmil cable protector would be used on a 500-kcmil conductor.

Cable protectors are available for cable-to-cable or cable-to-bus installation, for either aluminum or copper conductors. The type of cable protector illustrated in Fig. 17–20 is for use with a 500-kcmil copper conductor, and is rated at 600 volts with an interrupting rating of 200 000 amperes.

Fig. 17–21 is an example of how cable protectors may be installed on the service of the commercial building. A cable protector is installed at each end of each 500-kcmil conductor. Three conductors per phase terminate in the main switchboard where the service is then split into two 800-ampere bolted pressure switches.

Figure 17–22 shows three typical cable protector locations for the main service and Fig. 17–23 shows several typical cable protector types that may be used for various applications.

Fig. 17–20 Cable protector for 500-kcmil copper conductor.

Fig. 17–24 explains how cable protectors are used where more than one customer is connected to the same transformer.

Cable protectors are installed in the "hot" phase conductors only. They are not installed in the grounded neutral conductor.

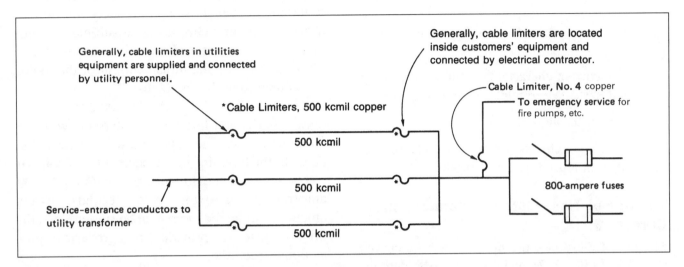

Fig. 17–21 Use of cable protectors.

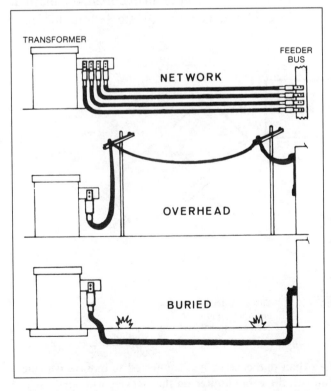

Fig. 17–22 Examples of typical cable protector locations. (Courtesy of Gould Shawmut.)

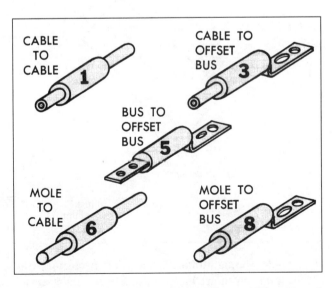

Fig. 17–23 Types of cable protectors. (Courtesy of Gould Shawmut.)

TIME–CURRENT CHARACTERISTIC CURVES AND PEAK LET-THROUGH CHARTS

The electrician must have a basic amount of information concerning fuses and their application to be able to make the correct choices and decisions for everyday situations that arise on an installation.

The electrician must be able to use the following types of fuse data:

- time–current characteristic curves, including total clearing and minimum melting curves.
- peak let-through charts.

Fuse manufacturers furnish this information for each of the fuse types they produce.

The Use of Time–Current Characteristic Curves

The use of the time–current characteristic curves shown in Figs. 17–25 and 17–26 can be demonstrated by considering a typical problem. Assume that an electrician must select a fuse to protect a motor that is governed by *Section 28*. This section indicates that the running overcurrent protection shall be based on not over 125% of the full-load running current of the motor. The motor in this example is assumed to have a full-load current of 24 amperes. Therefore, the size of the required protective fuse is determined as follows:

$24 \times 1.25 = 30$ amperes

Now the question must be asked: is this 30-ampere fuse capable of holding the inrush current of the motor (which is approximately four to five times the running current) for a sufficient length of time for the motor to reach its normal speed? For this problem, the inrush current of the motor is assumed to be 100 amperes.

Refer to Fig. 17–25. At the 100-ampere line, draw a horizontal line until it intersects with the 30-ampere fuse line. Then draw a vertical line down to the base line of the chart. At this point, it can be seen that the 30-ampere fuse will hold 100 amperes for approximately 40 seconds. In this amount of time, the motor can be started. In addition, the fuse will provide running overload protection as required.

If the time–current curve for thermal overload relays in a motor controller is checked, it will show that the overload element will open the same current in much less than 40 seconds. Therefore, in the event of an overload, the thermal overload elements will operate before the fuse

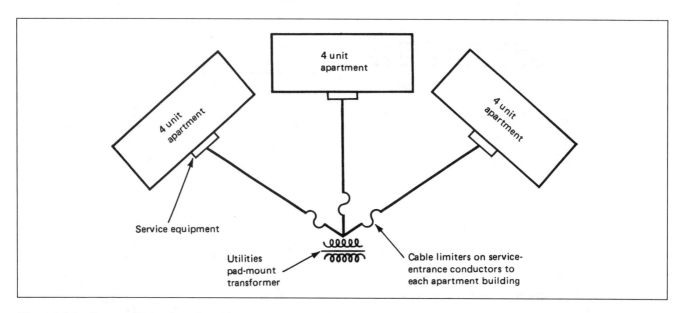

Fig. 17–24 Some utilities install cable protectors where more than one customer is connected to one transformer. Thus, if one customer has problems in the main service equipment, the cable limiter on that service will open, isolating the problem from other customers on the same transformer. This is common for multiunit apartments, condominiums, and shopping centres.

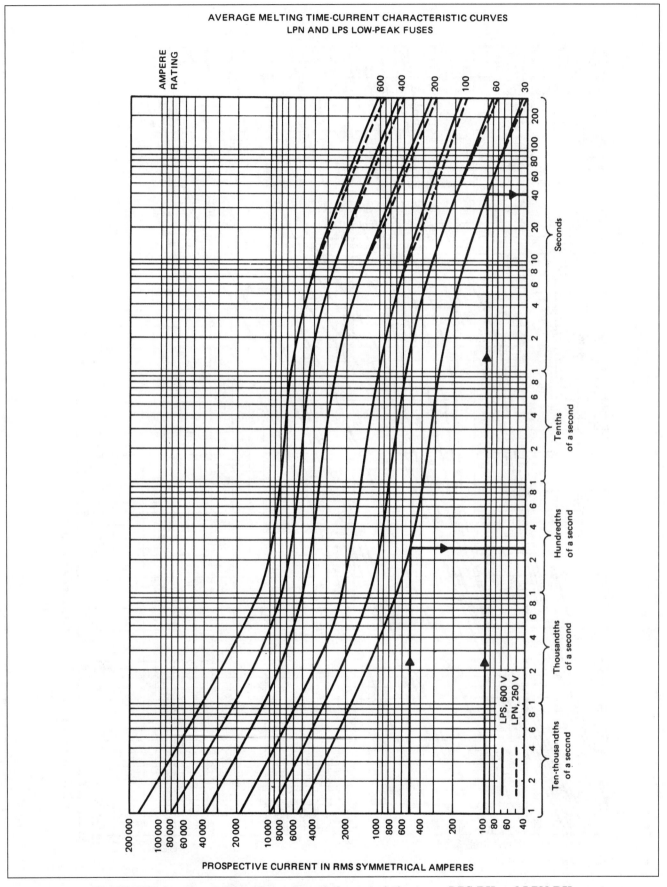

Fig. 17–25 Average melting time–current characteristic curves LPS-RK and LPN-RK.

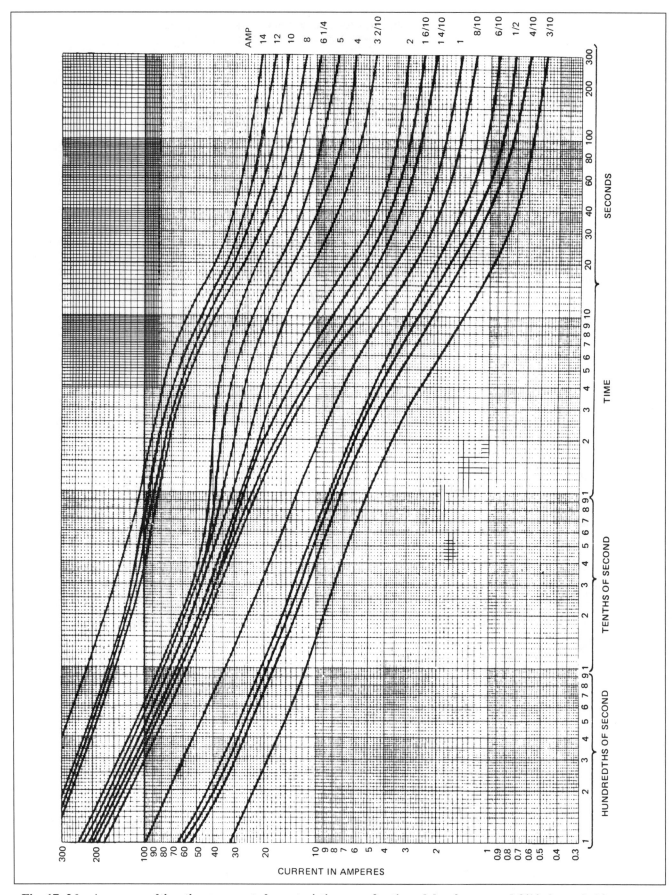

Fig. 17–26 Average melting time–current characteristic curve for time-delay fuses rated 3/10 through 14 amperes.

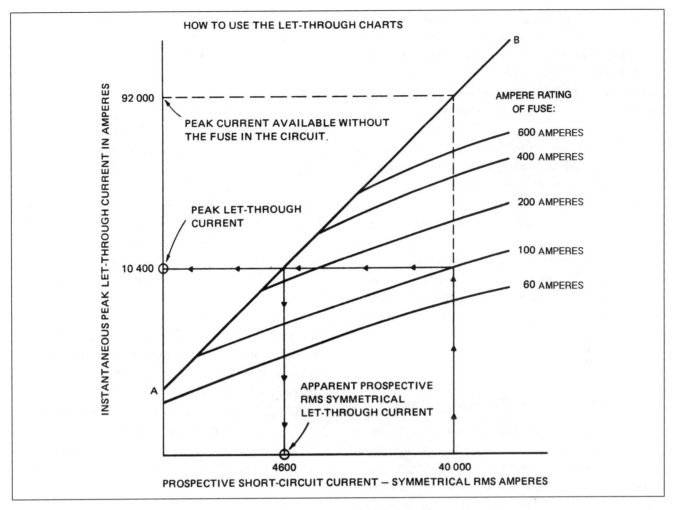

Fig. 17–27 Using the let-through charts to determine peak let-through current and apparent prospective RMS symmetrical let-through current.

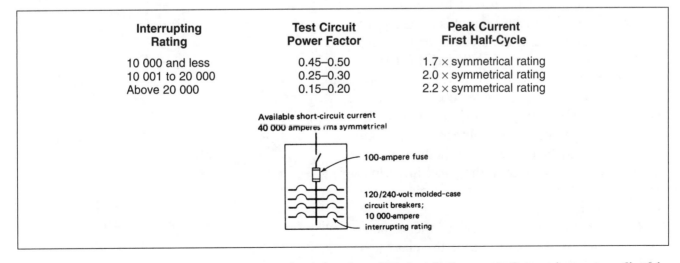

Fig. 17–28 Example of a fuse used to protect circuit breakers. This installation meets the requirements outlined in *Rule 14–014*.

opens. If the overload elements do not open for any reason, or if the contacts of the controller weld together, then the properly sized dual-element fuse will open, reducing the possibility of motor burnout. The above method is a simple way to obtain backup or double motor protection.

Referring again to Fig. 17–25, assume that a short circuit of approximately 500 amperes occurs. Find the 500-ampere line at the left of the chart and then, as before, draw a horizontal line until it intersects with the 30-ampere fuse line. From this intersection point, drop a vertical line to the base line of the chart. This value indicates that the fuse will open the fault in slightly over $\frac{2}{100}$ (0.02) seconds. On a 60-hertz system (60 cycles per second), one cycle equals 0.016 seconds. Therefore, a 30-ampere fuse will clear a 500-ampere fault in just over one cycle.

The Use of Peak Let-Through Charts

The withstand ratings of electrical equipment (such as bus duct, switchboard bracing, controllers, and conductors) and the interrupting ratings of circuit breakers are given in the published standards of the Canadian Standards Association, Underwriters Laboratories, NEMA, CEMA, and the Insulated Cable Engineering Association (ICEA). The withstand ratings may be based either on the peak current (I_p) or on the root mean square (RMS) current.

As an example of the use of the let-through chart in Fig. 17–27, assume that it is necessary to protect moulded case circuit breakers. These breakers have an interrupting rating of 10 000 RMS symmetrical amperes. The available short-circuit current at the panel is 40 000 RMS symmetrical amperes. The fuse protecting the circuit breaker panel is rated at 100 amperes. This 100-ampere fuse may be located either in the panel or at some distance away from the panel in a separate disconnect switch placed at the main service or distribution panel, Fig. 17–28.

Now refer to Fig. 17–27. Find the 40 000-ampere point on the base line of the chart and then draw a vertical line upward until it intersects the 100-ampere fuse curve. Move horizontally to the left until Line A–B is reached. Then move vertically downward to the base line. The value of this point on the base line is 4600 amperes and is the apparent RMS amperes. This means that the current-limiting effect of the fuse, when subjected to a 40 000 RMS symmetrical ampere fault, permits an apparent let-through current of only 4600 RMS symmetrical amperes. Therefore, the 100-ampere fuse selected readily protects the circuit breaker (with a 10 000-ampere interrupting rating) against the 40 000 RMS symmetrical ampere fault. Actual tests conducted by manufacturers have shown that the same fuse can be used to protect a circuit breaker with a 10 000-ampere interrupting rating when 100 000 amperes are available.

In Fig. 17–27, note that at a current of 40 000 RMS symmetrical amperes, the peak current, I_p, available at the first half cycle is 92 000 amperes if there is no fuse in the circuit to limit the let-through current. When the 100-ampere fuse is installed, the instantaneous peak let-through current, I_p, is limited to just 10 400 amperes.

The magnetic stresses that occur in the example given can be compared. When the current-lim-

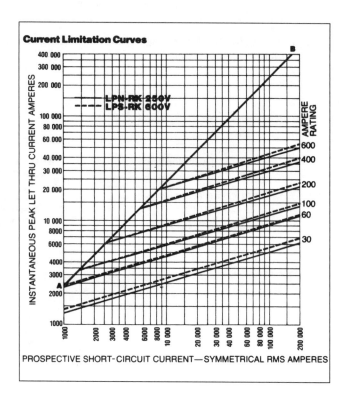

Fig. 17–29 Current-limiting effect of low-peak, dual-element, time-delay Class RK1 fuses.

iting fuse is installed, the magnetic stresses are only 1/80 of the stresses present when a noncurrent-limiting overcurrent device is used to protect the circuit.

Fig. 17–29 shows a peak let-through chart for a family of dual-element fuses. Fuse manufacturers provide peak let-through charts for the various sizes and types of fuses they produce.

CSA Standard C22.2 No. 5.1-M91 describes the testing of moulded case circuit breakers. One of the tests performed on these circuit breakers is a short-circuit test. The test circuit is calibrated for a given amount of fault current at a specific power factor. The lower the power factor, the higher will be the first half-cycle peak current.

Presently, there are three basic levels of short-circuit tests for moulded case circuit breakers.

Knowing how to use peak let-through charts to determine both instantaneous peak let-through current and apparent RMS let-through current of a current-limiting fuse enables the user to select current-limiting fuses that will protect the downstream circuit breakers. Manufacturers of circuit breakers also provide technical data that show fuse and breaker combinations for use at various available fault-current levels.

Some inspection authorities will no longer accept the use of the peak let-through charts. Only the manufacturers' technical data for series-tested combinations are accepted in some jurisdictions.

Peak let-through charts are also available for current-limiting circuit breakers. The current-limiting effect of a circuit breaker is different from that of a current-limiting fuse. The current-limiting effect of a fuse is much greater than that of a breaker having the same ampere rating. Therefore, refer to fuse peak let-through charts when applying current-limiting fuses. Use breaker peak let-through charts when applying current-limiting breakers.

Fig. 17–30 shows how a manufacturer might indicate the maximum size and type of fuse that will protect circuit breakers against fault currents that exceed the breakers' interrupting rating. Circuit breakers are tested according to the requirements of *CSA Standard C22.2 No. 5.1-M91*. Panelboards are tested according to the requirements of *CSA Standard C22.2 No. 29-M89*. Motor controllers are tested to *CSA Standard C22.2 No. 14-M91*. Switchboards are tested to *CSA Standard C22.2 No.31-M89*.

Therefore, to protect circuit breakers for fault levels that exceed their interrupting rating, refer to the manufacturers' data, or use acceptable engineering practices for the selection of current-limiting overcurrent devices as discussed in Units 18 and 19 of this text.

CIRCUIT BREAKERS

The *C.E.C., Part I* defines a circuit breaker as a device that is designed to open and close a circuit by nonautomatic means and to open the circuit automatically on a predetermined overcurrent without being damaged itself when properly applied within its rating, *Section 0, Definitions*.

Moulded case circuit breakers are the most common type in use today, Fig. 17–31. The tripping mechanism of this type of breaker is enclosed in a moulded plastic case. The *thermal-magnetic* type of circuit breaker is covered in this unit. Another type is the power circuit breaker, which is larger, of heavier construction, and more costly. Power circuit breakers are used in large industrial applications. However, these breakers perform the same electrical functions as the smaller moulded case breakers.

Circuit breakers are also available with solid-state adjustable trip settings. When compared to the standard moulded case circuit breaker, these breakers are very costly.

Moulded case circuit breakers are approved by CSA testing laboratories and are covered by NEMA and UL standards. *Rules 14–300* through *14–308* state the basic requirements for circuit breakers and are summarized as follows:

- Breakers shall be trip free so that the internal mechanism trips the breaker to the *Off* position even if the handle is held in the *On* position.

- The breaker shall clearly indicate if it is in the *On* or *Off* position.

Pow-R-Line 2: INTEGRATED EQUIPMENT RATINGS

Voltage	Branch Breakers	Max. Fault Level Available (kA sym.)	Main or Upstream Breaker Type				
			100A	150A	225A	400A	600A
208Y/120 and 240	GB, GHB, GBH	100	EDH	EDH	HJD	HKD	HLD
		200	FDC	FDC	JDC	KDC, LCL	—
480Y/277	GHB, GBH	25	FD, HFD	FD, HFD	JD, HJD	KD, HKD	—
		65	HFD	HFD	HJD(1)	HKD(1)	—
		100	FDC	FDC	JDC(1), LCL	KDC(1), LCL	—
		200	FDB+LFD	FDB+LFD	—	—	—
600Y/347	GBH	14	FDB, FD, HFD	FDB, FD, HFD	KD, HKD	KD, HKD	HLD(2)
		18	FD, HFD	FD, HFD	KD, HKD	KD, HKD	HLD(2)
		25	HFD	HFD	—	—	—
		50	FDB+LFD	FDB+LFD	—	—	—

NOTES: 1. Branch Breakers: 1 pole 15-100A, 2,3 pole 15-50A.
2. Branch Breakers: 2 and 3 pole only.

Pow-R-Line 2: INTEGRATED EQUIPMENT RATINGS

Voltage	Branch Breakers	Max. Fault Level Available (kA sym.)	Main or Upstream Fuse Type				
			100A	150A	225A	400A	600A
208Y/120 and 240	GHBS	65	J, T	J, T	J, T	—	—
	1P 15-20A	100	J, T	—	—	—	—
	GB, GHB, GBH	100	J, T, R	J, T, R	J, T, R	J, T	T(1)
		200	J, T, R	J, T	J, T	J, T	—
480Y/277	GHBS	65	J, T	J, T	J, T	—	—
	1P 15-20A	100	J, T	—	—	—	—
	GHB, GBH	100	J, T, R	J, T, R	J, T, R	J, T	T(1)
		2000	J, T, R	J, T	J, T	J, T	—
600Y/347	GBH	100	J	J	J	—	—

NOTES: 1. Branch Breakers: 2 and 3 pole only.

Fig. 17–30 Series-tested equipment ratings (Courtesy of Cutler-Hammer Canada.)

- The breaker shall be nontamperable; that is, it cannot be readjusted (to change its trip point or time required for operation) without dismantling the breaker or breaking the seal.
- The rating shall be durably marked on the breaker. For the smaller breakers with ratings of 100 amperes or less and 600 volts or less, this marking must be moulded, stamped, or etched on the handle or other portion of the breaker that will be visible after the cover of the panel is installed.
- Every breaker having an interrupting rating other than 5000 amperes shall have its interrupting rating shown on the breaker.
- *Rule 30–710(3)* and the CSA require that when a circuit breaker is to be used as a toggle switch for fluorescent lighting, the

Fig. 17–31 Moulded case circuit breakers.

breaker must be CSA approved for switching duty. Circuit breakers are used as toggle switches when it is desired to control 120- or 347-volt lighting circuits (turn them on and off) from the panel instead of installing separate toggle switches. Breakers approved for use as toggle switches will bear the letters "SWD" on the label. Breakers not marked in this manner must not be used as switches.

- A circuit breaker should not be loaded to more than 80% of its current rating for loads over 225 amperes that could be on for more than 3 hours in any 6 hour period, unless the breaker is marked otherwise, such as those breakers approved by the CSA for 100% loading (100% rated), *Rule 8–104(3,4)*.

- If the voltage rating on a circuit breaker is marked with a single voltage (example: 240 volts), the breaker may be used in grounded or ungrounded systems where the voltage between any two conductors does not exceed the breaker's marked voltage rating.

- If the voltage rating on a circuit breaker is marked with a slash voltage (example: 347/600 volts or 120/240 volts), the breaker may be used where the voltage to ground does not exceed the breaker's lower voltage marking, and the voltage between any two conductors does not exceed the breaker's higher voltage marking. This type of breaker is thus restricted to systems having a grounded neutral.

Thermal-Magnetic Circuit Breakers

A thermal-magnetic circuit breaker contains a bimetallic element. On a continuous overload, the bimetallic element moves until it unlatches the inner tripping mechanism of the breaker. Harmless momentary overloads do not move the bimetallic element far enough to trip the breaker. If the overload is heavy, or if a short circuit occurs, then the magnetic mechanism within the circuit breaker trips instantly. The time required for the breaker to open the circuit completely depends upon the magnitude of the fault current and the mechanical condition of the circuit breaker. This time may range from approximately one-half cycle to several cycles.

Circuit breaker manufacturers calibrate and set the tripping characteristic for most moulded case breakers. Breakers are designed so that it is difficult to alter the set tripping point, *Rule 14–304*. For certain types of breakers, however, the trip coil can be changed physically to a different rating. Adjustment provisions are made on some breakers to permit the magnetic trip range to be changed. For example, a breaker rated at 100 amperes may have an external adjustment screw with positions marked HI-MED-LO. The manufacturer's application data for this breaker indicates that the magnetic tripping occurs at 1000 amperes, 750 amperes, or 500 amperes respectively for the indicated positions. These settings usually have a tolerance of ±10%.

The *ambient-compensated* type of circuit breaker is designed so that its tripping point is not affected by an increase in the surrounding temperature. An ambient-compensated breaker has two elements: the first element heats due to the current passing through it and because of the heat of the surrounding air; the second element is affected only by the ambient temperature. These elements act in opposition to each other. In other words, as the tripping element tends to lower its

276 Unit 17 Overcurrent Protection: Fuses and Circuit Breakers

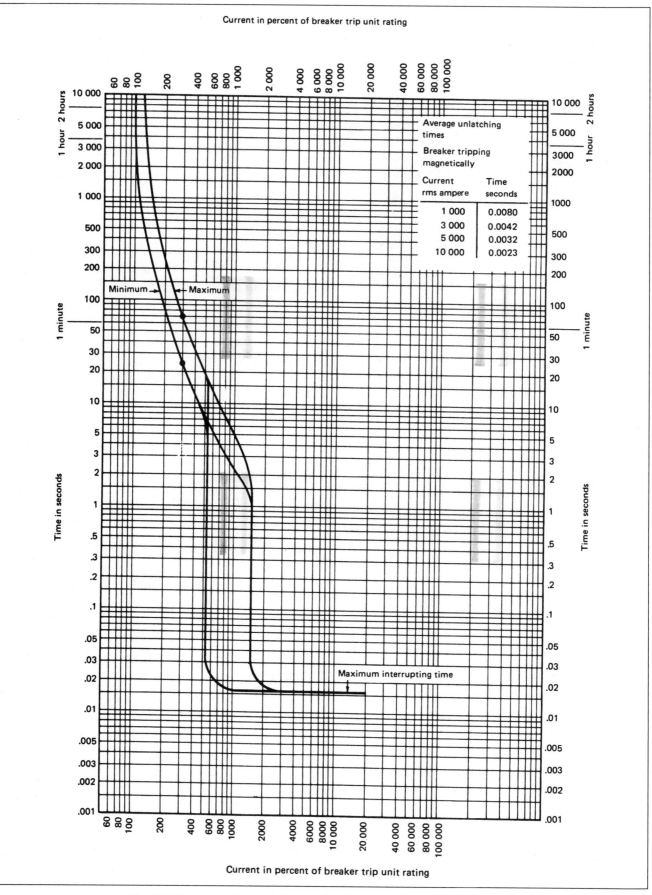

Fig. 17-32 Typical moulded-case circuit-breaker, time-current curve.

tripping point because of a high ambient temperature, the second element exerts an opposing force that stabilizes the tripping point. Therefore, current through the tripping element is the only factor that causes the element to open the circuit.

Factors that can affect the proper operation of a circuit breaker include moisture, dust, vibration, corrosive fumes and vapours, and excessive tripping and switching. As a result, care must be taken when locating and installing circuit breakers and all other electrical equipment.

The interrupting rating of a circuit breaker is marked on the breaker label. The electrician should check the breaker carefully for the interrupting rating since the breaker may have several voltage ratings with a different interrupting rating for each. For example, assume that it is necessary to select a breaker having an interrupting rating at 240 volts of at least 50 000 amperes. A close inspection of the breaker may reveal the following data:

Voltage	Interrupting Rating
240 volts	65 000 amperes
480 volts	25 000 amperes
600 volts	18 000 amperes

Recall that for a fuse, the interrupting rating is marked on the fuse label. This rating is the same for any voltage up to and including the maximum voltage rating of the fuse.

The standard full-load ampere ratings of nonadjustable circuit breakers are the same as those for fuses. As previously mentioned, additional standard ratings of fuses are 1, 3, 6, and 10 amperes.

The time–current characteristics curves for circuit breakers are similar to those for fuses. A typical circuit breaker time–current curve is shown in Fig. 17–32. Note that the current is indicated in percentage values of the breaker trip unit rating. Therefore, according to this graph, the 100-ampere breaker being considered will:

1. carry its 100-ampere (100%) rating indefinitely
2. carry 300 amperes (300%) for a minimum of 25 seconds and a maximum of 70 seconds
3. unlatch its tripping mechanism in 0.0032 seconds (approximately one-quarter cycle) with a current of 5000 amperes
4. interrupt the circuit in a maximum time of 0.016 seconds (one cycle) with a current of 5000 amperes (5000%)

The same time–current curve can be used to determine that a 200-ampere circuit breaker will:

1. carry its 200-ampere (100%) rating indefinitely.
2. carry 600 amperes (300%) for a minimum of 25 seconds and a maximum of 70 seconds.
3. unlatch its tripping mechanism in 0.0032 seconds (approximately one-quarter cycle) with a current of 5000 amperes.
4. interrupt the circuit in a maximum time of 0.016 seconds (one cycle) with a current of 5000 amperes (2500%).

This example shows that if a short circuit in the magnitude of 5000 amperes occurs on a circuit with both the 100-ampere breaker and the 200-ampere breaker installed, they will open together because they have the same unlatching times. In many instances, this action is the reason for otherwise unexplainable power outages (see Unit 19). This is rather common when heavy (high value) fault currents occur on circuits protected with moulded case circuit breakers.

Common Misapplication

A possible violation of *Rules 14–012* and *14–014* is the installation of a main circuit breaker (such as a 100-ampere breaker) that has a high interrupting rating (such as 50 000 amperes) while making the assumption that the branch-circuit breakers (with interrupting ratings of 10 000 amperes) are protected adequately against the 40 000-ampere short circuit, Fig. 17–33.

Standard circuit breakers with high interrupting ratings cannot protect a standard circuit breaker with a lower interrupting rating. Thus an interrupting rating on a circuit breaker included in a piece of equipment does not automatically qualify the equipment in which the circuit breaker is installed for use on circuits with higher available currents than the rating of the equipment itself.

Since the opening time of a 100-ampere circuit breaker generally exceeds one-half cycle, almost the full amount of current (40 000 amperes) will enter the branch-circuit breaker section of the panel where the breaker interrupting rating is 10 000 amperes. **The probable outcome of such a situation is an electrical explosion with risk of serious personal injury or property loss resulting from a fire.**

A similar situation exists when a fuse with an interrupting rating of 50 000 amperes is installed ahead of a plug or cartridge fuse having an interrupting rating of 10 000 amperes. As in the case of circuit breakers, a fuse having a high interrupting rating cannot protect a fuse having a low interrupting rating against fault currents in excess of the interrupting rating of the lower-rated fuse. Therefore, *all* fuses in a given panel must be capable of interrupting the maximum fault current available at that panel.

SERIES-TESTED COMBINATIONS

Series-rated circuit breakers are also called *series-connected* circuit breakers.

The CSA has tested circuit-breaker panels and load centres under the test conditions of two breakers (and, in a few cases, three) installed in series. Series combinations with current-limiting fuses ahead of circuit breakers are also tested. In the example shown in Fig. 17–34, both the 20-ampere and the 100-ampere breakers will trip if they are properly matched. The series arrangement of breakers results in a cushioning effect when the two breakers trip simultaneously. Both

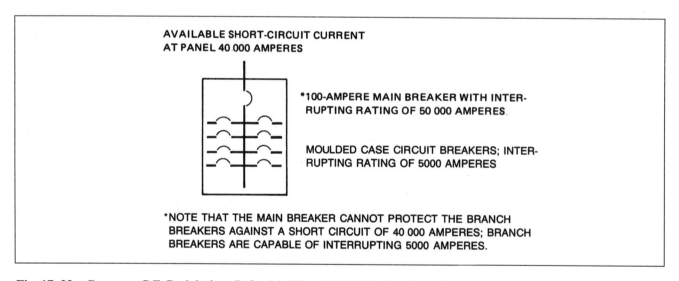

Fig. 17–33   Common *C.E.C.* violation, *Rules 14–012* and *14–014*, when installing standard-type moulded-case circuit breakers.

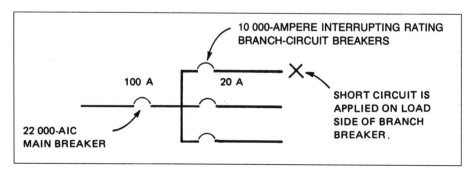

Fig. 17–34   Series-rated circuit breakers. In this example, both the 20-ampere breaker and the 100-ampere main breaker trip off under high-level fault conditions.

circuit breakers **must** trip in order for the series combination to obtain the CSA certification. A panel tested under these conditions may be rated as an entire unit. For the situation in Fig. 17–34, the panel is rated as suitable for connection to a system that can deliver no more than 22 000 amperes of short-circuit current.

The disadvantage of series-rated circuit breakers is that for heavy short circuits on any of the branch circuits, the main breaker also trips and causes the complete loss of power to the entire panel. Fig. 17-30, page 274, is an example of a manufacturer's series-tested combination datasheet.

On normal overload conditions on a branch circuit, the branch breaker will trip off without causing the main breaker to trip off. Refer to Unit 18 for more data on nonselectivity.

The advantage of installing series-rated breakers is that their cost is somewhat lower than installing breakers "fully rated" for the available fault current at the panelboard or load centre.

CAUTION: Be extremely careful when installing series-rated breakers. Do *not* "mix" different manufacturers' breakers. Do *not* use any breakers that have not been CSA approved as being suitable for use in combination with one another.

Series-connected breakers have been described as incomplete in certain constructional features or restricted in performance capabilities and are intended for use as components of complete equipment submitted for investigation rather than for direct separate installation in the field. The final acceptance of the component is dependent upon its installation and use in complete equipment submitted to the CSA.

Rule 14–014(d) requires that the manufacturer of a series-rated panel must legibly mark the equipment. The label will show the many ampere ratings and catalogue numbers of those breakers that are suitable for use with that particular piece of equipment. Fig. 17-30, page 274, is an example of a manufacturer's series-tested combination datasheet.

Rule 14–014(e) requires that there be "field marking" of series-related equipment. Most electrical inspectors expect the electrical contractor to attach this labelling to the equipment. For example, the main distribution equipment of the commercial building is located remote from the individual tenants' panelboards. The electrical contractor will affix legible and durable labels to each disconnect switch in the main distribution equipment, identifying the size and type of overcurrent device that has been installed for the proper protection of the downstream panelboards. See Fig. 17-35.

This "Caution—Series-Rated System" labelling will alert future users of the equipment that they should not indiscriminately replace existing overcurrent devices or install additional overcurrent devices with sizes and types of fuses or circuit breakers not compatible with the series-rated system as originally designed and installed. Otherwise, the series-rated system might not operate properly, creating a hazard to life and property.

CURRENT-LIMITING BREAKERS

A current-limiting circuit breaker will limit the let-through energy (I^2t) to something less than the I^2t of a one-half cycle wave.

The label on this type of breaker will show the words "Current-Limiting." The label will also indicate the breaker's let-through characteristics or will indicate where to obtain the let-through characteristic data. These let-through data are necessary to ensure adequate protection of downstream components (wire, breakers, controllers, bus bar bracing, and so on). Therefore, it is important when installing circuit breakers to ensure not only that the circuit breakers have the proper interrupting rating (capacity), but also that all of the components connected to the circuit downstream from the breakers are capable of withstanding the let-through current of the breakers. For the example on page 277, this means that the branch-circuit conductors must be capable of withstanding 40 000 amperes for approximately one cycle (the opening time of the breaker).

CAUTION—SERIES-RATED SYSTEM
_____AMPERES AVAILABLE
IDENTIFIED REPLACEMENT COMPONENT REQUIRED

Fig. 17–35 Example of label on series-rated breaker.

COST CONSIDERATIONS

As previously discussed, there are a number of different types of circuit breakers to choose from. The selection of the type to use depends upon a number of factors, including the interrupting rating, selectivity, space, and cost. When an installation requires something other than the standard moulded-case-type circuit breaker, the use of high-interrupting or current-limiting types may be necessary. The electrician and/or design engineer must complete a short-circuit study to ensure that the overcurrent protective devices provide proper protection against short-circuit conditions as required by the *C.E.C., Part I*.

The following list price cost comparisons are for various 100-ampere, 240-volt, panelboard-type circuit breakers.

Type	Interrupting Rating	List Price
Standard, plug-in	10 000	$ 165.00
Standard, bolt-on	22 000	$ 322.00
Standard, bolt-on	100 000	$ 390.00
Current-limiting	200 000	$1163.00
Electronic solid-state having adjustable trips and ground fault protection, 225 frame size	65 000	$3323.00

Moulded-case circuit breakers with an integral current-limiting fuse are available. The thermal element in this type of breaker is used for low overloads, the magnetic element is used for low-level short circuits, and the integral fuse is used for short circuits of high magnitude. The interrupting capacity of this breaker/fuse combination (sometimes called a *limiter*) is higher than that of the same breaker without the fuse.

Table 29 shows that on motor branch circuits, the maximum setting of an inverse-time circuit breaker should be sized at not over 250% of the full-load current of the motor. *Rule 28-210* indicates that instantaneous trip breakers should be set at not over 1300% of the motor full-load current or 215% LRC.

CAUTION: Carefully read the label on the equipment to be connected. When the label states "fuses," fuses must be used for the overcurrent protection. Circuit breakers are not permitted in this case. Refer to Fig. 17–36 and *Rule 28–202*.

Some labels might read "maximum size overcurrent device," in which case fuses or circuit breakers could be used for branch-circuit overcurrent protection.

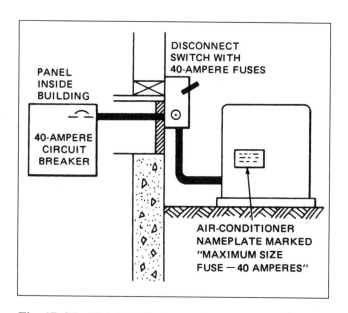

Fig. 17–36 This installation conforms to *Rules 28–202* and *28–604(5)*. The disconnect switch is within sight of and within 3 m of the unit and contains the 40-ampere fuses called for on the air-conditioner nameplate as the branch-circuit protection.

REVIEW

Note: Refer to the *C.E.C., Part I* or the plans as necessary.

1. What is the purpose of overcurrent protection? _____
_____
_____

2. List the four factors that must be considered when selecting overcurrent protective devices.

   (1) _____   (3) _____

   (2) _____   (4) _____

3. What sections of the *C.E.C., Part I* require that overcurrent protective devices have adequate interrupting ratings? _____

_____

4. Indicate the ampere rating of the fuse or circuit breaker that is selected to protect the following copper conductors. (Refer to *Table 13*.) No motors connected.

   a. No. 12 TW75   _____-ampere overcurrent device

   b. No. 8 T90 Nylon   _____-ampere overcurrent device

   c. No. 3 R90   _____-ampere overcurrent device

   d. No. 3/0 RW90 XLPE (600 V) _____-ampere overcurrent device

5. A single-phase motor draws a full-load current of 40 amperes. Complete parts (a), (b), and (c) for this motor.

   a. Install _____-ampere dual-element fuses in a _____-ampere switch.

   b. Install _____-ampere current-limiting fuses in a _____-ampere switch.

   c. Install a _____-ampere circuit breaker, standard inverse-time type.

6. The best current-limiting fuses generally contain _____ links surrounded by _____ arc-quenching filler. Current-limiting fuses can also contain _____ links.

7. In your own words, explain the meaning of the phrase "*interrupting rating*."

_____

_____

_____

_____

_____

8. In your own words, explain the meaning of the term "*current-limiting*."

_____

_____

_____

9. Define the following terms:

   a. I^2t _____

   b. I_p _____

   c. RMS _____

10. Provide the correct information for the fuse classes indicated:

	Ampere Range	Voltage Rating	Interrupting Rating
Class G fuses	_____A	_____V	_____A
Class H fuses	_____A	_____V	_____A
Class K fuses	_____A	_____V	_____A
Class J fuses	_____A	_____V	_____A
Class L fuses	_____A	_____V	_____A
Plug fuses	_____A	_____V	_____A
Class R fuses	_____A	_____V	_____A
Class T fuses	_____A	_____V	_____A
Class CC fuses	_____A	_____V	_____A

11. Class T and Class J fuses (will) (will not) fit into standard fuse clips. (Circle the correct answer.)

12. Using Fig. 17–25, determine the opening time for a 60-ampere fuse that is loaded to 300 amperes. _____ seconds.

13. A short-circuit current of 800 amperes will open a 60-ampere fuse of the type shown in Fig. 17–25 in ____ seconds. A short-circuit current of 5000 amperes will open the same fuse in ____ seconds.

14. Using the charts in Figs. 17–27 and 17–29, determine the approximate instantaneous peak let-through values and the apparent (equivalent) values of current for the following short-circuit currents:

Prospective Short-Circuit Current	Fuse	Instantaneous Peak Let-Through	Apparent RMS Current Amperes
a. 100 000 A	60 A, 250 V	_____A	_____A
b. 40 000 A	60 A, 250 V	_____A	_____A
c. 200 000 A	200 A, 600 V	_____A	_____A
d. 30 000 A	100 A, 250 V	_____A	_____A

15. A section of plug-in bus duct is braced for 14 000 RMS amperes. The available short-circuit current is 30 000 RMS symmetrical amperes. Using Fig. 17–29, determine if a 200-ampere, 250-volt, low-peak, dual-element fuse will limit the current sufficiently to protect the bus duct against a short circuit. Explain. _____

_____

_____

_____

16. If the same section of plug-in bus duct (question 15) is connected to a 200-ampere, standard moulded-case circuit breaker having an interrupting rating of 42 000 amperes, will the duct be protected properly? Explain. _____

_____

_____

_____

17. In a thermal-magnetic circuit breaker, overloads are sensed by the _____ element; short circuits are sensed by the _____ element.

18. Is it possible to install

    a. a 20-ampere Type D plug fuse in a 15-ampere fuseholder? _____

    b. a 30-ampere Class T fuse in a standard (Class H) 20-ampere fuseholder? _____

    c. a 100-ampere Class H fuse in a 100-ampere fuseholder for Class J fuses? _____

    d. a Class L fuse in a 400-ampere standard switch? _____

19. A cable limiter

    a. is a short-circuit device only.   T   F

    b. is not to be used for overload protection.   T   F

    c. is generally connected to both ends of large paralleled conductors
       so that if a fault occurs on one of the conductors, that faulted cable is
       isolated from the system.   T   F

    d. is rated at 600-volt, 200 000-ampere interrupting capacity.   T   F

20. When the label on equipment states "Maximum size fuse 50 amperes," is it permitted to connect the equipment to a 50-ampere circuit breaker? _____

21. Class CC fuses are control circuit fuses easily recognized by the _____ on one end. These fuses are rated at (125 V) (250 V) (480 V) (600 V). Fill in and circle the correct answers.

22. Current-limiting circuit breakers will limit the energy to something (less than) (equal to) (more than) the available fault current. Circle the correct answer.

23. The interrupting rating of a circuit breaker (automatically qualifies) (does not qualify) the equipment in which the breaker is installed to the same interrupting rating. Circle the correct answer.

24. It has been learned from the electric utility that the available fault current at the main service equipment will be approximately 55 000 RMS symmetrical amperes. Which of the following statements are TRUE and which are FALSE?

    It is permissible to select:

    a. a standard main breaker rated at 65 000 AIC and branch breakers rated at 10 000 AIC. _____

    b. Class L and Class RK5 fuses in equipment CSA approved at 20 000 amperes. _____

    c. main and branch-circuit breakers rated at 42 000 AIC. _____

    d. main fuses of the current-limiting type, then determine if, under fault conditions, the fuses will limit the let-through current to some value less than 10 000 amperes. Branch breakers can then be 10 000 amperes interrupting rating. _____

    e. "series-connected" breakers properly matched by the manufacturer in their CSA-recognized combination panel. The panel is to be rated at 65 000 RMS symmetrical amperes. It is understood that should a fault occur on a branch breaker, the branch breaker and the main breaker will trip off, thereby shutting off all power in the building. _____

    f. all breakers having a 65 000 AIC rating (be sure other components, such as the conductors, motor controllers, and wire can withstand 65 000 amperes). _____

25. What two sections of the *C.E.C., Part I* cover the marking of series-rated systems with the words "Caution—Series-rated system?"

    _____

26. You are asked to remove an existing fusible panel and replace it with a panel containing circuit breakers. List the information from the fusible panel that would be required to place an order for the new breaker panel.

_____   _____

_____   _____

_____   _____

_____

27. Magnetic forces vary as the square of the peak current. This is a very significant factor under short-circuit conditions. Recall that the peak for a normal sine wave is 1.4 times the RMS value. When a fault occurs, most of the resistance of the circuit has been bypassed (short circuited) so the major opposition to the flow of current is the reactance of the circuit. Thus, under fault conditions, the power factor of the faulted circuit becomes very low. This causes the peak current for the first half-cycle to become quite high.

The peak current for the first half-cycle for circuits having a power factor of 50% (0.50) is _____ times the RMS current.

The peak current for the first half-cycle for circuits having a power factor of 20% (0.20) is _____ times the RMS current.

28. Draw a line between the following statements and the correct percentage of FLA for dual-element time-delay fuses used for short-circuit and ground-fault protection of a motor branch circuit.

    a. Maximum percentage per *Table 29* is   125%

    b. Absolute maximum per *Rule 28–200* is   175%

    c. Fuse manufacturer's recommendations is   225%

29. In general, motor overload protection is based upon a percentage of the motor's full-load current. Draw a line between the motor type and the correct choice.

    a. Motors having a service factor of not less than 1.15   115%

    b. Motors having a service factor of less than 1.15   125%

    c. All other motors   175%

30. Referring to Fig. 17–30, it is necessary to install the feeder supplying a panelboard in which Type GB branch circuit breakers are installed. The available fault-current at the panelboard has been calculated to be approximately 75 000 amperes RMS symmetrical. The breakers in the panel have a marked interrupting rating of 10 000 amperes. What is the maximum size and type of fuse that must be used to protect this panelboard properly? _____

_____

31. For question 30, would the installation comply with the *C.E.C., Part I* if the main distribution board consisted of circuit breakers? The feeder to the panelboard is a 225-ampere circuit breaker having an interrupting rating of 75 000 amperes. Explain your answer.

_____

_____

UNIT 18

Short-Circuit Calculations and Coordination of Overcurrent Protective Devices

OBJECTIVES

After completing the study of this unit, the student will be able to
- perform short-circuit calculations using the point-to-point method
- calculate short-circuit currents using the appropriate tables and charts
- define the terms *coordination, selective systems,* and *nonselective systems*
- define the term *interrupting rating* and explain its significance
- use time–current curves

The student must understand the intent of *Rules 14–012, 14–014,* and *14–012 Appendix B.* These rules ensure that the fuses and/or circuit breakers selected for an installation are capable of interrupting—at the rated voltage—the current that may flow under any condition (overload, short circuit, or ground fault) with complete safety to personnel and without damage to the panel, load centre, switch, or electrical equipment in which the protective devices are installed.

An overloaded condition resulting from a miscalculation of load currents will cause a fuse to blow or a circuit breaker to trip in a normal manner. However, a miscalculation, a guess, or ignorance of the magnitude of the available short-circuit currents may result in the installation of breakers or fuses having inadequate interrupting ratings. Such a situation can occur even though the load currents in the circuit are checked carefully. **Breakers or fuses having inadequate interrupting ratings need only be subjected to a short circuit to cause them to explode, which could injure personnel and seriously damage the electrical equipment.** *The interrupting rating of an overcurrent device is its maximum rating and must not be exceeded.*

In any electrical installation, individual branch circuits are calculated as has been discussed previously in this text. After the quantity, size, and type of branch circuits are determined, these branch-circuit loads are then combined to determine the size of the feeder conductors to the respective panelboards. Most consulting engineers will specify that a certain number of spare branch-circuit breakers be installed in the panelboard, plus a quantity of spaces that can be used in the future.

For example, a certain computation may require a minimum of sixteen 20-ampere branch circuits. The specification might call for a 24-circuit panelboard with 16 active circuits, four spares, and four spaces. *Rule 8–108(2)* requires that two additional spaces be left in the panel in dwelling units, but this rule does not apply to commercial buildings.

The next step is to determine the interrupting rating requirements of the fuses or circuit breakers to be installed in the panel. *Rule 14–012* is an all-encompassing rule that covers the interrupting rating requirements for services, mains, feeders, subfeeders, and branch-circuit overcurrent devices. For various types of equipment, normal currents can be determined by checking the equipment nameplate current, voltage, and wattage ratings. In addition, an ammeter can be used to check for normal and overloaded circuit conditions.

A number of formulas can be used to calculate short-circuit currents. Manufacturers of fuses, circuit breakers, and transformers publish numerous tables and charts showing approximate values of short-circuit current. A standard ammeter must *not* be used to read short-circuit current, because this practice will result in damage to the ammeter and possible injury to personnel.

SHORT-CIRCUIT CALCULATIONS

The following sections will cover several of the basic methods of determining available short-circuit currents. As the short-circuit values given in the various tables are compared to the calculations, it will be noted that there are slight variances in the results. These differences are due largely to (1) the rounding off of the numbers in the calculations, and (2) variations in the resistance and reactance data used to prepare the tables and charts. For example, the value of the square root of three (1.732) is used frequently in three-phase calculations. Depending on the accuracy required, values of 1.7, 1.73, or 1.732 can be used in the calculations.

In actual practice, the available short-circuit current at the load side of a transformer is less than the values shown in Problem 1. However, this simplified method of finding the available short-circuit currents will result in values that are conservative.

The actual impedance value on a CSA-approved 25-kVA or larger transformer can vary ± 10% from the transformer's marked impedance. The actual impedance could be as low as 1.8% or as high as 2.2%. This will affect the available fault-current calculations.

For example, in Problem 1, the *marked* impedance is reduced by 10% to reflect the transformer's possible *actual* impedance. The calculations show this "worst case" scenario. All short-circuit examples in this text have the marked transformer impedance values reduced by 10%.

Another factor that affects fault-current calculations is voltage. Utility companies are allowed to vary voltage to their customers within a certain range. This might be ±10% for power services, and ±5% for lighting services. The higher voltage will result in a larger magnitude of fault current.

Another source of short-circuit current comes from electric motors that are running at the time the fault occurs. This is covered later on in this unit.

Thus, it can be seen that no matter how much data we plug into our fault-current calculations, there are many variables that are out of our control. What we hope for is to arrive at a result that is reasonably accurate so that our electrical equipment is reasonably safe insofar as interrupting ratings and withstand ratings are concerned.

In addition to the methods of determining available short-circuit currents that are provided below, there are computer programs that do the calculations. These programs are fast, particularly where there are many points in a system to be calculated.

Determining the Short-Circuit Current at the Terminals of a Transformer Using the Impedance Formula

PROBLEM 1:

Assume that the three-phase pad-mounted transformer installed by the utility company for the commercial building has a rating of 300 kVA at 120/208 volts with an impedance of 4% (from the transformer nameplate). The available short-circuit current at the secondary terminals of the transformer must be determined. To simplify the calculation, it is also assumed that the utility can deliver unlimited short-circuit current to the primary of the transformer. In this case, the transformer primary is known as an *infinite bus* or an *infinite primary*.

The first step is to determine the normal full-load current rating of the transformer:

$$I \text{ (at the secondary)} = \frac{kVA \times 1000}{E \times 1.73} = \frac{300 \times 1000}{208 \times 1.73}$$

$$= 834 \text{ amperes normal full load}$$

Using the impedance value given on the nameplate of the transformer, the next step is to find a multiplier that can be used to determine the short-circuit current available at the secondary terminals of the transformer.

The factor of 0.9 shown in the calculations below reflects the fact that the transformer's actual impedance might be 10% less than that marked on the nameplate and would be a worst case condition. In electrical circuits, the lower the impedance, the higher the current.

If the transformer is marked 4% impedance, then

$$\text{multiplier} = \frac{100}{4 \times 0.9} = 27.778$$

and short-circuit current = 834 × 27.778 = 23 167 amperes.

If the transformer is marked 2% impedance, then

$$\text{multiplier} = \frac{100}{2 \times 0.9} = 55.556$$

and short-circuit current = 834 × 55.556 = 46 334 amperes.

If the transformer is marked 1% impedance, then

$$\text{multiplier} = \frac{100}{1 \times 0.9} = 111.111$$

and short-circuit current = 834 × 111.111 = 92 667 amperes.

Determining the Short-Circuit Current at the Terminals of a Transformer Using Tables

Table 18–1 shows the short-circuit currents for a typical transformer. Transformer manufacturers publish short-circuit tables for many sizes of transformers having various impedance values. Table 18–1 provides data for a 300-kVA, three-phase transformer with an impedance of 2%.

According to the table, the symmetrical short-circuit current is 42 090 amperes at the secondary terminals of a 120/208-volt three-phase transformer (refer to the zero-foot row of the table). This value is on the low side because the manufacturer that developed the table did not allow for

Table 18-1 Symmetrical short-circuit currents at various distances from a liquid-filled transformer (300 kVA transformer, 2% impedance).

SYMMETRICAL SHORT-CIRCUIT CURRENTS AT VARIOUS DISTANCES FROM A LIQUID-FILLED TRANSFORMER (300 kVA TRANSFORMER, 2% IMPEDANCE)

WIRE-SIZE (COPPER)

	DIST. (FT.)	#14	#12	#10	#8	#6	#4	#1	0	00	000	2-000	0000	250 kcmil	2-250 kcmil	3-300 kcmil	350 kcmil	2-350 kcmil	3-350 kcmil	3-400 kcmil	500 kcmil	2-500 kcmil	750 kcmil	4-750 kcmil
208 VOLTS	0	42090	42090	42090	42090	42090	42090	42090	42090	42090	42090	42090	42090	42090	42090	42090	42090	42090	42090	42090	42090	42090	42090	42090
	5	6910	10290	14730	19970	25260	29840	34690	35770	36640	37340	39610	37930	38270	40100	40870	38840	40410	39870	40960	39300	40650	39650	41460
	10	3640	5610	8460	12350	17090	22230	29030	30760	32210	33410	37340	34420	35030	38270	39710	36040	38840	39870	40010	36850	39300	37480	40840
	25	1500	2360	3670	5650	8430	12150	18930	21170	23240	25000	31710	26750	27780	33590	36560	29550	34780	36930	37230	31020	35730	32190	39090
	50	760	1200	1890	2950	4530	6810	11740	13670	15610	17510	25090	19320	20520	27780	32250	22660	29550	32850	33340	24520	31020	26050	36480
	100	380	600	960	1510	2350	3610	6610	7920	9320	10810	17510	12330	13380	20520	26010	15400	22660	26850	27530	17250	24520	18860	32190
	200	190	300	480	760	1190	1860	3510	4280	5140	6090	10810	7110	7860	13380	18660	9360	15400	19590	20370	10820	17250	12150	26050
	500	80	120	190	310	480	760	1460	1800	2180	2630	4990	3130	3500	6510	10030	4290	7820	10770	11400	5100	9120	5870	16570
	1000	40	60	100	150	240	380	740	910	1110	1350	2630	1620	1820	3500	5650	2250	4290	6140	6560	2710	5100	3160	10310
	5000	10	10	20	30	50	80	150	180	230	280	550	330	380	740	1260	470	930	1380	1490	570	1130	670	2560
240 VOLTS	0	37820	37820	37820	37820	37820	37820	37820	37820	37820	37820	37820	37820	37820	37820	37820	37820	37820	37820	37820	37820	37820	37820	37820
	5	7750	11330	15810	20720	25260	28940	32560	33340	33960	34460	36080	34870	35120	36420	36960	35520	36640	37020	37070	35840	36800	36090	37370
	10	4140	6320	9400	13430	18040	22670	28230	29560	30660	31550	34460	32290	32730	35120	36140	33470	35520	36260	36350	34060	35840	34510	36930
	25	1720	2700	4180	6360	9360	13190	19640	21620	23380	24920	30240	26260	27090	31650	33860	28480	32530	34130	34340	29610	33230	30510	35680
	50	870	1380	2160	3360	5130	7620	12730	14630	16480	18230	24920	19850	20900	27090	30610	22740	28480	31060	31430	25570	29610	25570	33780
	100	440	700	1100	1730	2680	4100	7380	8770	10220	11720	18230	13220	14210	20900	25600	16150	22740	26280	26830	17860	24300	19310	30510
	200	220	350	550	880	1370	2130	3990	4830	5770	6790	11720	7880	8650	14240	19200	10190	16150	20030	20710	11650	17860	12960	25570
	500	90	140	220	350	560	870	1670	2050	2490	2990	5610	3540	3960	7230	10890	4820	8590	11620	12250	5700	9920	6520	17200
	1000	40	70	110	180	280	440	850	1050	1280	1550	2990	1850	2080	3960	6300	2560	4820	6820	7270	3070	5700	3570	11130
	5000	10	10	20	40	60	90	170	210	260	320	630	380	430	860	1440	540	1070	1580	1710	660	1290	770	2910
480 VOLTS	0	18910	18910	18910	18910	18910	18910	18910	18910	18910	18910	18910	18910	18910	18910	18910	18910	18910	18910	18910	18910	18910	18910	18910
	5	10450	12820	14750	16150	17080	17690	18200	18310	18400	18470	18690	18520	18550	18730	18800	18610	18760	18810	18810	18650	18780	18690	18850
	10	6750	9170	11630	13780	15400	16530	17540	17740	17910	18040	18470	18150	18210	18550	18690	18320	18610	18710	18720	18400	18650	18470	18800
	25	3180	4740	6770	9150	11520	13570	15690	16160	16540	16840	17830	17100	17250	18040	18380	17490	18180	18410	18440	17690	18280	17840	18630
	50	1680	2590	3900	5680	7840	10170	13190	13960	14600	15120	16840	15560	15820	17250	17870	16260	17490	17940	18000	16610	17690	16890	18360
	100	860	1350	2090	3180	4680	6600	9820	10810	11690	12460	15120	13130	13540	15820	16930	14240	16260	17060	17170	14810	16610	15260	17840
	200	440	690	1080	1680	2560	3810	6370	7320	8240	9110	12460	9930	10450	13540	15300	11370	14240	15530	15710	12150	14810	12780	16890
	500	180	280	440	700	1080	1670	3040	3640	4290	4960	8010	5560	6140	9360	11820	7050	10320	12190	12500	7880	11150	8600	14550
	1000	90	140	220	350	550	860	1620	1970	2370	2800	4960	3270	3610	6140	8520	4300	7050	8940	9290	4960	7880	5560	11830
	5000	20	30	40	70	110	180	340	420	510	620	1210	750	840	1610	2600	1040	1980	2820	3020	1250	2350	1450	4730
600 VOLTS	0	15130	15130	15130	15130	15130	15130	15130	15130	15130	15130	15130	15130	15130	15130	15130	15130	15130	15130	15130	15130	15130	15130	15130
	5	10210	11790	12920	13690	14180	14500	14770	14820	14870	14900	15010	14930	14940	15040	15070	14970	15050	15080	15080	15000	15060	15010	15100
	10	7270	9270	11010	12350	13280	13890	14410	14520	14610	14680	14900	14730	14770	14940	15020	14820	14970	15020	15030	14870	15000	14900	15070
	25	3740	5370	7280	9230	10920	12200	13410	13670	13870	14040	14570	14170	14250	14680	14850	14380	14750	14870	14890	14490	14800	14570	14980
	50	2040	3080	4500	6270	8170	9950	11940	12400	12770	13060	14040	13310	13460	14250	14590	13700	14380	14620	14650	13900	14490	14050	14840
	100	1060	1650	2510	3730	5290	7080	9650	10350	10930	11420	13060	11840	12090	13460	14080	12510	13700	14150	14210	13120	13900	13120	14570
	200	540	850	1330	2040	3040	4390	6840	7640	8390	9050	11420	9640	10010	12090	13160	10640	12510	13290	13390	11160	12850	11580	14050
	500	220	350	550	860	1330	2010	3550	4180	4830	5480	8180	6110	6530	9210	10960	7310	9900	11210	11400	7990	10470	8600	12700
	1000	110	170	280	440	680	1050	1950	2360	2800	3270	5480	3760	4110	6530	8540	4780	7310	8860	9120	5410	7990	5970	10940
	5000	20	40	60	90	140	220	420	520	640	770	1470	910	1030	1930	3030	1260	2340	3270	3480	1510	2750	1740	5180

the ± impedance variation allowed by the CSA standard. Problem 1 indicates that the available short-circuit current at the secondary of the transformer is 46 334 amperes at 2% impedance.

Determining the Short-Circuit Current at Various Distances from a Transformer Using Table 18–1

The amount of available short-circuit current decreases as the distance from the transformer increases, as indicated in Table 18–1. See Problem 2.

Determining Short-Circuit Currents at Various Distances from Transformers, Switchboards, Panelboards, and Load Centres Using the Point-to-Point Method

A simple method for determining the available short-circuit currents (also referred to as fault current) at various distances from a given location is the *point-to-point method*. Reasonable accuracy is obtained when this method is used with three-phase and single-phase systems.

The following procedure demonstrates the use of the point-to-point method:

Step 1: Determine the full-load rating of the transformer in amperes from the transformer nameplate, Table 18–2, or the following formulas:

a. For three-phase transformers:

$$I_{FLA} = \frac{kVA \times 1000}{E_{L-L} \times 1.73}$$

where E_{L-L} = line-to-line voltage

b. For single-phase transformers:

$$I_{FLA} = \frac{kVA \times 1000}{E_{L-L}}$$

Step 2: Find the percent impedance (Z) on the nameplate of the transformer.

Table 18–2 Short-Circuit Currents Available from Various Size Transformers

Voltage+ and Phase	KVA	Full Load Amps	% Impedance†† (Name plate)	†Short-Circuit Amps
120/240 1 ph.*	25	104	1.6	11 431
	37.5	156	1.6	16 961
	50	209	1.7	21 065
	75	313	1.6	32 789
	100	417	1.6	42 779
	167	695	1.8	60 038
120/208 3 ph.	150	417	2.0	23 166
	225	625	2.0	34 722
	300	834	2.0	46 333
	300	834	4.0	23 166
	500	1388	2.0	77 111
	750	2080	5.0	66 036
	1000	2776	5.0	88 127
	1500	4164	5.0	132 180
	2000	5552	5.0	123 377
	2500	6950	5.0	154 444
277/480 3 ph.	112.5	135	1.0	15 000
	150	181	1.2	16 759
	225	271	1.2	25 062
	300	361	1.2	33 426
	500	601	1.3	51 368
	750	902	5.0	28 410
	1000	1203	5.0	36 180
	1500	1804	5.0	57 261
	2000	2406	5.0	53 461
	2500	3007	5.0	66 822

† Three-phase short-circuit currents based on "infinite" primary.
* Single-phase values are L–N values at transformer terminals. These figures are based on change in turns ratio between primary and secondary, 100 000 KVA primary, zero feet from terminals of transformer, 1.2 (%X) and 1.5 (%R) for L–N vs. L–L reactance and resistance values, and transformer X/R ratio =3.
†† UL listed transformers 25KVA or greater have a ± 10% impedance tolerance. "Short Circuit Amps" reflect a worst case scenario.
+ Fluctuations in system voltage will affect the available short-circuit current. For example, a 10% increase in system voltage will result in a 10% increase in the available short-circuit currents shown in the table.

Notes: The interrupting rating for an overcurrent protective device is the device's *maximum* rating under standard test conditions. This interrupting rating must not be exceeded.
Three-phase line-to-ground (neutral) can vary from 25% to 125% of the L–L–L bolted fault-current value. Use 100% as typical.
For single-phase, centre-tapped transformers, it is common practice to multiply the L–L bolted fault current value by 1.5 to determine the approximate L–G (L–N) fault current value. Maximum short-circuit current will occur across the 120-volt transformer terminals.

PROBLEM 2:
For a 300-kVA transformer with a secondary voltage of 208 volts and 2% impedance, find the available short-circuit current at a main switch that is located 7.6 metres (25 feet) from the transformer. The main switch is supplied by four 750-kcmil copper conductors per phase in steel conduit.

Refer to Table 18–1 and read the value of 39 090 amperes in the column on the right-hand side of the table for a distance of 25 feet.

Unit 18 Short-Circuit Calculations and Coordination of Overcurrent Protective Devices

AWG or kcmil	Copper Conductors — Three Single Conductors						Copper Conductors — Three Conductor Cable					
	Steel Conduit			Nonmagnetic Conduit			Steel Conduit			Nonmagnetic Conduit		
	600V	5KV	15KV	600V	5KV	15KV	600V	5KV	15KV	600V	5KV	15KV
14	389	—	—	389	—	—	389	—	—	389	—	—
12	617	—	—	617	—	—	617	—	—	617	—	—
10	981	—	—	981	—	—	981	—	—	981	—	—
8	1557	1551	1557	1556	1555	1558	1559	1557	1559	1559	1558	1559
6	2425	2406	2389	2430	2417	2406	2431	2424	2414	2433	2428	2420
4	3806	3750	3695	3825	3789	3752	3830	3811	3778	3837	3823	3798
3	4760	4760	4760	4802	4802	4802	4760	4790	4760	4802	4802	4802
2	5906	5736	5574	6044	5926	5809	5989	5929	5827	6087	6022	5957
1	7292	7029	6758	7493	7306	7108	7454	7364	7188	7579	7507	7364
1/0	8924	8543	7973	9317	9033	8590	9209	9086	8707	9472	9372	9052
2/0	10755	10061	9389	11423	10877	10318	11244	11045	10500	11703	11528	11052
3/0	12843	11804	11021	13923	13048	12360	13656	13333	12613	14410	14118	13461
4/0	15082	13605	12542	16673	15351	14347	16391	15890	14813	17482	17019	16012
250	16483	14924	13643	18593	17120	15865	18310	17850	16465	19779	19352	18001
300	18176	16292	14768	20867	18975	17408	20617	20051	18318	22524	21938	20163
350	19703	17385	15678	22736	20526	18672	19557	21914	19821	22736	24126	21982
400	20565	18235	16365	24296	21786	19731	24253	23371	21042	26915	26044	23517
500	22185	19172	17492	26706	23277	21329	26980	25449	23125	30028	28712	25916
600	22965	20567	17962	28033	25203	22097	28752	27974	24896	32236	31258	27766
750	24136	21386	18888	28303	25430	22690	31050	30024	26932	32404	31338	28303
1000	25278	22539	19923	31490	28083	24887	33864	32688	29320	37197	35748	31959

AWG or kcmil	Aluminum Conductors — Three Single Conductors						Aluminum Conductors — Three Conductor Cable					
	Steel Conduit			Nonmagnetic Conduit			Steel Conduit			Nonmagnetic Conduit		
	600V	5KV	15KV	600V	5KV	15KV	600V	5KV	15KV	600V	5KV	15KV
14	236	—	—	236	—	—	236	—	—	236	—	—
12	375	—	—	375	—	—	375	—	—	375	—	—
10	598	—	—	598	—	—	598	—	—	598	—	—
8	951	950	951	951	950	951	951	951	951	951	951	951
6	1480	1476	1472	1481	1478	1476	1481	1480	1478	1482	1481	1479
4	2345	2332	2319	2350	2341	2333	2351	2347	2339	2353	2349	2344
3	2948	2948	2948	2958	2958	2958	2948	2956	2948	2958	2958	2958
2	3713	3669	3626	3729	3701	3672	3733	3719	3693	3739	3724	3709
1	4645	4574	4497	4678	4631	4580	4686	4663	4617	4699	4681	4646
1/0	5777	5669	5493	5838	5766	5645	5852	5820	5717	5875	5851	5771
2/0	7186	6968	6733	7301	7152	6986	7327	7271	7109	7372	7328	7201
3/0	8826	8466	8163	9110	8851	8627	9077	8980	8750	9242	9164	8977
4/0	10740	10167	9700	11174	10749	10386	11184	11021	10642	11408	11277	10968
250	12122	11460	10848	12862	12343	11847	12796	12636	12115	13236	13105	12661
300	13909	13009	12192	14922	14182	13491	14916	14698	13973	15494	15299	14658
350	15484	14280	13288	16812	15857	14954	15413	15490	15540	16812	17351	16500
400	16670	15355	14188	18505	17321	16233	18461	18063	16921	19587	19243	18154
500	18755	16827	15657	21390	19503	18314	21394	20606	19314	22987	22381	20978
600	20093	18427	16484	23451	21718	19635	23633	23195	21348	25750	25243	23294
750	21766	19685	17686	23491	21769	19976	26431	25789	23750	25682	25141	23491
1000	23477	21235	19005	28778	26109	23482	29864	29049	26608	32938	31919	29135

Ampacity	Plug-In Busway		Feeder Busway		High Imped. Busway
	Copper	Aluminum	Copper	Aluminum	Copper
225	28700	23000	18700	12000	—
400	38900	34700	23900	21300	—
600	41000	38300	36500	31300	—
800	46100	57500	49300	44100	—
1000	69400	89300	62900	56200	15600
1200	94300	97100	76900	69900	16100
1350	119000	104200	90100	84000	17500
1600	129900	120500	101000	90900	19200
2000	142900	135100	134200	125000	20400
2500	143800	156300	180500	166700	21700
3000	144900	175400	204100	188700	23800
4000	—	—	277800	256400	—

Table 18–3 Table of C values.

Step 3: Find the transformer multiplier M_1:

$$M_1 = \frac{100}{\text{transformer \% impedance} \times 0.9}$$

Note: Because the marked transformer impedance can vary ±10% per the CSA standard, the 0.9 factor above takes this into consideration to show "worst case" conditions.

Step 4: Determine the transformer let-through short-circuit current at the secondary terminals of transformer. Use tables or the following formula:

a. For three-phase transformers (L–L–L):

$$I_{SCA} = \text{transformer FLA} \times M_1$$

b. For single-phase transformers (L–L):

$$I_{SCA} = \text{transformer FLA} \times M_1$$

c. For single-phase transformers (L–N):

$$I_{SCA} = \text{transformer FLA} \times M_1 \times 1.5$$

At the secondary terminals of a single-phase centre-tapped transformer, the L–N fault current is higher than the L–L fault current. At some distance from the terminals, depending on the wire size and type, the L–N fault current is lower than the L–L fault current. The L–N fault current can vary from 1.33 to 1.67 times the L–L fault current. These figures are based on the different turns ratios between the primary and the secondary, infinite source current, zero distance from the terminals of the transformer, and $1.2 \times \%$ reactance (X) and $1.5 \times \%$ resistance (R) for the L–N vs L–L resistance and reactance values. For simplicity, in Step 4c we used an approximate multiplier of 1.5. See Fig. 18–2 (page 294) for example.

Step 5: Determine the f factor:

a. For three-phase faults:

$$f = \frac{1.73 \times L \times I_{SCA}}{N \times C \times E_{L\text{-}L}}$$

b. For single-phase, line-to-line (L–L) faults on single-phase, centre-tapped transformers:

$$f = \frac{2 \times L \times I_{SCA}}{N \times C \times E_{L\text{-}L}}$$

c. For single-phase, line-to-neutral (L–N) faults on single-phase, centre-tapped transformers:

$$f = \frac{2 \times L \times I_{SCA}}{N \times C \times E_{L\text{-}N}}$$

where

- L = the length of the circuit to the fault, in feet
- C = the constant derived from Table 18–3 for the specific type of conductors and wiring method
- $E_{L\text{-}L}$ = the line-to-line voltage
- $E_{L\text{-}N}$ = the line-to-neutral voltage
- N = the number of conductors in parallel

Step 6: After finding the f factor, refer to Table 18–4 and locate the corresponding value of the multiplier M_2, or, calculate it as follows:

$$M_2 = \frac{1}{1+f}$$

Step 7: Multiply the available fault current at the beginning of the circuit by the multiplier M_2 to determine the available symmetrical fault current at the fault.

$$I_{SCA} \text{ at fault} = I_{SCA} \text{ at beginning of circuit} \times M_2$$

Motor Contribution: All motors running at the instant a short-circuit occurs contribute to the short-circuit current. The amount of current from

CHART FOR M_2 MULTIPLIER

f	M_2	f	M_2
0.01	0.99	1.20	0.45
0.02	0.98	1.50	0.40
0.03	0.97	2.00	0.33
0.04	0.96	3.00	0.25
0.05	0.95	4.00	0.20
0.06	0.94	5.00	0.17
0.07	0.93	6.00	0.14
0.08	0.93	7.00	0.13
0.09	0.92	8.00	0.11
0.10	0.91	9.00	0.10
0.15	0.87	10.00	0.09
0.20	0.83	15.00	0.06
0.30	0.77	20.00	0.05
0.40	0.71	30.00	0.03
0.50	0.67	40.00	0.02
0.60	0.63	50.00	0.02
0.70	0.59	60.00	0.02
0.80	0.55	70.00	0.01
0.90	0.53	80.00	0.01
1.00	0.50	90.00	0.01
		100.00	0.01

$$M_2 = \frac{1}{1+f}$$

Table 18–4 Simple chart to convert f values to M_2 multiplier when using the point-to-point method.

the motors is equal approximately to the starting (locked rotor) current for each motor. This current value depends upon the type of motor and its characteristics. It is a common practice to multiply the full-load ampere rating of the motor by four or five to obtain a close approximation of the locked rotor current and provide a margin of safety. For energy-efficient motors, multiply the motor's full-load current rating by six to eight times for a reasonable approximation of the fault-current contribution. The current contributed by running motors at the instant a short circuit occurs is added to the value of the short-circuit current at the main switchboard prior to the start of the point-to-point calculations for the rest of the system. To simplify the following problems, motor contributions have not been added to the short-circuit currents.

SHORT-CIRCUIT CURRENT VARIABLES

Phase-to-Phase-to-Phase Fault

The three-phase fault current determined in Step 7 is the *approximate* current that will flow if the three "hot" phase conductors of a three-phase system are shorted together, in what is commonly referred to as a "bolted fault." This is the worst case condition.

Phase-to-Phase Fault

To obtain the *approximate* short-circuit current values when two "hot" conductors of a three-phase system are shorted together, use 87% of the three-phase current value. In other words, if the three-phase current value is 20 000 amperes when the three "hot" lines are shorted together (L–L–L

PROBLEM 3:

It is desired to find the available short-circuit current at the main switchboard of the commercial building. Once this value is known, the electrician can provide overcurrent devices with adequate interrupting ratings and the proper bus bar bracing within the switchboard (*Rules 14–012, 14–014,* and *Appendix B*). Figure 18–1 shows the actual electrical system for the commercial building. As each of the following steps in the point-to-point method is examined, refer to the entries given in Tables 18–3 and 18–4 to determine the necessary values of C, f, and M_2.

Step 1: $I_{FLA} = \dfrac{kVA \times 1000}{E_{L-L} \times 1.73} = \dfrac{300 \times 1000}{208 \times 1.73} = 834$ amperes

Step 2: Multiplier, $M_1 = \dfrac{100}{\text{Transformer \% impedance} \times 0.9} = \dfrac{100}{4 \times 0.9} = 27.78$

Step 3: $I_{SCA} = 834 \times 27.78 = 23\,168.5$

Step 4: $f = \dfrac{1.73 \times L \times I}{3 \times C \times E_{L-L}} = \dfrac{1.73 \times 80 \times 23\,168.5}{3 \times 22\,185 \times 208} = 0.2316$

Step 5: $M_2 = \dfrac{1}{1+f} = \dfrac{1}{1+0.2316} = 0.812$

Step 6: The short-circuit current available at the line-side lugs on the main switchboard is: $23\,168.5 \times 0.812 = 18\,812$ RMS symmetrical amperes.

value), then the short-circuit current to two "hot" lines shorted together (L–L) is approximately:

20 000 × 0.87 = 17 400 amperes

Phase-to-Neutral (Ground)

For solidly grounded three-phase systems, such as the 120/208-volt system that supplies the commercial building, the phase-to-neutral (ground) bolted short-circuit current can vary from 25% to 125% of the L–L–L bolted short-circuit current. Therefore, it is common practice to consider the L–N or L–G short-circuit current value to be the same as the L–L–L short-circuit current value.

EXAMPLE: If the three-phase L–L–L fault current has been calculated to be 20 000 amperes, then the L–N fault current is approximately:

20 000 × 1.00 = 20 000 amperes

In summary:
L–L bolted short-circuit current = 87% L-L-L
L–N bolted short-circuit current = 100% L-L-L

The main concern is to provide the proper interrupting rating for the overcurrent protective devices, and the adequate withstand rating for the equipment. Therefore, for most three-phase electrical systems, the line–line–line bolted fault-current value will provide the desired level of safety.

Arcing Fault Values (Approximate)

There are times when one wishes to know the currents in arcing faults. These currents vary considerably. The following multipliers give acceptable approximations.

Type of Fault	480 Volts	208 Volts
L–L–L	0.89	0.12
L–L	0.74	0.02
L–G	0.38*	0

*Some reference books indicate this value to be 0.19.

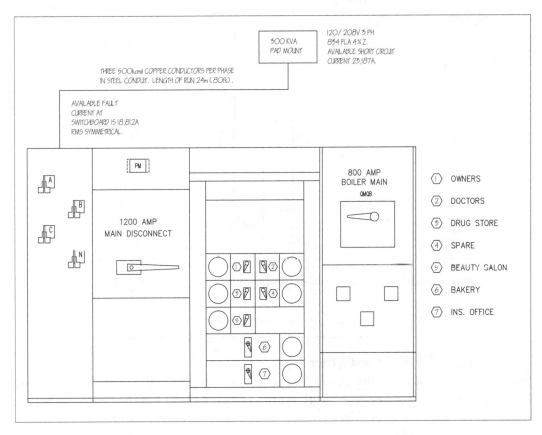

Fig. 18–1 Electrical system for the commercial building.

EXAMPLE: What is the approximate line-to-line arcing fault value on a 208/120-volt system where the line–line–line fault current has been calculated to be 30 000 amperes?

Solution:

30 000 × 0.02 = 600 amperes

As shown in Problem 3, the fuses or circuit breakers located in the main switchboard of the commercial building must have an interrupting capacity of at least 39 847 RMS symmetrical amperes. It is good practice to install protective devices having an interrupting rating at least 25% greater than the actual calculated available fault current. This practice generally provides a margin of safety to permit the rounding off of numbers, as well as compensating for a reasonable amount of short-circuit contribution from any electrical motors that may be running at the instant the fault occurs.

The fuses specified for the commercial building have an interrupting rating of 200 000 amperes (see the Specifications). In addition, the switchboard bracing is specified to be 50 000 amperes.

If current-limiting fuses are installed in the main switchboard feeders protecting the various panelboards, breakers having an interrupting rating of 10 000 amperes may be installed in the panelboards. This installation meets the requirements of *Rule 14-012*. (If necessary, the student should review the sections covering the application of peak let-through charts in Unit 17.)

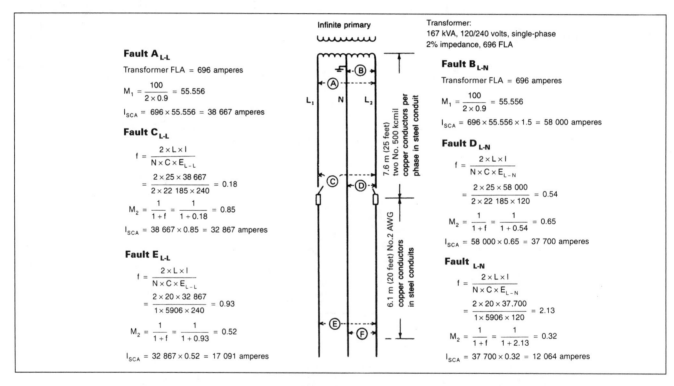

Fig. 18–2 Point-to-point calculation for single-phase, centre-tapped transformer. Calculations show L–L and L–N values.

PROBLEM 4: Single-Phase Transformer

This problem is illustrated in Figure 18–2. The point-to-point method is used to determine the currents for both line-to-line and line-to-neutral faults for a 167-kVA, 2% impedance transformer on a 240/120-volt, single-phase system. Note that the impedance has been reduced by 10%.

If noncurrent-limiting overcurrent devices (standard moulded-case circuit breakers) are to be installed in the main switchboard, breakers having adequate interrupting ratings must be installed in the panelboards. A short-circuit study must be made for each panelboard location to determine the value of the available short-circuit current.

The cost of circuit breakers increases as the interrupting rating of the breakers increases. The most economical protection system generally results when current-limiting fuses are installed in the main switchboard to protect the breakers in the panelboards or series rated breaker combinations are used. In this case, the breakers in the panelboards will have the standard 10 000-ampere interrupting rating.

Summary

1. To meet the requirements of *Rules 14–012* and *14–014*, it is absolutely necessary to determine the available fault currents at various points on the electrical system. If a short-circuit study is not done, the selection of overcurrent devices may be in error, resulting in a hazard to life and property.

2. For an installation using fuses only, the fuses must have an interrupting rating *not less* than the available fault current. The electrical equipment to be protected by the fuses must be capable of withstanding the let-through current of the fuse.

3. For an installation using fuses to protect panels that contain circuit breakers having inadequate interrupting ratings (less than the available fault current), use the peak let-through charts or fuse-breaker application tables from manufacturers to ensure the proper selection of fuses that will adequately protect the downstream circuit breakers.

4. For an installation using standard circuit breakers only, the breakers must have an interrupting rating *not less* than the available fault current at the point of application. It is recommended that the breakers used have interrupting ratings *at least 25% greater* than the available fault current.

To use current-limiting circuit breakers, refer to the manufacturer's application data.

To use series-connected (rated) circuit breakers, refer to the manufacturer's application data. (An example of series rated breakers may be found on page 274.) Remember that under heavy fault conditions the above installations will be nonselective. This situation is discussed in detail on the following pages.

COORDINATION OF OVERCURRENT PROTECTIVE DEVICES

While this text cannot cover the topic of electrical system coordination (selectivity) in detail, the following material will provide the student with a working knowledge of this important topic.

What Is Coordination? (*Rule 14–012, Appendix B*)

A situation known as *nonselective coordination* occurs when a fault on a branch circuit opens not only the branch-circuit overcurrent device, but also the feeder overcurrent device, Fig. 18–3. Nonselective systems are installed unknowingly and cause needless power outages in portions of an electrical system that should not be affected by a fault.

A *selectively coordinated system*, Fig. 18–4, is one in which *only* the overcurrent device immediately upstream from the fault opens. Obviously, the installation of a selective system is much more desirable than a nonselective system, in that all the other circuits are unaffected by a fault on an adjacent circuit.

The importance of selectivity in an electrical system is of primary concern in health-care facilities, where maintaining essential electrical systems is paramount. The unexpected loss of power in certain areas of hospitals, nursing care centres, and similar health-care facilities can be catastrophic.

The importance of selectivity (system coordination) is also critical to industrial installations where additional hazards would be introduced should a disorderly shutdown occur. Coordination may be defined as the proper localizing of a fault condition to restrict outages to the equipment affected.

Some local electrical codes require that all circuits, feeders, and mains in buildings such as schools, shopping centres, assembly halls, nursing homes, retirement homes, churches, restaurants, and any other places of public occupancy be selectively coordinated so as to minimize the dangers associated with total power outages.

Nonselectivity of an electrical system is not considered good design practice, and is generally accepted only as a trade-off for a low-cost installation.

It is advisable to check with the authority enforcing the *C.E.C., Part I* before proceeding too far in the selection of overcurrent protective devices for a specific installation.

By knowing how to determine the available short-circuit current and ground-fault current, the electrician can make effective use of the time–current curves and peak let-through charts (Unit 17) to find the length of time required for a fuse to open or a circuit breaker to trip.

What Causes Nonselectivity?

In Fig. 18–5, a short circuit in the range of 3000 amperes occurs on the load side of a 20-ampere breaker. The magnetic trip of the breaker is adjusted permanently by the manufacturer to unlatch at a current value equal to ten times its rating, or 200 amperes. The feeder breaker is rated at 100 amperes; the magnetic trip of this breaker is set by the manufacturer to unlatch at a current equal to ten times its rating, or 1000 amperes. This type of breaker generally cannot be adjusted in the field. Therefore, a current of 200 amperes or more will cause the 20-ampere breaker to trip instantly. In addition, any current of 1000 or more will cause the 100-ampere breaker to trip instantly.

For the breakers shown in Fig. 18–5, a momentary fault of 3000 amperes will trip (unlatch) both breakers. Since the flow of current in a series circuit is the same in all parts of the circuit, the 3000-ampere fault with trigger both magnetic trip mechanisms. The time–current curve shown in Fig. 17–32 (page 276) indicates that for a 3000-ampere fault, the unlatching time for both breakers is 0.0042 second and the interrupting time for both breakers is 0.016 second.

The term "interrupting time" refers to the time it takes for the circuit breakers' contacts to open, thereby stopping the flow of current in the circuit. Refer to Figs. 18–5 and 18–6.

This example of a nonselective system should make apparent to the student the need for a thorough study and complete understanding of time–current curves, fuse selectivity ratios, and unlatching time data for circuit breakers. Otherwise, a blackout may occur, such as the loss of exit and emergency lighting. The student must be able to determine available short-circuit currents (1) to ensure the proper selection of protective

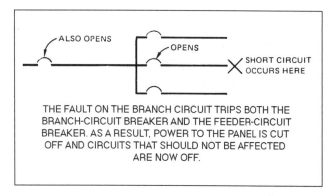

Fig. 18–3 Nonselective system.

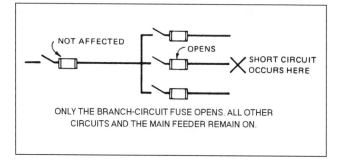

Fig. 18–4 Selective system.

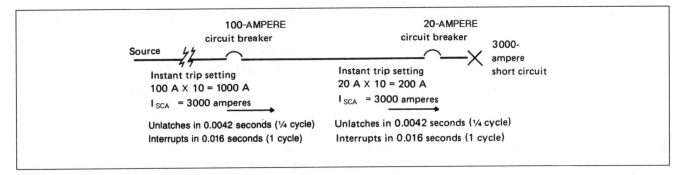

Fig. 18–5 Nonselective system.

devices with adequate interrupting ratings, and (2) to provide the proper coordination as well.

Fig. 18–6 shows an example of a selective circuit. In this circuit, the available fault current is only 500 amperes. This current trips the 20-ampere breaker instantly (the unlatching time of the breaker is approximately 0.0080 second and its interrupting time is 0.016 second). The graph in Fig. 17–32 (page 276) indicates that the 100-ampere breaker interrupts the 500-ampere current in a range from 7 to 20 seconds. This relatively lengthy trip time range is due to the fact that the 500-ampere fault acts upon the current thermal trip element only, and does not affect the magnetic trip element, which operates on a current of 1000 amperes or more.

Selective System Using Fuses

The proper choice of the various classes and types of fuses is necessary if selectivity is to be achieved, Fig. 18–7. Indiscriminate mixing of fuses of different classes, time–current characteristics, and even manufacturers may cause a system to become nonselective.

To ensure selective operation under low-overload conditions, it is necessary only to check and compare the time–current characteristic curves of fuses. Selectivity occurs when the curves do not cross one another.

Fuse manufacturers publish *selectivity guides,* similar to the one shown in Fig. 18–8, to be used for short-circuit conditions. When using these

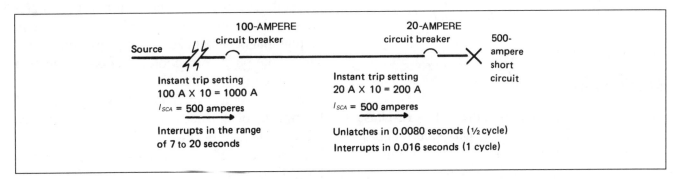

Fig. 18–6 Selective system.

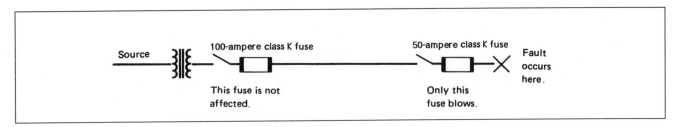

Fig. 18–7 Selective system using fuses.

guides, selectivity is achieved by maintaining a specific amperage ratio between the various classes and types of fuses. A selectivity chart is based on any fault current up to the maximum interrupting ratings of the fuses listed in the chart.

Selective System Using Circuit Breakers

Circuit-breaker manufacturers publish time–current characteristic curves and unlatching information, see Fig. 17–32 (page 276).

For normal overload situations, a circuit breaker having an ampere rating lower than the ampere rating of an upstream circuit breaker will trip. The upstream breaker will not trip. The system is selective for normal overload currents.

For low-level faults (short circuits) less than the instantaneous trip setting of an upstream circuit breaker, a circuit breaker having an ampere rating lower than the ampere rating of the upstream breaker will trip, and the upstream breaker will not trip. The system is selective, Fig. 18–6.

For fault-current levels above the instantaneous trip setting of the upstream circuit breaker, both the branch-circuit breaker and the upstream circuit breaker will trip off. The system is nonselective, Fig. 18–5.

There are no ratio selectivity charts for breakers as there are for fuses.

SINGLE PHASING

The *C.E.C., Part I* requires that all three-phase motors be provided with running overcurrent protection to each phase, *Rule 28–200*. A line-to-ground fault generally will blow one fuse. There will be a resulting increase of 173 to 200% in the line current in the remaining two connected phases. This increased current will be sensed by the motor fuses and overload relays when such fuses and relays are sized at 125% or less of the full-load current rating of the motor. Thus, the fuses and/or the overload relays will open before the motor windings are damaged. When properly matched, the overload relays open before the fuses. The fuses offer "back-up" protection if the overload relays fail to open for any reason.

A line-to-line fault in a three-phase motor will blow two fuses. In general, the operating coil of the motor controller will drop out, thus providing protection to the motor winding.

To reduce *single-phasing* problems, each three-phase motor must be provided with individual overload protection through the proper sizing of the overload relays and fuses. Phase failure relays are also available. However, can other equipment be affected by a single-phasing condition?

In general, loads that are connected line-to-neutral or line-to-line, such as lighting, receptacles, and electric heating units, will not burn out under a single-phasing condition. In other words, if one main fuse blows, then two-thirds of the lighting, receptacles, and electric heat will remain on. If two main fuses blow, then one-third of the lighting remains on, and a portion of the electric heat connected line-to-neutral will stay on.

Nothing can prevent the occurrence of single-phasing. What must be detected is the increase in current that occurs under single-phase conditions. This is the purpose of motor overload protection as required in *Rule 28–300*.

It is essential to maintain some degree of lighting in occupancies such as stores, schools, offices, and health-care facilities (such as nursing homes). A total blackout in these public structures has the potential for causing panic and extensive personal injury. A loss of one or two phases of the system supplying a building should not cause a complete power outage in the building.

Unit 18 Short-Circuit Calculations and Coordination of Overcurrent Protective Devices

RATIOS FOR SELECTIVITY

LINE-SIDE FUSE	LOAD-SIDE FUSE									
	KRP-C HI-CAP time-delay Fuse 601-6000A Class L	KTU LIMITRON fast-acting Fuse 601-6000A Class L	KLU LIMITRON time-delay Fuse 601-4000A Class L	KTN-R, KTS-R LIMITRON fast-acting Fuse Class K 0-600A	JJS, JJN TRON fast-acting Fuse Class T 0-600A	JKS LIMITRON quick-acting Fuse Class J 0-600A	FRN-R, FRS-R FUSETRON dual-element Fuse Class K 0-600A	LPN-R, LPS-R LOW-PEAK dual-element Fuse Class K 0-600A	LPJ LOW-PEAK time-delay Fuse Class J 0-600A	SC Type Fuse (Class G) 0-60A
KRP-C HI-CAP time-delay Fuse 601-6000A Class L	2:1	2:1	2.5:1	2:1	2:1	2:1	4:1	2:1	2:1	--
KTU LIMITRON fast-acting Fuse 601-6000A Class L	2:1	2:1	2.5:1	2:1	2:1	2:1	6:1	2:1	2:1	--
KLU LIMITRON time-delay fuse 601-4000A Class L	2:1	2:1	2:1	2:1	2:1	2:1	4:1	2:1	2:1	N/A
KTN-R, KTS-R LIMITRON fast-acting Fuse 0-600A Class RK1	N/A	N/A	N/A	3:1	3:1	3:1	8:1	3:1	3:1	4:1
JJN, JJS TRON fast-acting Fuse 0-600A Class T	N/A	N/A	N/A	3:1	3:1	3:1	8:1	3:1	3:1	4:1
JKS LIMITRON quick-acting Fuse 0-600A Class J	N/A	N/A	N/A	3:1	3:1	3:1	8:1	3:1	2:1	4:1
FRN-R, FRS-R, FUSETRON dual-element Fuse 0-600A Class RK5	N/A	N/A	N/A	1.5:1	1.5:1	1.5:1	2:1	1.5:1	1.5:1	1.5:1
LPN-R, LPS-R LOW-PEAK dual-element Fuse 0-600A Class RK1	N/A	N/A	N/A	3:1	3:1	3:1	8:1	2:1	2:1	4:1
LPJ Low-Peak Fuse 15-600A Class J	N/A	N/A	N/A	3:1	1.5:1	1.5:1	8:1	1.5:1	2:1	2:1
SC Type Fuse 0-60A Class G	N/A	N/A	N/A	2:1	2:1	2:1	4:1	3:1	3:1	2:1

N/A = NOT APPLICABLE

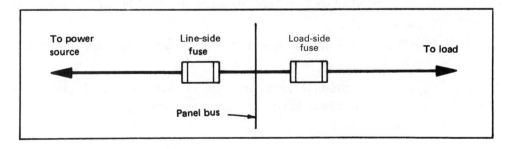

Fig. 18–8 Selectivity guide.

PROBLEM 5:
It is desired to install 100-ampere, low-peak time delay Class J fuses in a main switch, and a 50-ampere time delay Class J fuse in the feeder switch. Is this combination of fuses selective?

Refer to the Selectivity Guide in Fig. 18–8. The chart indicates that a 2:1 minimum ratio must be maintained for this type of fuse. Since 100:50 is a 2:1 ratio, the installation is selective. In addition, any fuse of the same type having a rating of less than 50 amperes will also be selective with the 100-ampere main fuse. That is, if a fault occurs on the load side of the 50-ampere fuse, only the 50-ampere fuse will open.

REVIEW

Note: Refer to the *C.E.C., Part I* or the plans as necessary.

1. Using Table 18–1, determine the available short-circuit currents on a 208-volt system for the following:

    a. 15.24 m (50 ft) of No. 1 conductor _____ A

    b. 7.62 m (25 ft) of No. 3/0 conductor _____ A

    c. 15.24 m (50 ft) of 500-kcmil conductor _____ A

    d. 15.24 m (50 ft) of two No. 3/0 conductors per phase _____ A

    e. 30.48 m (100 ft) of No. 4/0 conductor _____ A

2. a. Define *selectivity*. _____
   _____
   _____
   _____

   b. Define *nonselectivity*. _____
   _____
   _____
   _____

3. Indicate whether the following systems are selective or nonselective:

    a. a KRP-C 800-ampere fuse installed upstream of an LPS-R 200-ampere fuse _____

    b. an FRS-R 600-ampere Fusetron fuse installed ahead of an LPS-R 400-ampere low-peak fuse _____

    c. a KTS-R 400-ampere Limitron fuse installed ahead of an FRS-R 200-ampere Fusetron fuse _____

    d. a 225-ampere main circuit breaker in a panel that contains 20-ampere branch-circuit breakers, where the breakers have adequate interrupting capacity for the 20 000 amperes of available fault current _____

4. A series-connected panel contains a 1200-ampere main breaker with many branch-circuit breakers rated at 100 amperes and 225 amperes. The branch-circuit breakers have an individual interrupting rating of 10 000 amperes. The main circuit breaker has an individual interrupting rating of 42 000 amperes. In combination the panel is CSA listed as suitable for use at fault currents not to exceed 65 000 amperes.

   a. For a fault current above 10 000 amperes on the load side of one of the 225-ampere breakers, the main breakers (will trip) (will not trip). Circle the correct answer.

   b. For a fault current of 4000 amperes on the load side of one of the 100-ampere breakers, the main breaker (will trip) (will not trip). Circle the correct answer.

5. Refer to Fig. 18–9 and calculate the available short-circuit current at Panel "A." This panel is supplied by a 7.6 m (25 ft) run of No. 3/0 copper conductors in steel conduit. Use the point-to-point method and show all calculations.

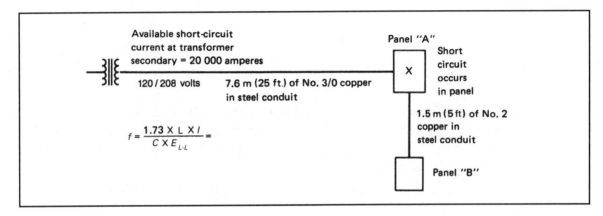

Fig. 18–9

(STUDENT CALCULATIONS)

6. Calculate the available fault current if a short circuit occurs in Panel "B", Fig. 18–9. Use the point-to-point method and show all calculations.

(STUDENT CALCULATIONS)

7. Calculate the available fault current if a short circuit occurs in the bakery panel of the commercial building covered in this text. Use the power riser diagram and building drawings to determine the size and length of conductor from the main switchboard to the bakery panel. Show all calculations.

(STUDENT CALCULATIONS)

8. A three-phase, L–L–L bolted short-circuit current has been calculated to be 40 000 RMS symmetrical amperes. What is the approximate value of:

   a. a line-to-line fault? _____

   b. a line-to-ground fault? _____

UNIT 19

Equipment and Conductor Short-Circuit Protection

OBJECTIVES

After completing the study of this unit, the student will be able to

- understand that all electrical equipment has a withstand rating
- discuss the withstand rating of conductors
- understand important *C.E.C., Part I* sections that pertain to interrupting rating, available short-circuit current, current-limitation, effective grounding, bonding, and temperature limitation of conductors
- understand what the term *ampere-squared seconds* means
- perform calculations to determine how much current a copper conductor can safely carry for a specified period of time before being damaged or destroyed
- refer to charts to determine conductor withstand ratings
- understand that the two forces present when short-circuits or overloads occur are *thermal* and *magnetic*
- discuss the 3 m (9.84 ft) "tap" conductor installation, *Rule 14–100*

All electrical equipment, including switches, motor controllers, conductors, bus duct, panelboards, load centres, switchboards, and so on, have an ability to withstand a certain amount of electrical energy for a given amount of time before damage to the equipment occurs. This gives rise to the term *withstand rating*.

CSA standards specify certain test criteria for the above equipment. For example, switchboards must be capable of withstanding a given amount of fault current for at least three cycles. Note that both the amount of current and the length of time for the test are specified.

Simply stated, withstand rating is the ability of the equipment to "hold together" for the time it takes the overcurrent protective device (fuse or circuit breaker) to respond to the fault condition. Where the equipment is intended to break current, such as a fuse or circuit breaker, the equipment is marked with its interrupting rating. See *Rule 14–012* for the definition of interrupting rating. Electrical equipment manufacturers conduct exhaustive tests to determine the withstand or interrupting ratings of their products.

Equipment tested in this manner is usually marked with the size and type of overcurrent protection required. For example, the label on a motor controller might indicate a maximum size fuse for different sizes of thermal overloads used in that controller. If the marking indicates that fuses must be used for the overcurrent protection, a circuit breaker would not be permitted.

The ability of a current-limiting overcurrent device to limit the let-through energy to a value less than the amount of energy that the electrical system is capable of delivering means that the equipment can be protected against fault-current values of high magnitude. Equipment manufacturers specify current-limiting fuses to minimize the potential damage that might occur in the event of a high-level short circuit or ground fault.

The electrical engineer and/or electrical contractor must perform short-circuit studies, and then determine the proper size and type of current-limiting overcurrent protective device that can be used ahead of the electrical equipment that does not have an adequate withstand or interrupting rating for the available fault current to which it could be subjected. The calculation of short-circuit currents is covered in Unit 18.

Study the normal-, overloaded-, and short-circuit diagrams (Fig. 19–1, A–C) and observe how Ohm's law is applied to these circuits. The calculations for the ground-fault circuit are not shown, because the impedance of the return ground path can vary considerably. *Rule 10–500* states that the ground path must

1. be permanent and continuous,
2. be able to safely carry any fault current that the path might be required to carry, and
3. have sufficiently low impedance to limit the voltage to ground and to facilitate the operation of the overcurrent devices in the circuit.

CONDUCTOR WITHSTAND RATING

Up to this point in the text, we have covered in detail how to compute conductor ampacities for branch circuits, feeders, and service-entrance conductors, with the basis being the connected load and/or volt-amperes per square metre. Then the demand factors and other diversity factors are applied. If we have followed the *C.E.C., Part I* rules for calculating conductor size, the conductors will not be damaged under overload conditions because the conductors will have the proper ampacity and will be protected by the proper size and type of overcurrent protective device.

Tables in the *C.E.C., Part I*, such as *Table 2*, cover the ampacity ratings for various sizes and insulation types of conductors. These tables consider normal loading conditions.

But what happens to these conductors under high-level short-circuit and ground-fault conditions? Small conductors often are the weak link in an electrical system.

When short circuits and/or ground faults occur, the connected loads are bypassed, thus the term "short circuit." All that remains in the circuit are the conductors and other electrical devices such as breakers, switches, and motor controllers. The impedance (ac resistance) of these devices is extremely low, so for all practical purposes, the most significant opposition to the flow of current when a fault occurs is the conductor impedance. Fault-current calculations are covered in Unit 18.

The following discussion covers the actual ability of a conductor to maintain its integrity for the time it is called upon to carry fault current, instead of burning away and causing additional electrical problems. It is important that the electrician, electrical contractor, electrical inspector, and consulting engineer have a good understanding of what happens to a conductor under medium- to high-level fault conditions.

The withstand rating of a conductor, such as an equipment grounding conductor, main bonding jumper, or any other current-carrying conductor, indicates how much current the conductor can withstand for a small amount of time. This is the short-time withstand rating of the conductor.

Let us review some *C.E.C., Part I* sections that focus on the importance of equipment and conductor withstand rating, available fault currents, and circuit impedance.

- *Rule 14–012, Interrupting Rating:* Equipment intended to break current at fault levels shall have an interrupting rating sufficient for the system voltage and the current that is

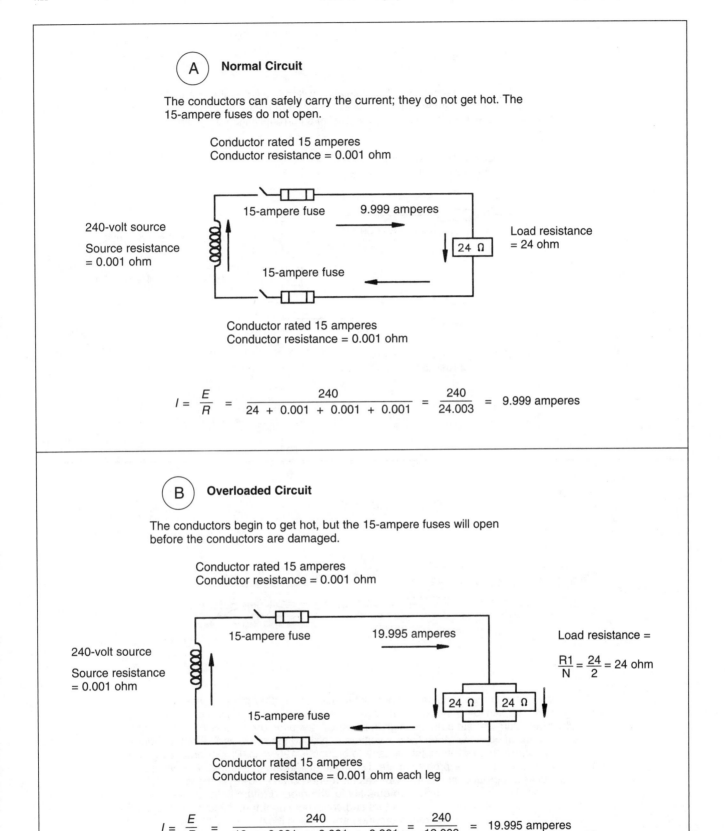

Fig. 19–1 These simple diagrams clearly illustrate the current flow in (A) a normal circuit, (B) an overloaded circuit, (C) a short circuit, and (D) a line-to-ground fault.

Short Circuit

The conductors get extremely hot. The insulation will melt off, and the conductor itself will melt unless the 15-ampere fuses open in a very short period of time and keep the fault current to a low value. This is the function of a current-limiting overcurrent device.

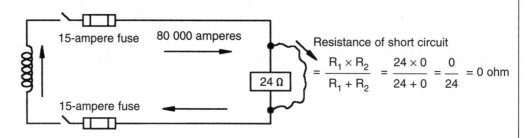

$$I = \frac{E}{R} = \frac{240}{0.001 + 0.001 + 0.001} = \frac{240}{0.003} = 80\ 000 \text{ amperes}$$

D Ground Fault

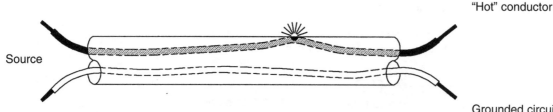

The "hot" conductor comes in contact with a metal raceway or other metal object. If the return ground path has low impedance, the overcurrent device protecting the circuit will clear the fault. If the return ground path has high impedance, the overcurrent device will not clear the fault. The metal object will then have a voltage to ground the same as the "hot" conductor has to ground. In house wiring, this voltage to ground is 120 volts. Proper grounding and ground-fault circuit interrupter protection are discussed elsewhere in this text. The calculation procedure for a ground fault is the same as for a short circuit, however the values of R can vary greatly because of the unknown quantity for the ground return path. Loose locknuts, bushings, and set screws on connectors and couplings, poor terminations, rust, etc. all contribute to the resistance of the return ground path, making it extremely difficult to calculate the actual ground-fault current values. Tremendous arcing damage can occur.

available at the line terminals of the equipment.

- *Section 0, Definition of Overcurrent Device:* An overcurrent device means any device capable of automatically opening an electric circuit, under both predetermined overload and short-circuit conditions, either by fusing of metal or by electro-mechanical means.

- *Rule 10–002:* System and circuit conductors are grounded to limit voltages due to lightning, line surges, or unintentional contact with higher voltage lines, and to stabilize the voltage to ground during normal operation. Equipment grounding conductors are bonded to the system grounding conductors to provide a low impedance path for fault current that will facilitate the operation of overcurrent devices under ground fault conditions.

- *Rule 10–500, Effective Grounding Path:* All grounding paths must:
  1. be permanent and continuous,
  2. be able to safely carry any current it might be required to carry, and
  3. have sufficiently low impedance to limit the voltage to ground and to facilitate the operation of the circuit protective devices.

- *Rules 10–400* and *10–402:* Metal raceways, cable trays, cable armour, cable sheath, enclosures, frames, fittings, and other metal noncurrent-carrying parts shall be effectively bonded where necessary to assure electrical continuity and the capacity to conduct safely any fault current likely to be imposed on them. Any nonconductive paint, enamel, or similar coating shall be removed at threads, contact points, and contact surfaces or the equipment shall be connected by means of fittings so designed as to make such removal unnecessary.

- *Rule 10–812, Grounding Conductor Size for AC Systems:* The size of the grounding conductor for a grounded system shall not be less than that given in *Table 17*.

For a system not grounded at the premises, the size of the grounding conductor shall not be less than that given in *Table 18*.

- *Rule 10–814, Bonding Conductor Size:* The size of copper or aluminum equipment bonding conductors shall not be less than that given in *Table 16*.

- *Rule 4–004(8), Ampacity of Wires and Cables:* Where the ambient temperature exceeds or is anticipated to exceed 30°C (86°F), the temperature correction factors from *Table 5A* shall apply.

CONDUCTOR HEATING

The amount of energy (heat) generated during a fault varies as the square of the RMS (root-mean-square) current multiplied by the time (duration of fault) in seconds. This value is expressed as I^2t.

The term I^2t is called "ampere-squared seconds."

Since watts = I^2R, we could say that the damage that might be expected under severe fault conditions can be related to (1) the amount of current flowing during the fault, (2) the time in seconds that the fault current flows, and (3) the resistance of the fault path. This relationship is expressed as I^2Rt.

Since the value of resistance under severe fault conditions is generally extremely low, however, we can simply think in terms of how much current (I) is flowing and for how long (t) it will flow.

It is safe to say that whenever an electrical system is subjected to a high-level short circuit or ground fault, less damage will occur if the fault current can be limited to a low value, and if the faulted circuit is cleared as fast as possible.

It is important to understand the time–current characteristics of fuses and circuit breakers in order to minimize equipment damage. Time–current characteristic curves were discussed in Unit 17.

CALCULATING AN INSULATED (75°C THERMOPLASTIC) CONDUCTOR'S SHORT-TIME WITHSTAND RATING

Copper conductors can withstand

- one ampere (RMS current)
- for five seconds
- for every 42.25 cmil (0.0214 mm^2) of cross-sectional area.

Note in the above statement that both current (how much) and time (how long) are included.

Let's take a real-world situation. If we wish to provide an equipment bonding conductor for a circuit protected by a 60-ampere overcurrent device, we find in *Table 16* that a No. 10 copper conductor is the *mimimum* size permitted.

Referring to Fig. 19–2, we find that the cross-sectional area of a No. 10 conductor is 10 380 cmil.

This No. 10 copper conductor has a 5-second withstand rating of:

$$\frac{10\ 380\ \text{circular cmil}}{42.25\ \text{cmil}} = 246\ \text{amperes}.$$

This means that 246 amperes is the maximum amount of current that a No. 10 copper insulated conductor can carry for 5 seconds without being damaged. This is the No. 10 conductor's short-time withstand rating.

Stating this information using the *thermal stress (heat)* formula, we have:

Thermal stress (heat) = I^2t

where

I = RMS current in amperes
t = time in seconds

Thus, the No. 10 copper conductor's 5-second withstand rating is:

$I^2t = 246 \times 246 \times 5 = 302\ 580$ ampere-squared seconds

With this basic information, we can easily determine the short-time withstand rating of this No. 10 copper insulated conductor for other values of time and/or current. The No. 10 copper conductor's 1-second withstand rating is:

I^2t = ampere-squared seconds

$I^2 = \dfrac{\text{ampere-squared seconds}}{t}$

$I = \sqrt{\dfrac{\text{ampere-squared seconds}}{t}}$

$= \sqrt{\dfrac{302\ 580}{1}}$

= 550 amperes

The approximate opening time of a typical moulded-case circuit breaker is one cycle, i.e., $\frac{1}{60}$ of a second or 0.0167 second. The No. 10 copper conductor's one-cycle withstand rating is:

$I = \sqrt{\dfrac{302\ 580}{0.0167}}$

= 4257 amperes

The typical clearing time for a current-limiting fuse is approximately one-fourth of a cycle or 0.004 second. The No. 10 copper conductor's one-quarter-cycle withstand rating is:

$I = \sqrt{\dfrac{302\ 580}{0.004}}$

= 8697 amperes

Therefore, a conductor can be subjected to large values of fault current if the clearing time is kept very short.

AWG	CM Area
14	4 110
12	6 530
10	10 380
8	16 510
6	26 240
4	41 740
3	52 630
2	66 360
1	83 690
0	105 600
00	133 100
000	167 800
0000	211 600

Fig. 19–2 Circular mil area of conductors.

Unit 19 Equipment and Conductor Short-Circuit Protection 309

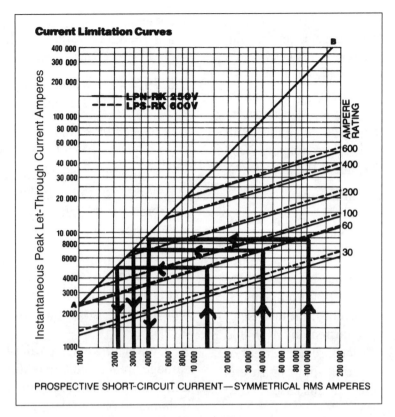

Fig. 19–3 Current-limiting effect of Class RK1 fuses. The technique required to use these charts is covered in Unit 17.

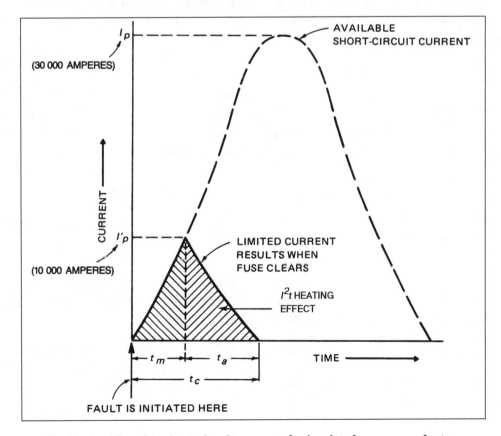

Fig. 19–4 Allowable short-circuit currents for insulated copper conductors.

When applying current-limiting overcurrent devices, it is important to use peak let-through charts to determine the apparent RMS let-through current before applying the thermal stress formula. For example, in the case of a 60-ampere Class RK1 current-limiting fuse, the apparent RMS let-through current with an available fault current of 40 000 amperes is approximately 3000 amperes. See Figs. 17–27 (page 271), and 19–3.

I^2t let-through of current-limiting fuse
= 3000 × 3000 × 0.004
= 36 000 ampere-squared seconds

Therefore, the No. 10 AWG copper equipment grounding conductor could be used where the available fault current is 40 000 amperes, if the circuit is protected by current-limiting fuses.

With an available fault current of 100 000 amperes, the apparent RMS let-through current of the 60-ampere Class RK1 current-limiting fuse is approximately 4000 amperes. In this case

I^2t = 4000 × 4000 × 0.004
= 64 000 ampere-squared seconds

Remember that, as previously discussed, the withstand I^2t rating of a No. 10 copper conductor is 302 580 ampere-squared seconds.

EXAMPLE: A 75°C thermoplastic insulated Type TW75 copper conductor can withstand 4200 amperes for one cycle.

a. What is the I^2t withstand rating of the conductor?

b. What is the I^2t let-through value for a non-current-limiting circuit breaker that takes one cycle to open? The available fault current is 40 000 amperes.

c. What is the I^2t let-through value for a current-limiting fuse that opens in 0.004 second when subjected to a fault of 40 000 amperes? The apparent RMS let-through current is approximately 4600 amperes. Refer to Fig. 17–27 (page 271).

d. Which overcurrent device (b or c) will properly protect the conductor under the 40 000-ampere available fault current?

Answers:
a. I^2t = 4200 × 4200 × 0.016
= 282 240 ampere-squared seconds

b. I^2t = 40 000 × 40 000 × 0.016
= 25 600 000 ampere-squared seconds

c. I^2t = 4600 × 4600 × 0.004
= 84 640 ampere-squared seconds

d. Comparing the I^2t withstand rating of the conductor to the I^2t let-through values of the breaker (b) and the fuse (c), the proper choice of protection for the conductor is (c).

The use of peak let-through, current-limiting charts is discussed in Unit 17. Peak let-through charts are available from all manufacturers of current-limiting fuses and current-limiting circuit breakers.

Conductors are particularly vulnerable to damage under fault conditions. Circuit conductors can be heated to a point where the insulation is damaged or completely destroyed. The conductor itself can actually burn away. In the case of equipment grounding conductors, if the conductor burns away under fault conditions, the equipment can become "hot," creating an electrical-shock hazard.

Even if the equipment grounding conductor does not burn, it can become so hot that it melts the insulation on adjacent conductors, shorting out the other circuit conductors in the raceway or cable. This results in further fault current conditions, damage, and hazards.

Of extreme importance are bonding jumpers, particularly the main bonding jumpers in service equipment. These main bonding jumpers must be capable of handling extremely high fault currents.

CALCULATING WITHSTAND RATINGS AND SIZING FOR BARE COPPER CONDUCTORS

A bare conductor can withstand higher levels of current than an insulated conductor of the same cross-sectional area. Do not exceed

- one ampere (RMS current)
- for 5 seconds

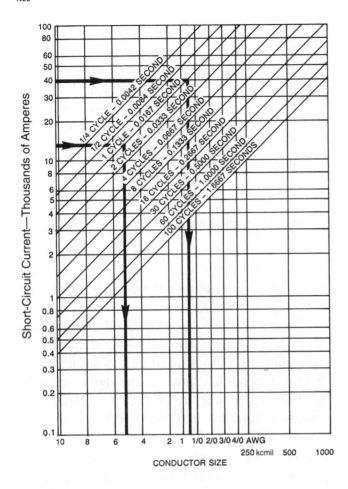

Fig. 19–5 Allowable short-circuit currents for insulated copper conductors.

Fig. 19–5 shows the withstand rating of 75°C thermoplastic insulated copper conductors. Many engineers will use these tables for bare grounding conductors because, in most cases, the bare equipment grounding conductor is in the same raceway as the phase conductors. An extremely hot equipment grounding conductor in contact with the phase conductors would damage the insulation on the phase conductors. For example, when nonmetallic conduit is used, the equipment grounding conductor and the phase conductors are in the same raceway.

To use the table, enter on the left side at the amount of fault current available. Then draw a line to the right, to the time of opening of the overcurrent protective device. Draw a line downward to the bottom of the chart to determine the conductor size.

EXAMPLE: A circuit is protected by a 60-ampere overcurrent device. Determine the minimum-size equipment grounding conductor for available fault currents of 14 000 amperes and 40 000 amperes. Refer to *Table 16*, the fuse peak let-through chart, and the chart showing allowable short-circuit currents for insulated copper conductors, Fig. 19-5.

Available Fault Current	Overcurrent Device	Conductor Size
40 000 amperes	Typical one-cycle breaker	No. 1/0 copper conductor
40 000 amperes	Typical RK1 fuse Clearing time 1/4 cycle or less	Using a 60-ampere RK1 current-limiting fuse, the apparent RMS let-through current is approximately 3000 amperes. *Table 16* shows a minimum No. 10 equipment bonding conductor. The allowable short-circuit current chart shows a conductor smaller than a No. 10. Therefore, No. 10 is the minimum-size equipment bonding conductor permitted.
14 000 amperes	Typical one-cycle breaker	No. 4 copper conductor
14 000 amperes	Typical RK1 fuse Clearing time 1/4 cycle or less	Using a 60-ampere RK1 current-limiting fuse, the apparent RMS let-through current is approximately 2200 amperes. *Table 16* shows a minimum No. 10 equipment bonding conductor. The allowable short-circuit current charts shows a conductor smaller than a No. 10. Therefore, No. 10 is the minimum-size equipment bonding conductor permitted.

- for every 30 cmil (0.0152 mm^2) of cross-sectional area of the conductor.

Because bare equipment grounding conductors are often installed in the same raceway or cable as the insulated circuit conductors, the "weakest link" of the system would be the insulation on the circuit conductors. Therefore, the conservative approach when considering conductor safe withstand ratings is to apply the **one ampere—for 5 seconds—for every 42.25 cmil** formula, or simply refer to a conductor short-circuit withstand rating chart, such as Fig. 19–5, and use the values for insulated conductors.

USING CHARTS TO DETERMINE A CONDUCTOR'S SHORT-TIME WITHSTAND RATING

The Insulated Cable Engineers Association, Inc. publishes much data on this subject. The graph in

You can see from the above examples that if the conductor size as determined from the allowable short-circuit current chart is larger than the size found in *Table 16*, then you must install the larger size. Installing a conductor too small to handle the available fault current could result in insulation damage or, in the worst case, the burning off of the equipment grounding conductor, leaving the protected equipment "hot."

If the conductor size as determined from the allowable short-circuit current chart is smaller than the size found in *Table 16*, then install an equipment bonding conductor *not smaller* than the minimum size required by *Table 16*.

There is another common situation in which it is necessary to install equipment bonding conductors larger than those shown by *Table 16*. When circuit conductors have been increased in size because of a voltage-drop calculation, the equipment bonding conductor ampacity should be increased proportionately. Voltage-drop calculations were covered in Unit 3.

Summary

When selecting equipment bonding conductors, bonding jumpers, or current-carrying conductors, the amount of short-circuit or ground-fault current available and the clearing time of the overcurrent protective device must be taken into consideration so as to minimize damage to the conductor and associated equipment. The conductor must not become a fuse. The conductor must remain intact under any values of fault current it is subjected to.

As stated earlier, the important issues are how much current will flow and how long the current will flow.

The choices are:

1. Install current-limiting overcurrent devices that will limit the let-through fault current and will reduce the time it takes to clear the fault. Then refer to *Table 16* to select the minimum-size equipment bonding conductor permitted by the *C.E.C., Part I,* and refer to the other conductor tables such as *Table 2* for the selection of circuit conductors.

2. Install conductors that are large enough to handle the full amount of available fault current for the time it takes a noncurrent-limiting overcurrent device to clear the short circuit or ground fault. Withstand ratings of conductors can be calculated or determined by referring to conductor withstand rating charts.

MAGNETIC FORCES

Magnetic forces acting upon electrical equipment (bus bars, contacts, conductors, and so on) are *proportional to the square of the **peak** current,* expressed as I_p^2.

Refer to Fig. 19–4 for the case in which there is no fuse in the circuit. The peak current, I_p (available short-circuit current), is indicated as 30 000 amperes. The peak let-through current, I_p, resulting from the current-limiting effect of a fuse, is indicated as 10 000 amperes.

EXAMPLE: Because magnetic forces are proportional to the square of the peak current, the magnetic forces (stresses) on the electrical equipment subjected to the full 30 000-ampere peak current are *nine times* that of the 10 000-ampere peak current let-through by the fuse. Stated another way, a current-limiting fuse that can reduce the available short-circuit peak current from 30 000 amperes to only 10 000 amperes will subject the electrical equipment to *only one-ninth* the magnetic forces.

Visual signs indicating that too much current was permitted to flow for too long a time include conductor insulation burning, melting and bending of bus bars, arcing damage, exploded overload elements in motor controllers, and welded contacts in controllers.

A current-limiting fuse (either a dual-element fuse or a straight current-limiting fuse) must be selected carefully. The fuse must have not only an adequate interrupting rating to clear a fault safely without damage to the fuseholder, but it also must be capable of limiting the let-through current (I_p) and the value of I^2t to the withstand rating of the equipment it is to protect.

The graph shown in Fig. 19–4 illustrates the current-limiting effect of fuses. The square of the area under the dashed line is energy (I^2t). I_p is the available peak short-circuit current that flows if there is no fuse in the circuit, or if the overcurrent device is not a current-limiting type. For a current-limiting fuse, the square of the shaded area of the graph represents the energy (I^2t), and the peak let-through current is I_p'. The melting time of the fuse element is t_m and the square of the shaded area corresponding to this time is the melting energy. The arcing time is shown as t_a; similarly, the square of the shaded area to this time is the arcing energy. The total clearing time, t_c, is the sum of the melting time and the arcing time. The square of the shaded area for time, t_c, is the total energy to which the circuit is subjected while the fuse clears. For the graph in Fig. 19–4, the area under the dashed line is six times greater than the shaded area. Since energy is equal to the area squared, the circuit is subjected to 36 times as much energy when it is protected by noncurrent-limiting overcurrent devices.

TAP CONDUCTORS

Branch-circuit conductors, feeder conductors, or service-entrance conductors must have an ampacity not less than the maximum load served, or, put another way, the load must not exceed the ampere rating of the circuit, *Rule 8–104(1,2)*.

The basic overcurrent protection requirement in *Rule 14–100* is that conductors shall be protected at the point where they receive their supply of current and at each point where the size of conductor is reduced. There are six exceptions allowed by *Rule 14–100*, each pertaining to specific applications of conductor taps.

These exceptions must be read and understood.

Of particular interest are feeder taps that are not over 3 m (10 ft), Fig. 19–6, and feeder taps that are not over 7.5 m (25 ft), Fig. 19–7. A number of requirements must be met when taps of this sort are made.

There are instances where a smaller conductor must be tapped from a larger conductor. For example, Fig. 19–6 shows a large feeder protected by a 400-ampere fuse. The tap conductor shall have an ampacity not less than either the load supplied by the tap or the rating of the switch or panel it feeds.

The above example of tapping a small conductor from a larger conductor that is protected

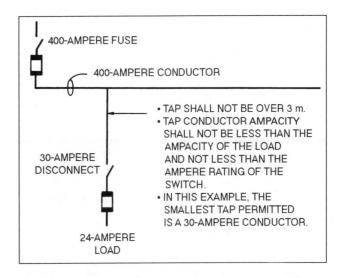

Fig. 19–6 *Rule 14–100(b)* requires that the tap be not more than 3 metres.

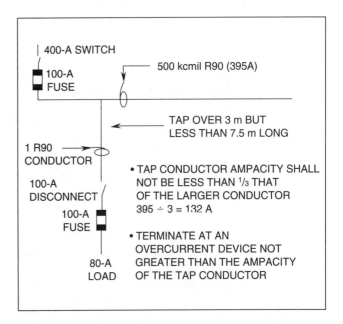

Fig. 19–7 *Rule 14–100(c)*, the 1/3 tap rule. This rule allows the tap to be over 3 m but not over 7.5 m long. It must have an ampacity not less than 1/3 that of the larger conductor.

by an overcurrent device much larger than the small tap conductor is an indication that the *C.E.C., Part I* is concerned with the protection of the smaller conductor. Note that it is the length of the conductor, not the length of the conduit run between the splitter and the switch, that is used for the calculation.

If the length of the tap from the splitter block to the terminal in the switch or panel is over 3 m, see Fig. 19–7. *Rule 14–100(c)* permits the use of a tap that does not exceed 7.5 m in length. This rule is often referred to as the "1/3 tap rule" because the ampacity of the tap conductor must be at least 1/3 the ampacity of the larger conductor feeding it. It also must terminate in an overcurrent device that has an ampacity equal to or less than the ampacity of the tap conductor. In Fig. 19–7, the tap must have an ampacity 1/3 that of the 395-ampere wire feeding the splitter. *Table 2* lists No. 1 Type R90 as wire suitable for 132 amperes.

Here again, the electrician must properly size the conductors for the load to be served, take into consideration the possibility of voltage drop, and then check the short-circuit withstand rating of the conductor to be sure that a severe fault will not cause damage to the conductor's insulation or, in the worst case, vaporize the conductor.

REVIEW

Note: Refer to the *C.E.C., Part I* or the plans as necessary.

1. Define *withstand rating*. _____
   _____
   _____
   _____

2. The label of a motor controller is marked "Maximum size fuse." To conform to *Rule 2–024*, which of the following is correct? Circle the letter of the best answer.

   a. Install fuses as stated on the label.

   b. Install a circuit breaker ahead of the controller.

   c. Either (a) or (b) is correct.

3. According to *Rule 10–500*, an effective grounding path must have the capacity to
   _____

4. What does the label "current-limiting" on an overcurrent device refer to?
   _____
   _____
   _____

5. What table of the *C.E.C., Part I* shows the minimum-size equipment bonding conductor?
   _____

6. The heating effect caused by current flowing through a conductor can be expressed by the term I^2t. What does this term mean? _____

_____

_____

_____

7. Copper 75°C insulated conductors can safely carry _____ amperes for _____ seconds for every _____ cmil of its cross-sectional area.

8. Calculate the maximum amount of current that a No. 12 copper 75°C insulated conductor can safely carry for 5 seconds.

9. What is the I^2t withstand rating of the No. 12 copper conductor in question 8?

10. Calculate the maximum amount of fault current that the No. 12 copper conductor in question 8 can withstand for:

    a. one cycle (0.016 second)

    b. one-quarter cycle (0.004 second)

11. Referring to Fig. 19–3, what are the instantaneous peak let-through current and apparent RMS current for the following 600-volt Class RK1 fuses?

Fuse Ampere Rating	Available Fault Current	Instantaneous Peak Let-Through Current	Apparent RMS Let-Through Current
600	200 000	_____	_____
400	100 000	_____	_____
200	40 000	_____	_____
100	15 000	_____	_____
60	10 000	_____	_____

12. What is the minimum-size 75°C thermoplastic-insulated copper conductor permitted for the following available short-circuit currents? Derive answers by using Fig. 19–5.

Fault Current	Clearing Time	Conductor Size
10 000 amperes	1 cycle	_____
10 000 amperes	$\frac{1}{2}$ cycle	_____
10 000 amperes	$\frac{1}{4}$ cycle	_____
30 000 amperes	1 cycle	_____
30 000 amperes	$\frac{1}{2}$ cycle	_____
30 000 amperes	$\frac{1}{4}$ cycle	_____

13. The available fault current is 50 000 amperes. A 200-ampere feeder is protected by a 200-ampere overcurrent device. The conductors are installed in a nonmetallic raceway (PVC) and therefore require that a separate equipment bonding conductor be installed. The inspector states that if current-limiting fuses are installed to protect the feeder, the proper equipment bonding conductor is found directly in *Table 16*.

    a. This would be a(n) _____ equipment bonding conductor.

    b. If a standard moulded case circuit breaker that has a one-cycle opening time is used as the feeder device, the inspector requires that a(n) _____ equipment bonding conductor be installed according to Fig. 19–5.

14. According to *Rule 14–100,* for taps not longer than 3 metres, the ampacity of the tap conductor shall be (not less than) (three times) the load and (greater than) (not less than) (equal to) the rating of the switch or panel it is feeding. Circle the correct answers.

UNIT 20

Low-Voltage Remote-Control Lighting

OBJECTIVES

After completing the study of this unit, the student will be able to

- list the components of a low-voltage remote-control wiring system
- understand the appropriate *C.E.C., Part I* rules governing the installation of a low-voltage remote-control wiring system
- demonstrate the correct connections for wiring a low-voltage remote-control system

LOW-VOLTAGE REMOTE-CONTROL LIGHTING

Conventional general lighting control is used in the major portion of the commercial building. For the drugstore, however, a method known as *low-voltage remote-control lighting* is selected because of the number of switches required and because it is desired to have extensive control of the lighting.

Relays

A low-voltage remote-control wiring system is relay operated. The relay is controlled by a low-voltage switching system and in turn controls the power circuit connected to it, Fig. 20–1. The low-voltage, split coil relay is the heart of the low-voltage remote-control system, Fig. 20–2. When the *on* coil of the relay is energized, the solenoid mechanism causes the contacts to move into the *on* position to complete the power circuit. The contacts stay in this position until the *off* coil is energized. When this occurs, the contacts are withdrawn and the power circuit is opened. The red wire is the *on* wire; the black wire is *off*, and the blue wire is common to the transformer.

The low-voltage relay is available in two mounting styles. One style of relay is designed to mount through a 16-mm ($\frac{1}{2}$-inch) trade size knockout opening, Fig. 20–3. For a 21-mm ($\frac{3}{4}$-inch) knockout, a rubber grommet is inserted to isolate the relay from the metal. This practice should ensure quieter relay operation. The second relay mounting style is the *plug-in relay*. This type of relay is used in an installation where several relays are mounted in one enclosure. The advantage of the plug-in relay is that it plugs directly into a bus bar. As a result, it is not necessary to splice the line voltage leads.

Single Switch

The switch used in the low-voltage remote-control system is a normally open, single-pole,

318 Unit 20 Low-Voltage Remote-Control Lighting

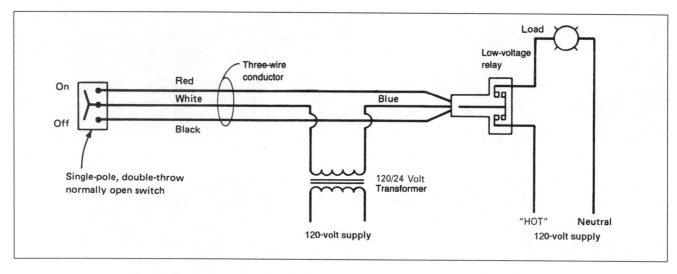

Fig. 20–1 Basic connection diagram for low-voltage remote-control system.

double-throw, momentary contact switch, Fig. 20–4. This switch is approximately one-third the size of a standard single-pole switch. In general, this type of switch has short lead wires for easy connections, Fig. 20–5.

These switch leads are colour-coded. The white wire is common and is connected to the 24-volt transformer source, the red wire connects to the *on* circuit, and the black wire connects to the *off* circuit.

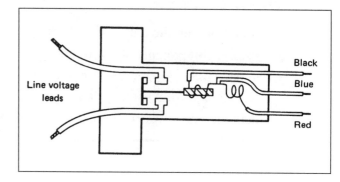

Fig. 20–2 Low-voltage relay.

Master Control

It is often desirable to control several circuits from a single location. Up to eight low-voltage switches, Fig. 20–6, will fit in the same space as that required for two conventional switches. This eight-switch master can be mounted on a $4\frac{11}{16}$-in (119-mm) box using an adapter provided with the master. Directory strips, identifying the switches' functions, can be prepared and inserted into the switch cover. An eight-switch master is used in the commercial building drugstore. The connection diagram is shown in Fig. 20–11.

If the control requirements are complex or extensive, a master sequencer, Fig. 20–7, can be installed. There is practically no limit to the number of relays that can be controlled, almost instantaneously, with this microprocessor-controlled electronic switch.

Fig. 20–3 Low-voltage relay. (Courtesy of General Electric Wiring Devices.)

Master Control with Rectifiers

Relays can be wired such that they can be operated both individually by separate switches and as a group by a master control. The principle of

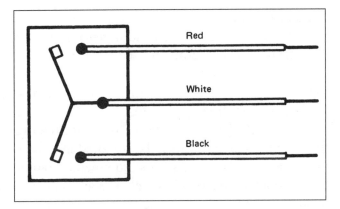

Fig. 20–4 Single-pole, double-throw, normally open, low-voltage control switch.

Fig. 20–5 Low-voltage switch. (Courtesy of General Electric Wiring Devices.)

Fig. 20–6 Eight-switch master control. (Courtesy of General Electric Wiring Devices.)

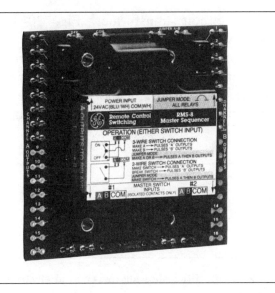

Fig. 20–7 Master sequencer. (Courtesy of General Electric Wiring Devices.)

operation of a master control with *rectifiers* is based on the fact that a rectifier permits current in only one direction, Fig. 20–8 and Fig. 20–12.

For example, if two rectifiers are connected as shown in Fig. 20–9, current cannot flow from A to B, or from B to A; however, current can flow from A to C, or from B to C. Thus, if a rectifier is placed in one lead of the low-voltage side of the transformer, and additional rectifiers are used to isolate the switches, a master switching arrangement can be achieved, Fig. 20–10. This method of master control is used in the drugstore.

Although a switching schedule is included in the specifications, the electrician may find it necessary to prepare a schematic diagram similar to that shown in Fig. 20–11. The relays and rectifiers (Fig. 20–12) for the drugstore master control are located in the low-voltage control panel.

WIRING METHODS

Section 16 governs the installation of a low-voltage system. *Section 16* provisions apply to remote-control circuits, low-voltage relay switching, low-energy power circuits, and low-voltage circuits.

The drugstore low-voltage wiring is classified as a Class 2 circuit. By definition, the power source of a low-voltage circuit is limited.

The circuit transformer, Fig. 20–13, is designed so that in the event of an overload, the output voltage decreases as the current increases. Any overload can be counteracted by this energy-limiting characteristic of the specially designed transformer core. If the transformer is not self-protected, a thermal device may be used to open the primary side to protect the transformer from overheating. This thermal device resets automatically as soon as the transformer cools. Other transformers are protected with nonresetting fuse links or externally mounted fuses.

Although the *C.E.C., Part I* does not require that the low-voltage wiring be installed in a raceway, the specifications for the commercial building do contain this requirement. The advantage of using a raceway for the installation is that it provides a means for making future additions at a minimum cost. A disadvantage of this approach is the initial higher construction cost.

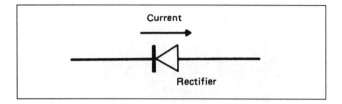

Fig. 20–8 Rectifier showing direction of electron flow (current).

Conductors

No. 18 conductors are used for low-voltage, remote-control systems. Larger conductors should be used for long runs to minimize the voltage drop. The cables for the installation usually contain two or three conductors. The insulation on the individual conductors may be either a double-cotton covering or plastic. Regardless of the type of insulation used, the individual conductors in the cables can be identified readily. To

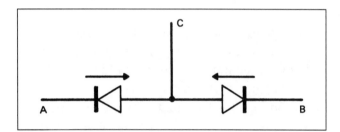

Fig. 20–9 Rectifier connected to allow current in one direction and block current in opposite direction.

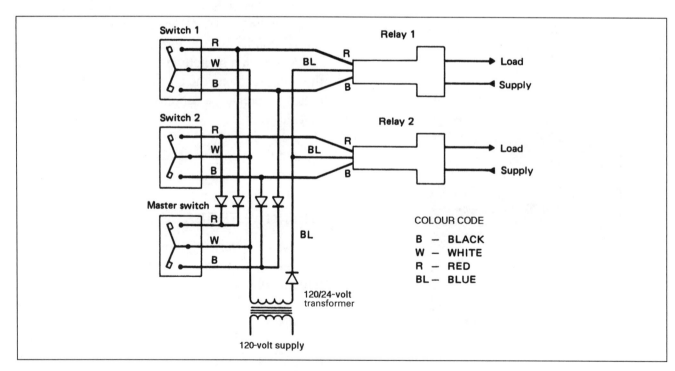

Fig. 20–10 Master control with rectifiers.

simplify connections, low-voltage cables are available with various conductor combinations, including blue–white, red–black, and black–white–red. These colour-coded wires are connected like-colour to like-colour throughout the installation. A cable suitable for outdoor use, either overhead or underground, is available for low-voltage, remote-control systems.

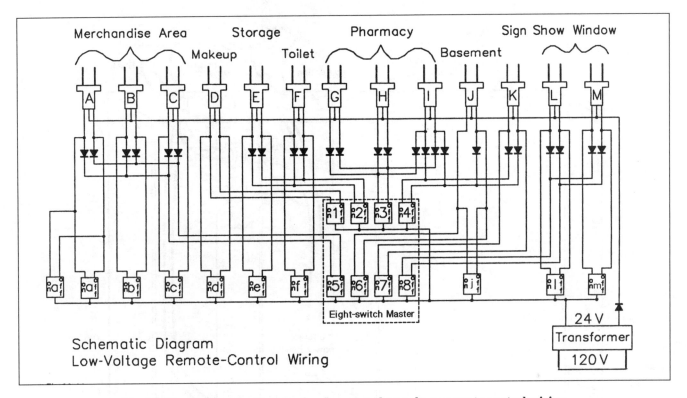

Fig. 20–11 Schematic diagram for drugstore low-voltage remote-control wiring.

Fig. 20–12 Rectifier.

Fig. 20–13 35-VA transformer.

Low-Voltage Panel

The low-voltage relays in the drugstore installation are to be mounted in an enclosure next to the power panel, Fig. 20–14. A barrier in this low-voltage panel separates the 120-volt power lines from the low-voltage control circuits, in compliance with *Rule 16–212*.

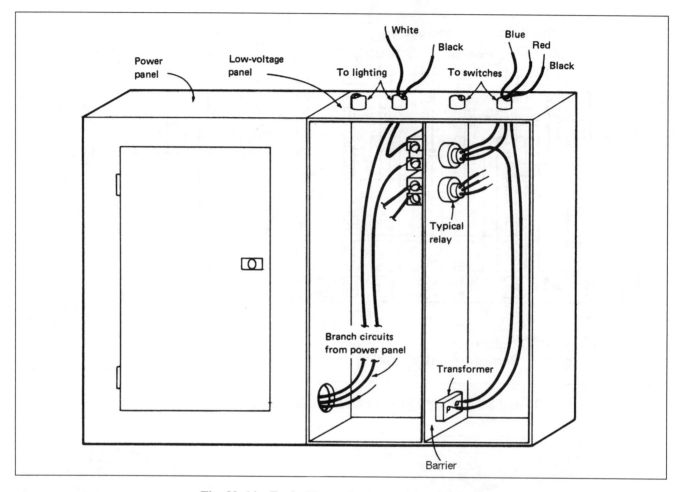

Fig. 20–14 Typical low-voltage panel installation.

REVIEW

Note: Refer to the *C.E.C., Part I* or the plans as necessary.

1. The low-voltage relay has a _____ coil design.

2. If the relay is in the *On* position and the *On* circuit is re-energized, the relay (will) (will not) change to the *Off* position. Circle the correct answer.

3. Match the following items by placing the letter of the correct wire in the appropriate blank of the column on the left.

   a. _____ Relay *On* circuit   A. Red wire
   b. _____ Relay common circuit   B. Black wire
   c. _____ Relay *Off* circuit   C. Blue wire

d. _____ Switch *On* circuit   D. White wire
e. _____ Switch *Off* circuit   E. Green wire
f. _____ Switch common circuit   F. Yellow wire

4. Indicate the current direction for the device shown by the symbol —▷|— (Diode)
   _____

5. What section of the *C.E.C., Part I* governs the installation of remote-control and signal circuits? _____

6. Low-voltage wiring (must) (need not) be installed in raceway to comply with the *C.E.C., Part I*. Circle the correct answer.

7. Complete the following connection diagram using rectifiers for the master control. Indicate wire colours.

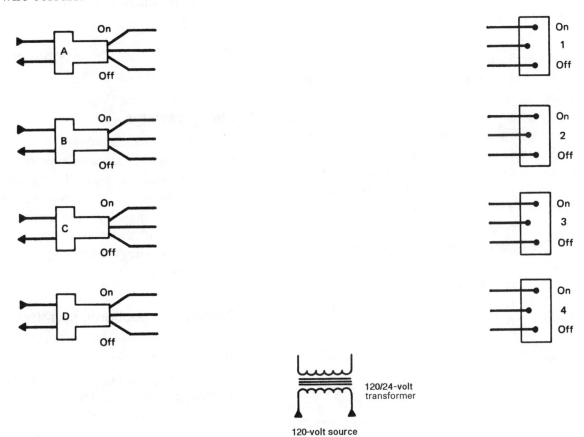

120/24-volt transformer

120-volt source

Switching schedule

Switch	Control relay
1	A
2	B
3	C & D
4	A & B

APPENDIX

Part A: Division 16 Electrical Specifications

SECTION 16010 ELECTRICAL GENERAL REQUIREMENTS

General

Comply with Division 1 — General Requirements. The specifications for this division are an integral part of the contract documents and shall be read accordingly.

Electrical drawings indicate only the general location of equipment and outlets. The contactor shall be responsible for the detail layout of equipment, outlets, raceways and wiring.

Outlets shown on architectural room elevations take precedence over positions or mounting heights on electrical drawings.

The owner reserves the right to move the location outlets 10' without additional charge providing the contractor is advised prior to installation.

Refer to architectural and structural drawings for accurate building dimensions.

Scope

The electrical contractor shall furnish all labour, tools, equipment and material to install, test and put into operation the electrical systems shown in the drawings or herein specified or both.

Submittals

Immediately after this contract is awarded the contractor shall submit a list of equipment, fixtures and materials to be incorporated into the work including:

- Manufacturer's name
- Equipment numbers
- Catalogue numbers
- Equipment description
- Delivery date

The list must have the approval of the architect before work is commenced.

Shop drawings

The contractor shall submit shop drawings, catalogue cuts and descriptive literature of all equipment installed as part of this contract. Shop drawings are particularly required for the following: detail drawings of the engine-generator set and base, and complete wiring diagrams and assembly drawings including the autostart and safety controls.

Permits, inspections, certificates and fees

The contractor shall deliver to the owner final certificates of inspection and approval by the inspection authority having jurisdiction when all work has been completed, tested and placed in operation.

Copies of electrical plans and specifications should be submitted as required by Electrical Inspection and Supply Authorities.

The contractor will notify inspection authorities in sufficient time to inspect work and pay all fees and costs.

Delivery, handling and storage

The contractor shall assume full responsibility to receive, handle, store and protect equipment, fixtures and materials covered under this contract.

Record drawings

The contractor will provide record drawings showing the locations of buried raceways and cables dimensioned to the centre line of building columns. Elevations should be shown with respect to finished grade or finished floor.

The location of all access panels for junction boxes, pull boxes and terminal cabinets should be shown.

All changes in the location and sizes of equipment, outlets, wiring and such other changes as may occur during the work should be shown.

Satisfactory record drawings must be filed with the architect before a final certificate of acceptance will be issued.

Relations with other trades

The work of the electrical trade shall be carried out in co-operation with other trades in such a manner as to avoid unnecessary delays and ensure the work of all trades is installed to the best advantage. The contractor shall assume full responsibility for laying out the work and for any damage to the owner or other trades caused by improper layout.

Workmanship

All work shall be in accordance with the Canadian Electrical Code, Provincial and Local Codes. Workmanship shall be of a uniformly high quality in regards to durability, efficiency, safety and neatness of detail. Inferior work shall be replaced without cost when so ordered by the architect.

Cutting patching

The contractor shall bear all costs of cutting and patching resulting from the work of this division.

Cleaning

The contractor shall clean up all refuse caused by the work at the end of the day and remove waste on a weekly basis. On completion of the work the contractor shall remove all surplus materials and waste.

Supports and hangers

The contractor shall provide all necessary supports, hangers, inserts and sleeves necessary to properly execute the work.

Testing

The contractor will test and check electrical and communication systems for correct operation in the presence of the engineer.

An insulation resistance test will be conducted using a "megger" (500 volt instrument on circuits up to 350 volts) on all lighting and power circuits prior to connecting devices and equipment. All test results should be recorded and submitted to the engineer for reference. The contractor shall repair or replace at no cost all circuits that do not meet the minimum requirements of the inspection authority.

In co-operation with mechanical trades the contractor will take and record clamp-on ammeter readings of mechanical equipment motors while the equipment is operating under load.

Electrical system voltages should be checked after the facility has been in operation for 60 days and tap settings adjusted as required.

SECTION 16050 BASIC MATERIALS AND METHODS

Products

The contractor will obtain from the engineer a list of acceptable equipment manufactures.

All material shall be new and bear the label of the Canadian Standards Association or other approval agency recognized by the inspection authority having jurisdiction.

Unless otherwise indicated, factory finish all equipment with ANSI/ASA #61 grey paint.

Section 16110 Raceways

The contractor shall install exposed conduits symmetrical to the building construction. Conduits should be routed to avoid beams columns and other obstructions. Conduits should cause minimum interference in spaces through which they pass.

Conduits should be grouped together whenever possible. Exposed conduits should be securely attached in place at intervals not exceeding the requirements of the C.E.C.

Three spare $\frac{3}{4}$-in (27 mm) trade size conduits shall be installed from each flush-mounted panel into the ceiling joist space above the panel.

Feeders shall be installed in either rigid metal conduit or rigid non-metallic conduit. The supply to the boiler shall be rigid metal conduit.

In the doctor's office, all branch-circuit conductors will be installed in electrical metallic tubing. In all other areas of the building, branch-circuit conductors may be installed in either electrical metallic tubing or electrical non-metallic tubing.

Section 16120 Conductors and cables

The conductor shall provide wires of the number and size shown in the drawings or described herein with spare conductors as indicated for the complete electrical system.

Pulling tensions and minimum bending radii should be limited to manufacturer's recommendations.

The contractor shall utilize cable lubricant when pulling cables through conduits and ducts.

Conductors shall have a 90°C rating, types RW90XLPE or T90 Nylon unless specified otherwise. All conductors shall be copper.

The minimum size conductors will be No. 12 AWG.

The neutral conductor of any feeder circuit shall not be smaller than the other circuit conductors.

Do not connect more than three lighting circuits from a three-phase panel or two circuits from a single-phase panel to a common neutral.

Do not exceed the C.E.C. requirements for voltage drop on any branch circuit or feeder.

Colour code feeder and branch circuits maintaining phase and colour sequence throughout.

Underground installation

Direct buried cables shall be installed in sifted sand free of rock, stone and other sharp objects with a 3" layer above and below.

Where direct buried cables pass under a roadway or area subject to vehicular traffic, install in suitable concrete encased ducts.

Section 16130 Boxes

Access panels shall be provided in ceilings where pull or junction boxes are not readily accessible.

The contractor shall neatly make cutouts for outlet boxes recessed in walls.

Boxes shall be supported independently of conduit and raceways.

Section 16140 Wiring devices

Switches shall be AC general-use, specification grade, 20 amperes, 120-277 volts, either single-pole, three-way or four-way, as shown on the plans.

When two or more switches are required at the same location they shall share a common wall plate unless otherwise noted.

The electrical contractor shall furnish and install, as indicated in blueprint E-1 (Electrical Symbol Schedule), receptacles meeting CSA standards and approved by the CSA. No more than eight 5-15R receptacles to be on one 2-wire circuit.

Wall plates for switches and receptacles shall be stainless steel unless otherwise noted on the drawings. Install wall plates after the painting of room surfaces has been completed.

Section 16-200 Power generation

The electrical contractor will furnish and install a motor-generator plan capable of delivering 12 KVA at 120/208 volts, wye-connected three-phase. Motor shall be for use with diesel fuel, liquid cooled, complete with 12-volt batteries and battery charger, mounted on anti-vibration mounts with all necessary accessories, including mufflers, exhaust piping, fuel tanks, fuels lines and remote derangement annunciator.

SECTION 16400 LOW VOLTAGE ELECTRICAL DISTRIBUTION

Service entrance board

The electrical contractor will furnish and install a service entrance as shown in the electrical power distribution diagram, which is found in the plans. The supply authority will install transformers and all primary service. The electrical contractor will provide a concrete pad and conduit rough in as required by the supply authority.

The electrical contractor will furnish and install a service lateral consisting of three sets of 4 500-kcmil conductors each set in rigid metallic or nonmetallic conduit. Conductors will be RW90-XLPE (600V) copper.

The electrical contractor will furnish and install service entrance equipment, as shown in the plans and detailed herein. The equipment will consist of nine switches, and the metering equipment will be fabricated in three type NEMA 1 sections. A continuous neutral bus will be furnished for the length of the equipment and shall be isolated except for the main bonding jumper to the grounding bus, which shall also be connected to each section of the service-entrance equipment and to the water main. The switchboard shall be braced for 25 000 RMS symmetrical. The metering will be located in one section and shall consist of seven meters. Five of these meters shall be for the occupants of the building and two shall serve the owner's equipment.

The switches shall be as follows:

1. Bolted pressure switch, three pole, 600 amperes with three 600-amp fuses.
2. Bolted pressure switch, three pole, 800 amperes with three 700-amp fuses.
3. Quick make quick break switch, three pole, 100 amperes with three 90-amp fuses.
4. Quick make quick break switch, three pole, 200 amperes with three 175-amp fuses.
5. Quick make quick break switch, three pole, 200 amperes with three 125-amp fuses.
6. Quick make quick break switch, two pole, 100 amperes with two 100-amp fuses.
7. Quick make quick break switch, three pole, 100 amperes with three 90-amp fuses.
8. Quick make quick break switch, three pole, 60 amperes with three 45-amp fuses.
9. Quick make quick break switch, three pole, 100 amperes.

The bolted pressure switches shall be knife-type switches constructed with a mechanism that automatically applies a high pressure to the blade when the switch is closed. Switch shall be rated to interrupt 200 000 RMS symmetrical amperes when used with current limiting fuses that have an equal rating.

The quick make quick break switches shall be constructed with a device that assists the operator in opening or closing the switch to minimize arcing. Switches shall rated to interrupt 200 000 RMS symmetrical amperes when used with current limiting fuses that have an equal rating. They shall have fuse clips for Class J fuses.

Fuses

Fuses 601 amps and larger shall have an interrupting rating of 200 000 RMS symmetrical amperes. They shall provide a time delay of not less than 4 seconds at 500 percent of their ampere rating. They shall be current limiting Class L.

Fuses 600 amps and less shall have an interrupting rating of 200 000 RMS symmetrical amperes. They shall provide a time delay of not less than 10 seconds at 500 percent of their ampere rating. They shall be current limiting Class J.

All fuses shall be selected to assure selective co-ordination.

Spare fuses shall be provided in the amount of 20 percent of each size and type installed, but in no case shall less than three spare fuses be supplied. These spare fuses shall be delivered to the owner at the time of acceptance of the project, and shall be placed in a spare fuse cabinet mounted on the wall adjacent to or located in the switchboard.

Fuse identification labels, showing the size and type of fuses installed, shall be placed inside the cover of each switch.

Bonding and grounding

The electrical contractor shall install and test a complete bonding and grounding system as shown in the drawings or herein specified or both.

Transformers, panelboards, metal enclosures, exposed building steel and metallic piping should be bonded to the main ground bus. The contractor shall provide ground connections to building steel with thermit welds or 9 mm silicon bronze alloy bolts, and peen ends of bolts after installation.

Grounding metal contact points should be cleaned of paint rust and other contaminates.

Exposed grounding conductors should be protected from mechanical damage metal with rigid PVC conduit.

The contractor shall test ground continuity and resistance prior to energizing electrical circuits and the grounding system efficiency for compliance with the Canadian Electrical Code. Ohmic resistance shall be verified in the presence of the inspection authority. Two copies of the test report shall be supplied to the architect.

Panelboards

The electrical contractor shall furnish panelboards as shown on the plans and detailed in the panelboard schedules. Panelboards shall be approved by the CSA. All interiors will have 225-ampere bus with lugs, rated at 75°C, for incoming conductors. Boxes will be of galvanized sheet steel and will provide wiring gutters, as required by the *C.E.C., Part I* (2002). Fronts will be suitable for either flush or surface installation and shall be equipped with a keyed lock and a directory card holder. All panelboards shall have an equipment grounding bus bonded to the cabinet. A typewritten directory identifying each breaker shall be provided in each panelboard. See Sheet E1 for a listing of the required poles and overcurrent devices.

Moulded case circuit breakers

Moulded case circuit breakers shall be installed in branch-circuit panelboards as indicated on the panelboard schedules on sheets E2 to E4. Each breaker shall provide inverse time delay under overload conditions and magnetic tripping for short circuits. The breaker operating mechanism shall be trip-free and multipole units will open all poles if one pole trips. Breakers shall have sufficient interrupting rating to interrupt 10 000 RMS symmetrical amperes.

Transfer switch

The electrical contractor will furnish and install a complete automatic load transfer switch capable of handling 12 KVA at 120/208 volts, three-phase. The switch control will sense a loss of power on any phase and signal the motor-generator to start. When emergency power reaches 80 percent of voltage and frequency, the switch will automatically transfer to the generator source. When the normal power has been restored for a minimum of 5 minutes, the switch will reconnect the load to the regular power and shut off the motor-generator. The switch shall be a NEMA 1 enclosure.

SECTION 16500 LIGHTING SYSTEMS

Luminaries

The electrical contractor will furnish and install luminaries, as described in the schedule shown in the plans. Luminaries shall be complete with all necessary supports, hangers' diffusers and lamps. Ballast, when required, shall be as shown on the drawings or as specified. Magnetic ballasts shall have a sound rating of A. Alternative luminaries other than those specified shall only be used with the written approval of the architect.

Luminaries shall be adequately supported.

SECTION 16700 COMMUNICATIONS

Time clock

A time clock shall be installed to control the lighting in the front and rear entries. The clock will be connected to Panel EM circuit No. 1. The clock will be 120 volts, one circuit, with astronomic control and a spring-wound carry-over mechanism.

Telephone system

The contractor shall supply and install a complete system of empty conduits, terminal cabinets, pull boxes and outlets for the telephone wiring as shown on the drawings or described herein. Minimum size conduit shall be $\frac{3}{4}$". Conduit runs shall have not more than two quarter bends and bending radius shall not be less than ten times the conduit diameter.

SECTION 16900 CONTROLS AND INSTRUMENTATION ELECTRIC

Low-voltage, remote-control switching

A low-voltage, remote-control switch system shall be installed in the drugstore as shown on the plans and detailed herein. All components shall be specification grade and constructed to operate on 24-volt control power. The transformer shall be a 120/24-volt, energy-limiting type for use on a Class II signal system.

Cabinet. A metal cabinet matching the panelboard cabinets shall be installed for the installation of relays and other components. A barrier will separate the control section from the power wiring.

Relays. Relays shall be 24-volt AC, split-coil type rated to control 20 amperes of tungsten or fluorescent lamp load.

Switches. The switches shall be complete with wall plate and mounting bracket. They shall be normally open, single-pole, double-throw, momentary contact switches with on-off identification.

Rectifiers. Heavy-duty silicon rectifiers with 7.5-ampere continuous duty rating shall be provided.

Wire. The wiring shall be in two- or three-conductor, colour-coded, No. 20 AWG wire.

Rubber grommet. An adapter will be installed on all relays to isolate the relay from the metal cabinet to reduce noise. Contractor is responsible to ensure the bonding of all metal parts.

Switching schedule. Connections will be made to accomplish the lighting control as shown in the switching schedule (Figure A–1).

Switch	Relay	Area Served	Branch Circuit #
RCa	A	Main area lighting	1
RCb	B	" " "	3
RCc	C	" " "	3
RCd	D	Makeup area	5
RCe	E	Storage	5
RCf	F	Toilet	5
	G	Pharmacy	7
	H	"	7
	I	"	7
RCj	J	Stairway and Basement	9
RCl, RCm	L,M	Show window	11,13
	K	Sign	15
RCM-1	D		
RCM-2	E,F		
RCM-3	G,H,I		
RCM-4	I,J,K		
RCM-5	A,B,C		
RCM-6	J		
RCM-7	K		
RCM-8	L,M		

Fig. A–1 Drugstore low-voltage, remote-control switching schedule.

Division 15 Mechanical

HEATING AND AIR-CONDITIONING SPECIFICATIONS

Only those sections that pertain to the electrical work are listed here.

Boiler

The heating contractor will furnish and install a 200-kW electric hot water heating boiler completely equipped with safety, operating and sequencing controls for 208-volt, three-phase electric power.

Hot water circulating pumps

The heating contractor will furnish and install five circulating pumps. Each pump will serve a separate rental area and be controlled from a thermostat located in that area as indicated on the electrical plans. Pumps will be 1/6 horsepower, 120 volts, single-phase. Overload protection will be provided by a manual motor starter.

Air-conditioning equipment

Air-conditioning equipment will be furnished and installed by the heating contractor in four of the rental areas indicated as follows.

Drugstore. A split-system packaged unit will be installed with a rooftop compressor-condenser and a remote evaporator located in the basement. The electrical characteristics of this system are:

Voltage:
208 volts, three-phase, 3 wire, 60 hertz

Hermetic refrigerant compressor-motor:
Rated load current 20.2 amperes, 208 volts, three-phase

Evaporator motor:
Full load current 3.2 amperes, 208 volts, single-phase

Condenser motor:
Full load current 3.2 amperes, 208 volts, single-phase

Minimum circuit ampacity:
31.65 amperes

Overcurrent protection:
50 amperes, time-delay fuse

Insurance Office. A single package unit will be installed on the roof with electrical data identical to that of the unit specified for the drugstore.

Beauty Salon. A single package unit will be installed on the roof and will have the following electrical characteristics:

Voltage:
208 volts, three-phase, 3 wire, 60 hertz

Hermetic refrigerant compressor-motor:
Rated load amperes 14.1, 208 volts, three-phase

Condenser-evaporator motor:
Full load amperes 3.3 at 208 volts, single-phase

Minimum circuit ampacity:
20.92 amperes

Overcurrent protection:
35 amperes, time-delay fuse

Doctor's Office. A single package will be installed on the roof and will have the following electrical characteristics:

Voltage:
208 volts, single-phase, 2 wire, 60 hertz

Hermetic refrigerant compressor-motor:
Rated load amperes 16.8 at 208 volts, single-phase

Condenser-evaporator motor:
Full load amperes 3.7 at 208 volts, single-phase

Minimum circuit ampacity:
24.7 amperes

Overcurrent protection:
40 amperes, time-delay fuse

Heating control

In cooled areas, the heating and cooling will be controlled by a combination thermostat located in the proper area as shown on the electrical plans.

PLUMBING SPECIFICATIONS

Only those sections that pertain to the electrical work are listed here.

Motors

All motors will be installed by the contractor furnishing the motors. All electrical power wiring to the motors and the connection of the motors shall be made by the electrical contractor. All control wiring and control devices will be the responsibility of the contractor furnishing the equipment to be controlled.

Sump pump

The plumbing contractor will furnish and install an electric motor-driven, fully automatic sump pump. Motor will be 1/2 horsepower, 120 volts, single-phase. Overload protection will be provided by a manual motor starter.

Part B: Useful Formulas

TO FIND	SINGLE PHASE	THREE PHASE	DIRECT CURRENT
AMPERES when KVA is known	$\dfrac{kVA \times 1000}{E}$	$\dfrac{kVA \times 1000}{E \times 1.73}$	not applicable
AMPERES when horsepower is known	$\dfrac{hp \times 746}{E \times \% \text{ eff.} \times pf}$	$\dfrac{hp \times 746}{E \times 1.73 \times \% \text{ eff.} \times pf}$	$\dfrac{hp \times 746}{E \times \% \text{ eff.}}$
AMPERES when kilowatts are known	$\dfrac{kw \times 1000}{E \times pf}$	$\dfrac{kW \times 1000}{E \times 1.73 \times pf}$	$\dfrac{kW \times 1000}{E}$
KILOWATTS	$\dfrac{I \times E \times pf}{1000}$	$\dfrac{I \times E \times 1.73 \times pf}{1000}$	$\dfrac{I \times E}{1000}$
KILOVOLT-AMPERES	$\dfrac{I \times E}{1000}$	$\dfrac{I \times E \times 1.73}{1000}$	not applicable
HORSEPOWER	$\dfrac{I \times E \times \% \text{ eff.} \times pf}{746}$	$\dfrac{I \times E \times 1.73 \times \% \text{ eff.} \times pf}{746}$	$\dfrac{I \times E \times \% \text{ eff.}}{746}$
WATTS	$E \times I \times pf$	$E \times I \times 1.73 \times pf$	$E \times I$

I = amperes E = volts kW = kilowatts kVA = kilovolt-amperes
hp = horsepower % eff. = percent efficiency pf = power factor

Fig. A–2 Useful formulas.

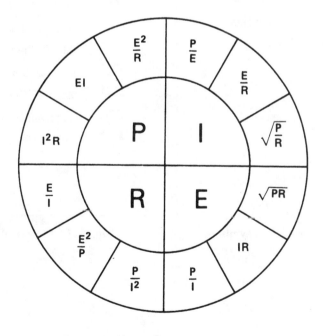

P = POWER IN WATTS
I = CURRENT IN AMPERES
R = RESISTANCE IN OHMS
E = ELECTROMOTIVE FORCE IN VOLTS

Fig. A–3 Equations based on Ohm's law.

Part C: Fire Protection Symbols

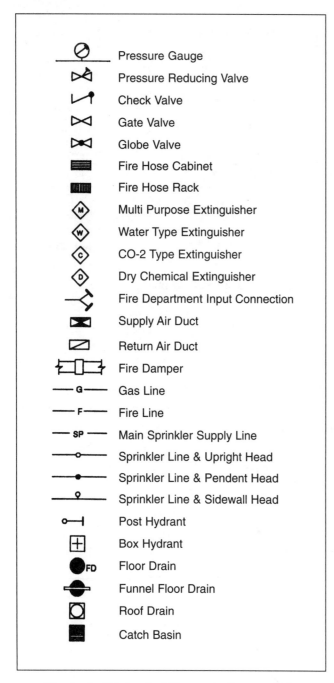

Fig. A–4 Mechanical fire protection symbols.

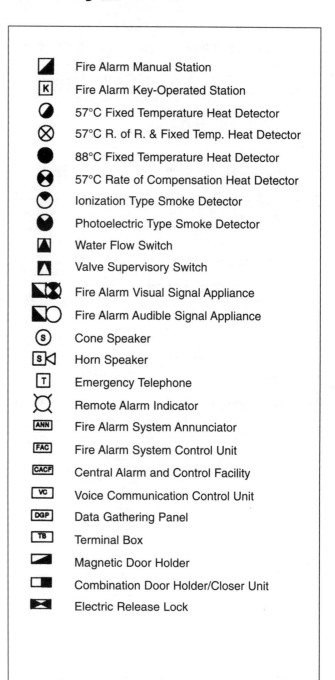

Fig. A–5 Electrical fire protection symbols.

GLOSSARY

ELECTRICAL TERMS COMMONLY USED IN THE APPLICATION OF THE *CANADIAN ELECTRICAL CODE, PART I*

Many of these definitions are quotations or paraphrases of definitions given in the *Canadian Electrical Code*.

Ampacity: The current carrying capacity of a conductor expressed in amperes.

Bonding: The joining together of all of the non-current carrying metal parts of the electrical system to assure electrical continuity between them.

Bonding conductor: A conductor that joins all of the non-current carrying metal parts of the electrical system and connects them to the service equipment or system grounding conductor.

Branch circuit: All of the wiring from the final overcurrent device protecting the circuit to the outlet(s).

Branch-circuit selection current (BCSC): A value used to select over-current protection, conductor sizes etc. for some refrigeration equipment. BCSC will be found on the equipment nameplate. BCSC is 64.1% of the maximum continuous current rating of the compressor.

Circuit ampacity: The ampacity of overcurrent device protecting the circuit or the ampacity of the conductors, whichever is less.

Circuit breaker: An electromechanical device designed to manually open and close a circuit and to automatically open a circuit under predetermined overload and short-circuit conditions.

Conductor: A wire or other form of metal used as part of an electrical system to convey current from one point to another.

Continuous load: A load that in normal operation is energized for more than a total of one hour in any two-hour period for loads of 225 amps or less and more than a total of three hours in any six-hour period for loads greater than 225 amps.

Current limiting fuse: A fuse that limits the amount of fault current let through the fuse under short circuit conditions by opening the circuit in less than one-quarter cycle.

Derated ampacity: The ampacity of a conductor after the correction factors of tables 5A to 5C of the C.E.C. have been applied.

Fault: An inadvertent connection between any two conductors of an electrical system or any conductor and ground.

Feeder: All of the wiring between the service box or other source of supply and the branch circuit overcurrent devices.

Ground: A connection to earth obtained by a ground electrode.

Ground fault circuit interrupter: A device designed to open an electric circuit within a specified time when a current to ground exceeds a predetermined value that is less than the setting of the overcurrent device protecting the circuit.

A class A ground fault circuit interrupter is designed to open a circuit when the current to ground is 6 mA or more in a time of $T = (20/I)^{1.43}$ where T = time in seconds and I = ground fault current between 4 mA and 260 mA or a maximum of 25 ms if the fault current is more than 260mA.

Ground fault protection: A device installed to provide protection of equipment by detecting line to ground faults. Required in solidly grounded circuits of 2000 A or more that operate at 150 volts or less to ground and 1000 A circuits operating at less than 750 volts phase to phase and more than 150 volts to ground.

Grounded circuit conductor: A current carrying conductor of the system that is intentionally connected to ground in order to limit the maximum voltage to ground of the system.

Interrupting rating: A rating given to protective devices to indicate the highest current that the device can interrupt without damage to associated equipment.

Maximum continuous current (MCC): A value determined by the manufacturers of hermatically sealed refrigerant compressors under high load (high refrigerant pressure) conditions.

Neutral conductor: the conductor of polyphase system or single phase 3 wire system that has a voltage that is equal in magnitude and phase spacing to the other conductors of the system.

Outlet: A point on a branch circuit where current is taken to supply utilization equipment.

Overcurrent device: a device that provides protection from current that exceeds the rating of the equipment or the ampacity of the conductors. Overcurrent devices are designed to provide protection under both overload and short-circuit conditions.

Overload device: A device designed to respond to overload conditions but not short circuits that is used to provide protection from current in excess of the ratings of the equipment or conductors.

Panelboard: An assembly consisting of overcurrent devices, busses, connections, and other related equipment enclosed in a cabinet.

Rated load current (RLA): A value determined by the manufacturer of the refrigeration equipment by actual operation (testing) at rated pressure and temperature conditions and voltage. RLA is usually 64.1% of the compressors maximum continuous current. When the same hermetic refrigerant motor-compressor is designed into different models the RLA for different products will most likely be different.

Selective co-ordination: The selection of overcurrent devices with time–current characteristics such that the effects of a fault will be localized.

Service: All of the wiring ahead (upstream) of the service box. The consumer's service is all of the wiring from the service box to the point where the supply authority makes connection. A supply service is one set of conductors run by supply authority from its mains to a consumers service.

Service box: An approved metal box or cabinet containing a circuit breaker or service switch and fuses, that may be locked or sealed and has provisions for the circuit breaker or service switch to be operated when the box or cabinet is locked and sealed.

Short circuit: A connection between any two or more conductors of an electrical system in such a way as to significantly reduce the impedance of the circuit. When inadvertent, a short circuit is referred to as a fault.

Switchboard: An assembly of large busses, switches, circuit breakers, fuses and other protective, controlling and measuring devices.

Ungrounded conductor: A conductor that has not been grounded; it is often referred to as a "hot" or line conductor. A potential difference (voltage) may exist between an ungrounded conductor and the grounded conductor or between two ungrounded conductors of an electrical system.

Unit substation: A unit substation is an assembly of busses, switches, fuses, circuit breakers, transformers and other protective measuring and controlling devices. A unit substation consists of a high voltage section that contains the high voltage switches and fuses, a transformer section that contains the transformer and a low voltage section that contains metering and distribution equipment.

INDEX

A
Air conditioning
    circuit requirements for, 134
    disconnect, 137
Aluminum conductors, 67
    connection problems of, 67
Ambient temperature, 60
Ampacity
    defined, 334
Appliances, 180
    grounding of, 186
    motor circuits for, 181
    overcurrent protection, 187
Approved, defined, 18
Arching fault values, 293
Architectural drafting symbols, 9, 10
Assemblies, multioutlet, 234

B
Bake oven, 189
Ballasts, 216
    electronic, 220
    fluorescent fixtures and lamps, 216
    power factor, 219
    sound rating, 219
Battery powered lighting, 82
Beauty salon, 227
    electrical drawings, 227
Blueprints, See Drawings
Bonding, 38
    service-entrance equipment, 33
Boiler control, 77
Boxes
    box fill and, 170
    dimensions, 170, 175
    sizing, 169
    switch (device), 167
Branch circuit
    defined, 51
    load calculations, 51
Breaker, branch circuits, 62

C
Canadian Electrical Code, Part I, 17
Canadian Standards Association (CSA), 17
Canadian Fire Alarm Association (CFAA), 105
Circuit breakers, 273
    cost considerations, 280
    current limiting, 279
    definition, 273
    interrupting ratings for, 274
    misapplication of, 277
    selective system, 298
    series-rated, 278
    thermal magnetic, 275
    time-current curve, 276
Circuit requirements, *See* specific piece of equipment
Circuits
    branch, defined, 51
    computer room, 238
Code use of metric (SI) measurements, 18
Communication systems, 237
Conductors
    aluminum, 47
    colour coding, 119
    connection problems, 67
    correction factors and, 60
    grounding electrode, 38
    grounding, sizes of, 42
    identification of, 119
    maximum number in conduit, 166
    neutral, 122
    remote control systems and, 320
    selecting boxes and, 167
    telephone, 236
    installation of, 68
    temperature effects on, 60
    voltage drop, 55, 63
    voltage drop and conductor size, 55, 66
    withstand rating (thermoplastic-covered TW75), 308
Conduits
    conductors in, 166
    derating factors, 60
    electrical metallic tubing (EMT), 158
    electrical nonmetallic tubing (ENT), 159, 165
    flexible metal, 159
    liquidtight, 160
    PVC, 162
    rigid, 156
Connections
    aluminum conductors and, problems of, 67
    wire, 61
Connectors
    flexible, 159
    liquidtight, 160
Conversion of imperial to metric units, 19
Contractor specification, 1, 324
Cooling system,
    compressor, 129
    condenser, 129
    control, 132
    electrical requirements for, 134
    evaporation, 128
    expansion valve, 130
    hermetic, 131
    installation, 132
    refrigeration, 127
Coordination of overcurrent device, 295
Copper conductors, rating/voltage loss, 63
Current-limiting circuit breakers, 279

D
Delta system
    four-wire, transformer connections, 27

open, transformer connection, 27
three-wire, transformer connections, 29
Derangement signal, 88
Derating
examples, 60
factors, conduits and, 58
Disconnect, 33
air conditioning and, 137
motors and, 181
service-entrance equipment and, 33
Doughnut machine, 188
Drawings
building drawings, 5
Drugstore, 245
Dual-element fuses, 252
Ductwork symbols, 12

E
Electric lamps, 193
Electric service
disconnect, 33
grounding, 33
metering, 30
transformers, 27
Electrical symbols, 13
Electrical work, specifications of, 324
Emergency power systems, 82
automatic transfer equipment, 88
derangement signals, 88
engine types, 84
sources of, 82
wiring for, 88
Energy savings consideration, 66
Evaporation, cooling system, 128
Exhaust fan (bakery), 180
Expansion valve, cooling system, 130

F
Feeder, panelboards and, 62, 121
Feeder loading schedule
bakery, 142
beauty salon, 227
doctor's office, 125
drugstore, 54
insurance office, 226
main service, 33
owner's circuits, 74
Feeders, 62
Fire
classes of, 93
detectors, 96
protection symbols, 332

pumps, 106
stages of, 93
triangle, 94
Flexible connections, 159
Floor boxes, 238, 241
installation and trimming, 242, 243
Floor outlets, 238
Fluorescent lamps, 214
ballasts, 216
compact, 214
Four-way switch, 151
Fuseholders, 256
Fuses
cable protectors, 266
cartridge, 255
classes, 257
current-limiting, 253–254
dual-element, 253
peak let-through charts, 249, 270
selective system, 297
testing, 266
time-current characteristic curves, 268

G
General conditions, specifications, 3
Generator, emergency power system, 82
Grounded neutral conductor, 36
Ground-fault protection, 45
Grounding, 186
Grounding the service, 33

H
Hazardous materials, 109
Hermetic compressors, cooling system, 131
High-frequency circuit, fluorescent lamps, 218
High-intensity discharge lamps, 221
High-pressure sodium lamps, 222
High-voltage metering, 32

I
Illuminance, 210
Incandescent lamps, 211
Instant-start circuit, fluorescent lamps, 217
Insurance office, electrical drawings, 226
Interrupting rating, 250

J
Junction boxes, 170

K
Kelvin, 210

L
Lamps, 193
fluorescent, 214
high-intensity discharge, 221
high-pressure sodium, 222
incandescent, 211
mercury, 221
metal halide, 222
sodium, low-pressure, 222
Lighting
battery powered, 80
remote-control low voltage, 317
Lighting fixture, See Luminaires
Lighting loads, 52
Loading schedule, See Feeder loading schedule
Low-voltage metering, 32
Lumen, defined, 209
Luminaires
lamp schedule, 202
styles A–Q, 202–208

M
Masonary boxes, 167
Master control, remote lighting, 318
Mercury lamps, 221
Metal conduits, rigid, 151
Metal halide lamps, 222
Metal raceways, installation of, 157
Metal tubing, electrical, 158
Metering electric service, 52
Motor loads, 181
Motors
branch circuits, 181
conductors supplying, 181, 186
feeder and, 186
overload protection, 182
short circuit, 182
Multioutlet assemblies, 234

N
Neutral current, 122
Neutral sizing, 121
Nonselectivity, overcurrent protection, 295
Notice to bidders, specifications, 1

O

Octagonal boxes, 170
Open delta system, transformer connections, 27
Outlets, floor, 238
Overcurrent protection, 248
   circuit breakers, 273
   circuits, 62
   conductor, 250
   continuous current rating, 250
   fuses, 250
   interrupting rating, 250
   response speed, 251
   transformers, 27
   voltage rating, 250
Overload protection, motors, 182
Owner's circuits, 73

P

Pad-mounted transformers, 24, 25, 26
Panelboards, 118
PE, *See* Registered Professional Engineers (PE), 16
Phase-to neutral short circuit, 293
Phase-to-phase fault, 292
Phase-to-phase-to-phase fault, 292
Piping symbols, 11
Plumbing symbols, 11
Power factor, ballasts, 219
Preheat circuit, fluorescent, 217
Proposal, specifications, 2

R

Raceways
   metal, 157
     multiple conductors, 60
     sizing of, 166
     support for, 176
     surface metal, 233
     underfloor, 238
Rapid-start circuit, fluorescent lamps, 217

Receptacles, 143
   CSA nonlocking, 145
   CSA locking, 146
   electronic equipment, 146
   ground-fault circuit interrupt, 147
   hospital grade, 145
Recessed luminaires, 196
Refrigeration, 127
Registered Professional Engineers (PEng), 16
Relays, remote-control lighting and, 317
Remote control lighting, 317

S

Schedule of drawings, specifications, 5
Series-rated circuit breakers, 278
Service conductors, 33
Service entrance, 30
Service equipment, bonding/grounding, 33
Sheet metal duct work symbols, 12
Short-circuit
   calculations, 286
   current protection, 285
Show window lighting load, 53
SI metric measurements, 19
Single-phase system, transformer connections, 27
Single-phasing, 298
Single-switch, remote-control lighting, 317
Single-pole switch, 150
Smoke detector, 99
Sodium lamps, low-pressure, 222
Sound rating, ballasts, 219
Specifications, 1
Standards Council of Canada, 16
Sump pump control, 75
Supplier label, 111
Switches, 149
System grounding, 42

T

Tap conductors, 313
Telephone system, 237
Thermal protection, recessed luminaires, 198
Thermal-magnetic circuit breakers, 275
Three-way switch, 151
Toggle switches, 149
Transformers
   connections, 27
   dry-type, 22
   liquid-filled, 23
   overcurrent protection, 27
   pad-mounted, 24, 25, 26
Tubing, electrical metallic, 158

U

Underfloor raceways, 238
Underground service, 24, 25, 26
Underground vault, main service, 25
Underwriters Laboratories of Canada (ULC), 17
Unit substation, installation, 23

V

Valve symbols, 11
Voltage drop, 55, 63

W

Water heater, 225, 228
WHMIS, 109
Wire
   connectors, 69
   size selection, 55
Wires, conduits/tubing and, 60
Wiring
   multioutlet assemblies, 235
   recessed luminaries, 201
   remote-control lighting, 321
Workplace label, 115